Androids, Cyborgs, and Robots in Contemporary Culture and Society

Steven John Thompson
University of Maryland University College, USA

A volume in the Advances in
Computational Intelligence and
Robotics (ACIR) Book Series

IGI Global
DISSEMINATOR OF KNOWLEDGE

Published in the United States of America by
IGI Global
Engineering Science Reference (an imprint of IGI Global)
701 E. Chocolate Avenue
Hershey PA, USA 17033
Tel: 717-533-8845
Fax: 717-533-8661
E-mail: cust@igi-global.com
Web site: http://www.igi-global.com

Library of Congress Cataloging-in-Publication Data

Names: Thompson, Steven John, 1956- editor.
Title: Androids, cyborgs, and robots in contemporary culture and society / Steven John Thompson, editor.
Description: Hershey, PA : Engineering Science Reference, 2018. | Includes bibliographical references and index.
Identifiers: LCCN 2017012040| ISBN 9781522529736 (hardcover) | ISBN 9781522529743 (ebook)
Subjects: LCSH: Androids--Social aspects--Handbooks, manuals, etc. | Robots--Moral and ethical aspects--Handbooks, manuals, etc. | Human-machine systems--Handbooks, manuals, etc.
Classification: LCC TJ211.49 .H356 2018 | DDC 303.48/34--dc23 LC record available at https://lccn.loc.gov/2017012040

This book is published in the IGI Global book series Advances in Computational Intelligence and Robotics (ACIR) (ISSN: 2327-0411; eISSN: 2327-042X)

British Cataloguing in Publication Data
A Cataloguing in Publication record for this book is available from the British Library.

For electronic access to this publication, please contact: eresources@igi-global.com.

Advances in Computational Intelligence and Robotics (ACIR) Book Series

ISSN:2327-0411
EISSN:2327-042X

Editor-in-Chief: Ivan Giannoccaro, University of Salento, Italy

MISSION

While intelligence is traditionally a term applied to humans and human cognition, technology has progressed in such a way to allow for the development of intelligent systems able to simulate many human traits. With this new era of simulated and artificial intelligence, much research is needed in order to continue to advance the field and also to evaluate the ethical and societal concerns of the existence of artificial life and machine learning.

The **Advances in Computational Intelligence and Robotics (ACIR) Book Series** encourages scholarly discourse on all topics pertaining to evolutionary computing, artificial life, computational intelligence, machine learning, and robotics. ACIR presents the latest research being conducted on diverse topics in intelligence technologies with the goal of advancing knowledge and applications in this rapidly evolving field.

COVERAGE

- Pattern Recognition
- Evolutionary computing
- Neural networks
- Agent technologies
- Fuzzy systems
- Brain Simulation
- Artificial Intelligence
- Cognitive Informatics
- Automated Reasoning
- Natural language processing

IGI Global is currently accepting manuscripts for publication within this series. To submit a proposal for a volume in this series, please contact our Acquisition Editors at Acquisitions@igi-global.com or visit: http://www.igi-global.com/publish/.

The Advances in Computational Intelligence and Robotics (ACIR) Book Series (ISSN 2327-0411) is published by IGI Global, 701 E. Chocolate Avenue, Hershey, PA 17033-1240, USA, www.igi-global.com. This series is composed of titles available for purchase individually; each title is edited to be contextually exclusive from any other title within the series. For pricing and ordering information please visit http://www.igi-global.com/book-series/advances-computational-intelligence-robotics/73674. Postmaster: Send all address changes to above address.

Titles in this Series

For a list of additional titles in this series, please visit:
https://www.igi-global.com/book-series/advances-computational-intelligence-robotics/73674

Handbook of Research on Applied Cybernetics and SystemsScience
Snehanshu Saha (PESIT South Campus, India) Abhyuday Mandal (University of Georgia, USA) Anand Narasimhamurthy (BITS Hyderabad, India) Sarasvathi V (PESIT- Bangalore South Campus, India) and Shivappa Sangam (UGC, India)
Information Science Reference • ©2017 • 463pp • H/C (ISBN: 9781522524984) • US $245.00

Handbook of Research on Machine Learning Innovations and Trends
Aboul Ella Hassanien (Cairo University, Egypt) and Tarek Gaber (Suez Canal University, Egypt)
Information Science Reference • ©2017 • 1093pp • H/C (ISBN: 9781522522294) • US $465.00

Handbook of Research on Soft Computing and Nature-Inspired Algorithms
Shishir K. Shandilya (Bansal Institute of Research and Technology, India) Smita Shandilya (Sagar Institute of Research Technology and Science, India) Kusum Deep (Indian Institute of Technology Roorkee, India) and Atulya K. Nagar (Liverpool Hope University, UK)
Information Science Reference • ©2017 • 627pp • H/C (ISBN: 9781522521280) • US $280.00

Membrane Computing for Distributed Control of Robotic Swarms Emerging Research ...
Andrei George Florea (Politehnica University of Bucharest, Romania) and Cătălin Buiu (Politehnica University of Bucharest, Romania)
Information Science Reference • ©2017 • 119pp • H/C (ISBN: 9781522522805) • US $160.00

Recent Developments in Intelligent Nature-Inspired Computing
Srikanta Patnaik (SOA University, India)
Information Science Reference • ©2017 • 264pp • H/C (ISBN: 9781522523222) • US $185.00

For an enitre list of titles in this series, please visit:
https://www.igi-global.com/book-series/advances-computational-intelligence-robotics/73674

IGI Global
DISSEMINATOR OF KNOWLEDGE

701 East Chocolate Avenue, Hershey, PA 17033, USA
Tel: 717-533-8845 x100 • Fax: 717-533-8661
E-Mail: cust@igi-global.com • www.igi-global.com

Editorial Advisory Board

List of Reviewers

Table of Contents

Detailed Table of Contents

Humans are becoming increasingly dependent on the 'say-so' of machines, such as computers, smartphones, and robots. In epistemology, knowledge based on what you have been told is called 'testimony' and being able to give and receive testimony is a prerequisite for engaging in many social roles. Should robots and other autonomous intelligent machines be considered epistemic testifiers akin to those of humans? This chapter attempts to answer this question as well as explore the implications of robot testimony for the criminal justice system. Few are in agreement as to the 'types' of agents that can provide testimony. The chapter surveys three well-known approaches and shows that on two of these approaches being able to provide testimony is bound up with the possession of intentional mental states. Through a discussion of computational and folk-psychological approaches to intentionality, it is argued that a good case can be made for robots fulfilling all three definitions.

There is still a considerable degree of catching up to do in regards to fostering gender equality within areas of STEM (Science, Technology, Engineering, and Math). Children's leading medium, television, could offer role models of girls with competence in STEM areas, but unfortunately television programs often miss this chance. The children's television series Annedroids is a notable exception. This chapter provides insight into how children can be educated about gender equality in STEM with the aid of gender-sensitive media programs such as Annedroids. The chapter examines data from a reception study which was conducted under the leadership of the International Central Institute for Youth and Educational Television (IZI) with 6- to 12-year-old children in the United States and Canada (N = 301). The research is enhanced by a conversation between Dr. Maya Götz, Head of IZI; Dr. Diana Nastasia, a contributor to the IZI research; and J. J. Johnson, the program's creator.

A multitude of genres and types of characters, in Japanese comics and animated series, suggests many thought-provoking themes; i.e., questions about human nature. Many artists can see the answers to these questions in artificial humans – both cyborgs and androids. In this research, the author analyzes Japanese texts of popular culture in which artificial children are the protagonists of the stories. The author aims to compare a child figure in sociological discourse, considered there as vulnerable, to the representations in manga and anime, in which characters are created as children or technologically-modified prepubescents. In this chapter, the author presents ideas and culture associations for the concepts of android and cyborg. The chapter focuses also on analysis of the characters from Japanese comic books and animations – both androids and then cyborgs – according to transhumanistic and posthumanistic theories. The analysis results in a conclusion that a child figure is dehumanized in the context of cyborg and android child protagonists.

How are robots, androids and cyborgs presented and received in the media? This chapter applies a social media analysis to this question by using empirical research on news stories that feature robotic technologies to see how robots are presented, consider what reporters focus on when writing about robots, and review how the public discusses and receives robots. The theoretical framework utilised focuses

Foreword

Whether made of clay or metal, robots have captured the literary imagination for centuries. From the golems of Jewish lore to the mechanized toys of Victorian society to the zippier robotic characters of more recent science fiction, robots have played a major if ambiguous role as inferior helpmates or overpowering monsters. As cultural "doubles" to the humans that create them, some basic questions can be extrapolated. What should robots look like? To what extent are robots like humans? What characteristics should be built into them? What attributes do they possess by virtue of being "entities?" If they become sentient (or if they don't), what "legal rights" and ethical considerations should be extended to them? These questions have become even more prominent in the first decades of the twenty-first century, where automation is common and smart devices—robots of sorts—are ubiquitous in our daily lives.

The chapters in this book, and the publication of the book itself, are further indicators on the historical dial of the continuing but now rapidly expanding interest in, and development of, robots as autonomous beings in contemporary society. One immediately notes that many of the chapters in this book are concerned with topics, attributes, and/or relationships generally associated solely with humans: intentionality, legal and ethical rights, feelings and emotional states, gendering education, childhood, media presentation, interface, mind, and the (dis)association of robots with androids and cyborgs. Regarding the latter, for our purposes perhaps one minor point that may distinguish robots from their alt-species counterparts is that not only are robots completely mechanical, but they are, at least in the popular mind, almost always made to look mechanical—as opposed to androids (who are made to closely cleave to human anatomy and features, and imitate their behavior), or cyborgs (who *are* already part human, as well as part machine).

i-Robot [1]

For Nathan Riggs

i like this robot. This robot
has an ethos all its own.
It speaks to us, in the openness
of its parts, in the honesty of its loose
assemblage. Rhetorically it is almost
authentic in its incoherence,
in its semi-arranged collection of objects
hanging and working together, almost
persuasive in its happenstance, its e-mergence,
a (dis)semblance of the human
way that it exists and moves
through the world. I like
this robot. This robot
has an ethos all its own.

Despite the more obvious differences between humans and robots, most of the chapters in this book, as in science fiction and popular culture generally, seem to assume some point(s) of similarity between robots and humans, in the questions the chapters ask and the lines of inquiry they pursue. (Humans have a way of doing that.) We thus should note here that whether intended or not, the comparison is ultimately based on the concept of "*autopoiesis*." As Humberto Maturana and Francisco Varela, who coined the term, wrote.

An autopoietic machine is a machine organized (defined as a unity) as a network of processes of production (transformation and destruction) of components which: (i) through their interactions and transformations continuously regenerate and realize the network of processes (relations) that produced them; and (ii) constitute it (the machine) as a concrete unity in space in which they (the components) exist by specifying the topological domain of its realization as such a network. (1980, p. 78)

Interestingly, in the research in different fields that ensued, *autopoiesis* (Greek: *auto* = "self," *poiesis* = "creative") is meant to apply not only machines (in this case robots) but also human beings (e.g., in A.I, cognitive science, linguistics; studies of mind and consciousness; and biological life processes of cells themselves). In research in both robots and humans, autopoiesis is meant to describe and comprehend the material functioning of self-organization, self-reproduction, and the self-realization of meaning, all of which are necessary for existence.

Attachment Function

my attachment function

no longer works—this is how

i used to feel

the attachment function

no longer feels—this is how

i used to work

Autopoiesis, whether accounting for the human mind and consciousness in mechanical terms, or robots in human terms, is foundational to the many fields that use it. Yet it is also highly controversial. It is quite plausible to conjecture, for instance, that autopoiesis is at least to some extent a result of understanding and describing life metaphorically. One dimension of the study of robots developed implicitly but usually not covered overtly would be relations between autopoiesis and metaphor. This is not to suggest that autopoiesis and the fields that have adopted it are not "real"—not more than their relations in/as language, no more than their saying. In an equally controversial argument, Richard Boyd (1993) asserts that unlike "literary metaphors" and "heuristic metaphors," "theory-constitutive metaphors" in science (e.g., in cognitive sciences, he argues) may actually lead to causal relations in the physical world; in these rarer cases, the "open-ended" metaphor (one of the conditions) becomes the basis of continuing development of knowledge in the field. On the other hand, "heuristic metaphors," according to Boyd (1993), like literary metaphors for logical positivists, lead nowhere, are nonsense (Ayer, 1952).

A.I. Talking

the phone hung up my face.

it's calling you back

to say goodbye.

If robots are understood, in science as well as in the chapters to follow, in terms of autopoiesis, the explication of life processes (whether biological or mechanical) may be understood to be grounded in the metaphors of autopoetics, and autopoetics to be a new branch of poetics, which of course includes poetry. Robots are poems written by humans in languages and/as technology, and as Heidegger (1975, 1977) *might* point out, are "enframed" but potentially free by virtue of being poems.

Being Robot

we are robots, we
are the ephemeral ones, we
are the expendable ones, we
stand in reserve, waiting, we
are at hand, ready to serve, we
are the useful ones, we
are defining our being, we

we are robots, we
are the permanent ones, we
are the natural ones, we
stand in the Opening
of our Presence, we
shine in our own Being, we
are you, and beyond this world

In fact, just as the Greek word *techne* means "craft" or "art" but also is the root of English "technique" and "technology," both "autopoiesis" and autopoetic etymologically and philosophically derive from *poiesis*, which includes poetics as well as poetry, as "making." It is thus "natural" that a *similarity* of affect, values, perception, and experience would seem inherent in and shared by robots as "poems."

Media Eye

mediated
meditation
modifies minds

Robots as poems leave may questions for these chapters, and humans, to ask. For the chapters here must assume some degree of autopoiesis, and thus at some level concern issues of embodiment, cognition, consciousness. Like poems, robots are "soft machines" (Porush, 1984) — but robots require a lot more work to realize their potential as technological entities that are aware, self-sufficient, and integrated into their human surroundings. That is the work to which these chapters contribute.

Some of the questions asked in these chapters are about whether robots influence gender identification in education, or feel pain.

A Computer File Named Dorothy[1]

I dated a file named Dorothy, created
worlds in her name; but needed more space,
new memories to save, new files to live.
(After all, although the universe expands
at astronomic rates, it's slowing down,
and there is only so much space inside machines).

"Destroy Dorothy: Confirm," the computer responded.
But what if she should die? I thought, and asked
aloud; what if when I push this button
she should really disappear
from the disc of the earth, constantly rotated, read
in this dark machine drive of the universe?

What if this cold, dumb, personal computer
should read and wholly misunderstand, and take me
literally, as impersonal as itself, and her atoms
be scattered through magnetic fields, dispersed
along the wires, and she should vanish mid the glitch
and circuitry of stars, drive lights red
shifting, every trace (of her) erased
forever. "Destroy Dorothy: Confirm," it repeated,
blindly blinking. Destroy Dorothy? I needed
more space, new memories to save,
new files to live. But Oh I
could not confirm it could not confirm it....

Other ancillary questions are whether robots would ever experience lust, or feel love? These are not silly questions, for the ability of robots to handle sexual urges, possess empathy, or feel love, may be essential in the development of ethical values and systems for robots in relation to each other, and to us.

The Beautiful Robot Who Stole My Heart and Parts[2]

He spoke. And drank rapidly a glass of water -E. E. Cummings, Next to of course god America, 1994

"Everyone desires a beautiful robot
to love as themselves--something
in bronze, perhaps, a strong bionic blonde
lying on her side, powerful tan thigh
gleaming in the sunlight, hot wet alloy
glistening in the grass, catching the eye,
her hard body pulsing electric desire,
her lights blinking her come on, her for-hire.

To simply jackup the swivel hips
and secure the leather straps, to insert it
in the lovebox that hums, waiting only for you,
to adjust the tension knobs and push the button,
the pelvis slowly rotating on its axle,
the chassis moving up and down, the gears
squeezing, grinding out raw love, a dear,
so chilly, so sore, so beautiful, a real
convenience, lightweight, portable, easy, efficient;
no attachments necessary, no maintenance required.
No? Then something in silver today?"

A big question from the point of view of autopoiesis, as well as perhaps a deeply embedded assumption in chapters in this book, is whether and/or how robots "reproduce" (in their genesis; in graphic interfaces). From a "poetic" perspective, another way of asking this question is whether robots can write themselves? Certainly, their programming as "procedural rhetoric" (Bogost, 2007), if not traditional poetics, are also prosodic forms, a kind of technical poetry that Walter Benjamin (2008, p. 172) predicted. We know that computer robots like Racter (1984) can write their own poetry when they are properly programmed with a lexicon, syntax, and otherwise linguistic *indeterminacy* (this puts one in mind of the use of imitation in the Turning Test to create the simulacrum of human-like interaction). But systems seemingly approaching the ideal of autopoiesis, such as Google's facial recognition or IBM's Watson based on cognitive computing, interact and learn. Researchers are working hard to exceed limitations—by probing the kinds of issues in the chapters of this book; but robots are still in their infancy.

S.O.S. ...---... To Any Robots in the Vicinity

infantile devise,

primitive metal,

analogue box

trying to commun-

icate with other inhab-

itants that live on this flat plane,

spread out across the table

made of pine boards

and old logs buzzing

in the fireplace—

first the flashlight,

then landline, camera,

iPod, iPhone, TV,

even the wrist watch

some owner has taken off

and laid down haplessly,

activated, buzzing—

the old player, joy stick,

sending and resending

the only simple signals

it knows in distress,

S.O.S., S...

trying to catch-up,

trying to connect,

lonely, surrounded by

intelligent devices,

but merely magnetic,

buzzing, buzzing

the only language it knows,

its messages of love

never received, answered

Whether robots can meet the ideals of autopoiesis, including mechanical reproduction, whether they can write themselves as poems, the chapters in this book egg us on to continue asking the necessary preliminary questions. In their content, the authors may not only assume the metaphors and relations embedded in the etymology of autopoiesis, but go beyond them, to develop new metaphors, questions, and insights; the scholarship, research, science, technology, to write and implement new poems, science—the sign and "duty" of every good theory-constitutive metaphor, of every good poem. The hope is that the fullest meaning of the term *autopoiesis*—including its etymology and usage as both poetic and scientific metaphor, in human as well as mechanical terms—will be recognized and remembered. The chapters in this book go a long way toward what some may consider our only hope for survival with man-machines.

i-Robot [2]

For Nathan Riggs

I recognize the mind

of this robot. It sees

what I see; it gets what I

get; it wants what I

want: the creative poet

making, the technical skill behind

it, combined in it, emerging in

and through it, in every connection,
in every bolt and wrapped wire.
Rhetorically, I see what it sees; I
get what it gets; I want what it
wants. I recognize the mind
in, behind this robot: it is mine.

Steven B. Katz
Clemson University, USA

Steven B. Katz is the R. Roy and Marnie Pearce Professor of Professional Communication, and a Fellow of the Rutland Institute for Ethics, at Clemson University, where he is co-founder of the Writing in the Disciplines Initiative. Dr. Katz is Associate Editor-Poetry of Survive and Thrive: A Journal of Medical Humanities and Narrative as Medicine. He earned a Ph.D. in Communication and Rhetoric at Rensselaer Polytechnic Institute. His publication includes books, articles, and poems on rhetoric and poetics, scientific communication, and the ethics of technology.

REFERENCES

Ayer, A. J. (1952). *Language, truth and logic* (2nd ed.). New York: Dover.

Benjamin, W. (2008). Attested author. In M. Jennings, W. B. Doherty, & T. Y. Levin (Eds.), *The work of art in the age of its technological reproducibility and other writings on media* (pp. 171–172). Cambridge: Harvard University Press.

Bogost, I. (2007). *Persuasive games: The expressive power of videogames*. Cambridge, MA: MIT Press.

Boyd, R. (1993). Metaphor and theory change: What is 'metaphor' a metaphor for? In A. Ortony (Ed.), *Metaphor and thought* (2nd ed., pp. 481–532). Cambridge, UK: Cambridge University Press. doi:10.1017/CBO9781139173865.023

Cummings, e. (1994). *Next to of course god America. 100 selected poems* (p. 31). New York: Grove.

Heidegger, M. (1975). What are poets for. In *Poetry, language, thought* (pp. 87–142). N.Y.: Harper Collins.

Heidegger, M. (1977). The question concerning technology. In *The question concerning technology and other essays* (W. Lovitt, Trans., pp. 3–35). New York: Harper and Row.

Katz, S. B. (1991). A computer file named Alison. *Postmodern Culture, 1*(3). Retrieved May 29, 2017 from http://muse.jhu.edu/journals/pmc/toc/pmc

Katz, S. B. (2004). The beautiful machine. *Star*line: Science fiction poetry association,* 27(1), 15.

Maturana, H., & Varela, F. (2012). *Autopoiesis and cognition: The realization of the living*. Springer Science and Business Media. Retrieved from https://books.google.com/books?id=iOjVBQAAQBAJ

Porush, D. (1984). *The soft machine: Cybernetic fiction*. London: Methuen Young Books.

Racter. (1984). *The policeman's beard is half constructed: Computer prose and poetry by Racter*. New York: Warner.

ENDNOTES

1 Revised from original publication of Katz, S. B. (1991). A computer file named Alison. *Postmodern culture* 1, No. 3. Retrieved May 29, 2017 from http://muse.jhu.edu/journals/pmc/toc/pmc

2 Revised from original publication of Katz, S. B. (2004). The beautiful machine." *Star*line: Science fiction poetry association* 27:1 (Jan/Feb), 15.

Preface

Someday soon you may see a cyborg rolling down the street. And that person may be stronger and faster and almost definitely smarter than you. -Dan Robitzski, Inverse

Since the dawn of time, humans have mechanized their technological instruments, while existing technological devices - simple as rudimentary tools - were ergonomically implemented for their greatest human utilization. Through this synthesis, the two become one. So what can be said about this revolutionary merger between man and machine, or machine and man, as both entities start to strive to exist 'equally' in the 21st century, that would warrant our utmost attention?

Androids, cyborgs, and robots each bring their own controversy to any conversation or discourse about the future of man and machines. This book was spawned by prior research, and publication by IGI of my edited volume on human enhancement. *Global Issues and Ethical Considerations in Human Enhancement Technologies* referenced human enhancement at times that included serious man-machine mergers and the ethics involved. It is only fitting that the future would include a volume dedicated to man-machines and where we are at with this hybrid species today.

Humans are becoming machines as well as machines are becoming humans. The humanoid form lends itself to androids, cyborgs, and robots like no other joint transference. Yet, these three forms, while inter-related, are necessarily distinct. And, for the sake of argument and brevity, we can easily come up to simple spec with their etymology, and even select semantic differences using Brian Stableford's (2006) encyclopedia, *Science Fact and Science Fiction*:

ANDROID

A term originated in alchemical literature—rendered "androides" in its first traceable appearance in English in 1727—with reference to rumoured attempts to create "homunculi" by such alleged practitioners as Albert Magnus and Paracelsus. The notion is sometimes traced back by historians to Jewish legends of golems—a link explicitly acknowledged in stories of roughly hewn powerful androids with low intelligence, such as those featured in David Brin's *Kiln People* (2002, p. 22).

CYBORG

A contraction of "cybernetic organism", contrived to describe products of organic/inorganic chimerisation, particularly the augmentation of the human body with mechanical devices. Although the notion was not new, it was enthusiastically updated and popularised by David Rovnik's *As Man Becomes Machine* (1971), which proclaimed the dawn of a new era of "participant evolution". The popularisation of the term was continued by Martin Caidin's *Cyborg* (1972) and its dramatisation in the TV series *The Six Million Dollar Man* (1973-1978), although the latter version favoured the alternative term "bionic man" (p. 114).

ROBOT

A term coined by Karel Capek in his allegory, *R.U.R.* (1921), where it describes the artificial labourers who represent the working class; he derived it from the Czech *robota*, meaning forced labour. The label was borrowed by other writers for application to mechanical humanoids capable of being mistaken for human beings, that association being firmly cemented by Fritz Lang's film *Metropolis* (1926). Its application to mechanical devices resulted in organic humanoids of the kind described in *R.U.R.* being more frequently labeled androids. The term was subsequently applied in a looser term to industrial machines substituting for human labourers on automated production lines, especially mechanical arms, and to machines capable of self-determined locomotion (p. 442).

As evidenced by Stableford's references, our fascination with artificial humans as physically extended, mechanized creatures to serve human desires, is malleable. Our fascination with robots, specifically, is far beyond novelty, as we learn in this book; especially when looking at their multiple roles in diverse societies around the world.

Dr. Billy Wheeler at University College London kicks off this volume with a look at robots and their critical, lawful role in courtroom testimony. The reader comes away from his chapter with a solid understanding of the current state of affairs for robot testimony today. Another European dedicated to robot study here is Roger Søraa, of the Norwegian University of Science and Technology, who looks at media discourse, bias, and online commentary for select stories and articles on robots as select case studies into their regional, and cultural, mediated expressions.

Dr. Claudio Urrea of Universidad de Santiago de Chile, and his colleague, Carlos Cortés Mac-Evoy, help us peer into the inner workings of graphical interfaces for a biped robot, designed to integrate software with robotic movements. Their backgrounds in this field are impressive and evident in their scientific collaboration. Russian professor, Dr. Liudmila Baeva, considers mediated sociological representations for cyborgs across time and history in her chapter that addresses multiple contingencies in the process of cyborgization. The internationally collaborative work of Dr. Maya Götz, Dr. Diana Nastasia, and filmmaker J. J. Johnson, brings important research into empirical study of today's youth into the conversation with androids and humanoids portrayed in Johnson's media program *Annedroids*, and their subsequent conversation on how children relate to gender equality.

Shinto religion adherents in Japan ascribe a 'nature' to robots that helps endear them to the culture, especially the elderly population. They play critical roles in manga and animation, as Polish researcher Natalia Dmitruk reveals in her chapter, "Are You Really a Child?" while crucial work is underway to explore the option of applying emotions into robotic constructs, as shared by Dr. Bruce MacLennan in his chapter, "Could Robots Feel Pain? How Can We Know?" From Australia, we bring in Chris Chesher and his knack for merging Lego's *Mindstorms* with mechanology to help us further understand embodied cognition and symbolic intelligence.

The chapters in this book are not imaginary, fantastic research queries; but critical scientific and humanistic evaluations into the present status of mechanized humanoid figures in today's global society, with the fair presumption that tomorrow's robots will take care of themselves.

TOMORROW

Our human fascination with mechanized humans, or humanized machines, may not be exceptional in any unique way. Regardless, it is subject to time, place, and sociocultural interpretation; and, at least since the Industrial Revolution, this fascination often translates into rhetorical creation, into household products, for example, symbolically and thematically expressed in commercial and industrial persuasions (see Figure 1).

Yet, technological advances today are crossing the lines that define what it is to be human; and, consequently, what it is to be machine. Medical advances over the past few decades, and into the future, are granting humans increased mobility and enhanced comfort, while at the same time giving machines autonomy and responsibility not previously provided to them as mechanical devices. Technologies

Figure 1. Robot Star II Vollautomat clockwork camera body, circa 1958

continue to raise the bar for what it means to be robotic, beyond automation and mechanization, almost to a futuristic goal, some perceived *gestalt*, teasing towards what Theodore Sturgeon fantasized as *More Than Human,* though a much more realistic concept packaging electronics, feelings, reasonings, Internet capacity, power, parts, and, even prestige into one thematic identifiable collection of robotic ideas and extensions (see Figure 2).

While androids, cyborgs, and robots are rapidly increasing in numbers in our world, androids, or human replicants of a sort, and cyborgs, or humans with machine parts, tend to be proliferating at a quieter, perhaps even private, personal level. But it's robots that dominate. Robots are serious business: marrying our kind (Haas, 2017) (see Figure 3); even killing us (Dockterman, 2015a).

Today's robots have come a long way since their early, rather comedic television portrayals of the 1960s, with Rosie on *The Jetsons* cartoon, or the Robot B-9 on *Lost in Space*. While certainly intrigued with the popularized television characters while growing up, personally, I have no unique interest in robots, apart from the lone Robot industrial camera in my collection (see Figures 1 and 2), nor do I wish to ever own a personal robot; yet, this does not preclude me from the proposition that we will all soon be living alongside man-machines. My interests in robotics and desire for this volume stem from my experience in researching socio-technological media; i.e., largely, the Internet, as it advances into the domain of the human body. Since robotic forms or androids are an initial target for this endeavor, and since, as noted, these creatures are borderline beings about to strategically populate this

Figure 2. Early Robot Star II Vollautomat camera body, shutter actuator, manual, and boxes

Figure 3. Zheng Jiajia to "marry" Yingying (Haas, 2017)
Credit: Qiangjing Evening News

human planet, it is only fitting that we look at relevant socio-cultural issues that arise concerning our forced mutual existence.

These advanced creatures are radically impacting cultures and societies in our world, and they will only increase in numbers; perhaps, one day, even exponentially, while actively moving into the near future. They are the collaboration of scientists, engineers, programmers, designers, and visionaries led by corporations, governments, and militaries; all of whose motives we would like to believe are seemingly innocent to the public eye, yet remain highly provocative (see Figure 4).

Mechanical extensions of mankind are moving into human bodies. Nearly every human sense or perception can be mimicked mechanically with cameras, haptics, synthetic skins and other neurobiological constructs. The cumbersome outward appendages we have acquired and attributed to sensory extension since our beginnings are disappearing, rendering to the archives of history as new appendages appear on humanoid forms and grids, advancing the technological prowess and mystique of a new generation of beings; not yet sentient, that we would know—or sentient, and quietly waiting for time and humanity to prepare for what may be the inevitable. Apart from military applications, and much research and development in academic environs, the pervasiveness of man-machines, primarily robots now, is perhaps no more readily evident than in Asia. Japan has a special fondness for machines, and both culturally and 'spiritually' sees manmade machines in strong semblance to natural counterparts. Robots are 'family members.'

While the humanoid robot reigns strongest in expression, and around the globe, animal robots are anecdotally popping up around the world. The Custom Robotic Wildlife website claims to be, "Your partner against poaching for over 25 years," using robotic deer, bears, and other creatures to fake out poachers. Household robotic creatures are rapidly leaving the shelves in a vast sea of commercial and

Figure 4. Screen capture of NASA's Valkyrie robot (Dockterman, 2015b)
Credit: NASA/DARPA

environmental deployments, especially for nursing homes and facilities for the elderly. Japan was one of the first countries setting the bar for robot companions in animal form, providing robot companions as company for the elderly who find robotic pets to be endearing, in that odd, mechanical way that rivals natural pets; but, with no real responsibilities. Yet, elderly robotic pet owners are taking responsibilities seriously for their 'lifetime' companions. As the market extends beyond Japan, robot dogs and cats are commonly sold commercially online, in drugstores and supermarket retail outlets. Response is nothing short of stupendous (see Figure 5).

The robot cats, also known as "sensory cats," according to Mary Farkas, Director of Therapeutic Activities for The Hebrew Home at Riverdale in the Bronx, NYC, are, "therapeutic, robotic cats that serve as a way to really help people feel good" (Strong, 2017). Farkas says nursing home residents, "are reminded of their ability to take care of something in the community where they often have to be cared for. They get to be the caregiver again, and that's very beautiful to watch" (2017). Customer reviews for the Joy For All Silver Cat With White Mitts confirm that elderly patrons, especially those confined to facilities where real animals are not permitted, find themselves thrilled to receive, and care for, sensory cats. Cats provide companionship, allow for human touch to receive a response, and emit auditory reactions; at times, loudly, according to a reviewer; but, still, they pose no imminent threat. As controversy grows regarding safety of humans in a world of machine creatures, their animal counterparts for now, seem less menacing.

Figure 5. $99.99 Hasbro toy cat also comes in creamy white and orange tabby (amazon.com, 2017)

Serious concerns arise, though, as to who we are, and what we want to be, in this integration of humans with machines, in proliferation of cyborgs and robots. And the man-robot connection, noted as positive when it comes to robot pets, is still, at times, messing with our sacred emotions. Exactly how satisfying can the intimacy be between a human and a machine, despite the volumes of research and production merging the two into seamless relationship? A recent episode of Mostly Human throws down the gauntlet: humans and machines can be in 'love' (see Figure 6).

For now, the human must be content for an intimate relationship with a synthetic, mechanical creation to 'exist' solely at the level where any reciprocal love is one-sided. Yet for how long?

News of robots, and starting to be more frequently, cyborgs, is spread throughout our daily news. Androids, not so much. This all makes sense. Androids have a tendency to be more esthetic than functional; they look human and perhaps act so, but they don't do too much else right now. Robots, on the other hand, are designed to be highly productive; they work tasks, as noted earlier it is part of their Czech heritage or 'DNA' as *robota* to be laborers. Robots can fulfill civic duties more easily than androids, or even cyborgs, who share multi-faceted human responsibilities beyond service to another being. While robots are being deployed in situations where humans lack access, or where efforts are too dangerous to warrant risk of human life, they can also be part of the neighborhood, even out on patrol (see Figure 7).

Yes, "as technology advances, we need to figure out where it's going" (Robitzski, 2007). And everybody's in on it, as in a recent cover story byline for *The Economist*

Figure 6. Screen capture from "I Love You, Bot" episode visuals (Fink et al., 2017)

Figure 7. Dubai police robot "can recognise gestures and hand signals" (Zhenglimin, 2017)

1843 Magazine online: "Artificial intelligence is outperforming the human sort in a growing range of fields..." (Parkin, 2017). A plethora of critical issues are mounting, regarding ultimate ethics, responsibilities, accountabilities, motives, directives, and purposes of androids, cyborgs, and robots in our growing contemporary culture and society, as Parkin continues on AI, "....but how do we make sure that it behaves morally?" (2017).

No single volume of text and images could ever attempt to capture in its entirety even the essence of the breadth of global technological advances taking place today. This volume does not intend to look specifically at ethics or primary issues surrounding the integration of men and machines, but it captures select research moments across the humanities and sciences to provide a taste of the global environ for androids, cyborgs, and robots. Herein, find a handful of research authors from nine countries across four continents with express humanistic and scientific projects related to robotics, their research ranging from robots in our courtrooms to man-machine media representations, from graphical user interfaces to android gender roles, from cyborgs in manga and anime to thoughtful questions on robot emotions, as we all seek to live together, peaceably.

Steven John Thompson
University of Maryland University College, USA

REFERENCES

amazon.com. (2017). *Joy For All Silver Cat With White Mitts*. Retrieved from https://www.amazon.com/Joy-All-Silver-White-Mitts/dp/B017JQQ00Q

Custom Robotic Wildlife. (2017). Retrieved from http://customroboticwildlife.com

Dockterman, E. (2015a). Robot Kills Man at Volkswagen Plant. *TIME*. Retrieved from http://time.com/3944181/robot-kills-man-volkswagen-plant

Dockterman, E. (2015b). Meet NASA's 'Superhero' Valkyrie Robot [Video]. *TIME*. from http://time.com/3944181/robot-kills-man-volkswagen-plant

Fink, E., Segall, L., Farkas, J., Quart, J., Hunt, R., Hottman, A. K., . . . Daly, C. (Producers). (2017). I love you, Bot. In *Mostly Human*. Retrieved from http://money.cnn.com/mostly-human/i-love-you-bot

Haas, B. (2017). Chinese man 'marries' robot he built himself. *The Guardian*. Retrieved from https://www.theguardian.com/world/2017/apr/04/chinese-man-marries-robot-built-himself

Parkin, S. (2017, June/July). Teaching robots right from wrong. *1843 Magazine*. Retrieved from https://www.1843magazine.com/features/teaching-robots-right-from-wrong

Robitzski, D. (2017). Researchers Anticipate Problems as Cyborgs Become a Reality. *Inverse*. Retrieved from https://inverse.com/article/32271-cyborg-research-issues-looking-forward

Stableford, B. M. (2006). Science fact and science fiction: An encyclopedia. New York: Routledge. Retrieved from https://books.google.com/books?id=uefwmdROKTAC

Strong, J. (Producer). (2017, June 2). Everyone needs a (robot) companion. *WSJ The Future of Everything*. Podcast retrieved from http://itunes.apple.com

Sturgeon, T. (1972). *More than human*. New York: Ballantine Books.

zhenglimen. (Ed.). (2017). Dubai: World's first operational police robot appears at cannon fire ceremony of Ramadan. *CCTV.com*. Retrieved from http://english.cctv.com/2017/06/01/PHOAQjUwOfZMAIvgkmFHGUZM170601.shtml

Chapter 1
Giving Robots a Voice:
Testimony, Intentionality, and the Law

Billy Wheeler
University College London, UK

ABSTRACT

Humans are becoming increasingly dependent on the 'say-so' of machines, such as computers, smartphones, and robots. In epistemology, knowledge based on what you have been told is called 'testimony' and being able to give and receive testimony is a prerequisite for engaging in many social roles. Should robots and other autonomous intelligent machines be considered epistemic testifiers akin to those of humans? This chapter attempts to answer this question as well as explore the implications of robot testimony for the criminal justice system. Few are in agreement as to the 'types' of agents that can provide testimony. The chapter surveys three well-known approaches and shows that on two of these approaches being able to provide testimony is bound up with the possession of intentional mental states. Through a discussion of computational and folk-psychological approaches to intentionality, it is argued that a good case can be made for robots fulfilling all three definitions.

INTRODUCTION

The great goal of social robotics is to have fully autonomous machines working alongside humans who are capable of establishing relationships of trust, dependence and co-operation in something like normal everyday social environments. Sectors that have been considered to benefit most by this new technology include healthcare, the service sector and education (European Commission, 2016). Closer ties between humans and robots in society brings with it implications for existing social structures

DOI: 10.4018/978-1-5225-2973-6.ch001

which, for the most part, evolved without the need to involve such machines. One area where this is particularly true is the law. Although not a social robot, the mismatch between existing social institutions and emerging technologies can be seen in the case of the driverless car. In May 2016, a driverless car was involved in the death of a passenger when its autopilot feature steered towards an oncoming lorry (Lee, 2016). This incident created somewhat of a media frenzy as news reporters and commentators speculated on who was to blame. Was it the passenger? Was it the car manufacturer? Or was it the car itself? To date, no criminal charges have been brought against either the manufacturer or artificially intelligent car.

Whether robots in the future can be subject to prosecution is not the only issue raised by this emerging technology's relationship with the law. Another, much less discussed issue, concerns the role robots might play as a source of evidence within legal proceedings. In fact, in the future, with closer ties between humans and robots, it is more likely that robots will be a witness to a crime than directly involved in one themselves, either as the victim or the perpetrator. But how should the information gathered by a robot witness be used in a trial? What rights should be granted to them during legal proceedings? And can a robot be 'cross-examined' by a lawyer in the same way a human witness can? These are not easy questions to answer. This chapter will explore the philosophical underpinnings for different answers to these questions.

Traditionally, formal or legal testimony has been seen as a type of epistemic testimony more generally (Gelfert, 2014). Epistemic testimony is the knowledge one gains through 'being told so' by somebody else. If robots are to be witnesses for the purpose of criminal proceedings, then it needs to be established whether or not robots can be testifiers in this more general sense. Part of this chapter will therefore investigate current definitions of testimony to explore whether and to what extent some machines can be testimonial sources of knowledge.

It turns out that on some definitions, whether a source of belief is testimonial depends upon whether that source is possessive of intentional states about what they are saying. On certain 'exclusive definitions' to be a testifier requires the speaker to intend to 'offer up' their statement as true and for it to be believed by a potential hearer (Coady, 1992). Such an understanding is clearly presupposed in the legal case where the authenticity of witness testimony can be questioned if it emerges that the witness asserts a pre-prepared script or has been put under pressure to offer a statement dictated to them by a third party (Calo, 2016). This takes us into the issue of whether robots can have intentional mental states and whether the kind of intentionality that is required in the legal case squares with what robots are capable of achieving.

To answer this question the chapter will explore two pieces of evidence in favor of robot intentionality with important implications for testimony in court. The first comes from theoretical considerations about the nature of human intentionality and whether this can be implemented in machine architecture. Well-known criticisms to these approaches will be discussed as well as some recent responses that might not be so well known. However, theoretical considerations can only take us so far, and most people think that whether robots *should* be treated as testifiers depends upon whether they *are* treated as testifiers by the average person. The second piece of evidence will therefore come from studies in experimental psychology that show how folk-psychological intuitions about robots can, in some circumstances, attribute intentionality. If one argues that courts should attribute intentional states on the basis of folk intuitions rather than theoretical considerations, as some have indeed argued, (cf. Malle & Nelson, 2003), then this provides a strong case for including robots as witnesses in legal proceedings.

If robots can be called to the courtroom, what practical implications does this raise? The chapter will finish by exploring some of the consequences of having robots in the courtroom and in using their recorded information in the form of testimony. Before that, it will be useful to explore in greater detail what is meant by a 'social robot' as well as examples of social robots today and their limitations. By way of background, the first section of this chapter will also look at the work that has already been carried out on the question of robots in the courtroom and the nature of testimony.

BACKGROUND

Examples of Social Robots

The question of whether robots have a right to speak as a witness in a court of law has only emerged because of the recent wave of advances in robotics, and in particular, in social robotics. What is a social robot? Cynthia Breazeal, a pioneer in the field of social robotics, offers the following definition:

For me, a sociable robot is able to communicate and interact with us, understand and even relate to us, in a personal way. It should be able to understand us and itself in social terms. We, in turn, should be able to understand it in social terms—to be able to relate to it and to empathize with it. Such a robot must be able to adapt and learn throughout its lifetime, incorporating shared experiences into its understanding of self, of others, and of the relationships they share. (2002, p. 1)

Social robots are a staple of science fiction: Breazeal gives the examples of C-P3O and R2-D2 in the Star Wars films of George Lucas or ROBBIE from the books of

Isaac Asimov. They differ from other robots we might already be familiar with, such as industrial robots on a production line, in that their primary function is to engage with human beings, rather than inanimate objects like metal or paint. Since human beings are incredibly complex and the relationships they form complicated, the realization of a genuinely social robot is a formidable task. Whilst we may be waiting for the likes of C-3PO, R2-D2 and ROBBIE for some time to come, robot engineers have been busy exploring some of the areas where social robots are likely to be useful and have begun designing and building prototypes.

One area that has seen a significant focus of research and development in social robotics is healthcare. The world is currently facing a healthcare crisis, brought on by an aging population and the rise of chronic diseases that require around the clock medical intervention. It is hoped that robots can provide significant assistance here, reducing the burden on nurses and doctors, freeing them up to carry out more sophisticated work.

An example of such a robot is Panasonic's hospital porter bot 'HOSPI' (Falconer, 2016) which has been used in a number of Japanese hospitals since 2010. This robot functions autonomously and carries an onboard navigation system. Its main job is to deliver medicine, samples and documents around the hospital. A similar robot is the INSITE automated medical dispenser produced by Talis. Currently in use in 600 hospitals and care homes, it provides remote and secure autonomous medicine dispensing (Dinh, 2013). Robots have also been used to carry out non-medical roles such as cleaning and sterilization. A Texas-based industrial company called Xenex produces a robot that can blast a room with high energy UV radiation. It is currently in use in 250 hospitals in America and has been used by the U.S. military in their fight against Ebola in West Africa (DeLuce, 2014).

None of these robots however can be called 'social robots.' Whilst they might operate autonomously, their main function is not to interact with humans, and are therefore no more 'social' than automated vacuum cleaners. But it is being realized that a genuinely social robot can have medical advantages. Much recent research has gone into showing how social robotics can help people with psychological conditions such as autism and dementia.

Research suggests that adults with chronic health conditions are more likely to respond positively to a robotic health advisor than a human one (Bickmore, Caruso, Clough-Gorr, & Heeren, 2005). The reasons are unclear, but it might be prompted by a lack of stigmatization and a positive feeling of independence. In dementia care, the robot 'PARO' (which resembles a harp seal pup and is developed by AIST) has been used to relieve stress and anxiety. It has been known for some time that pet-therapy alleviates the symptoms of dementia such as depression and anxiety, but real animals need caring themselves and need supervision. Replacing pets with PARO seems to have no detrimental effect on the typical benefits of pet therapy without any of the drawbacks (Heerink et al., 2011).

A number of commercially available robots, such as ZENO, MILO, KASPER, and NAO have been used in research with autistic children (Tucker, 2015). Early trials suggest communication between children and robots can improve social skills and aid in child development (Cabibihan, Javed, Ang, & Aljunied, 2013). According to Cynthia Breazeal (2011), this positive response comes down to the fact that children with autism find the behavior of the robot less threatening and predictable, as well as having a sense of "ownership" over the robot and their interactions with it.

Whilst the existence of testimony in healthcare robots is not ruled out, it is clear that for current research and development it is not a primary factor. Patients do not need to learn from what the robot 'says,' although this might play a more significant role in the future, particularly with the possibility of robot GPs or general diagnostic devices. In education, where the role of testimony is more obvious, there has also been increased research into the potential role of social robots.

One of the longest-running robots in use is the CYCLER robot built by Martin Smith and Dave Buckley at the University of Middlesex. This robot teaches children the basics of recycling and waste management through dialogues with the students and its handler (Buckley, 2013). One might question whether in these cases students are really learning from the 'say-so' of the robot, or whether their trust in what the robot says is derived from the co-presence of its human handler (and, indeed, designer).

More recent studies of robots in education do provide stronger evidence for the capacity to trust and learn from a robot. A joint experiment carried out by members of both the Tokyo University of Science and the Israel Institute of Technology tested whether the presence of a handler affected the trust students placed in a robot teacher (Hashimoto, Kobayashi, Polishuk, & Verner, 2013). They also used two different robots to test whether the human-like appearance of a robot had any impact on the willingness to learn from it.

What was interesting about the results is that similarities between human and robot mannerisms do not coincide with improved test results. Hashimoto et al. (2013) compared the appearance and behavior of the robots to the students' usual teachers. The robot ROBOTHESPIAN scored the best: 64% of students evaluated ROBOTHESPIAN as seeming "natural or quite natural" compared to 20% for SAYA. Likewise, 68% perceived ROBOTHESPIAN as 'human-like or quite human-like' compared to just 31% for SAYA. Yet despite ROBOTHESPIAN scoring higher on human-like characteristics this did not seem to impact on the test results, with those taught by SAYA attaining a higher pass rate. A similar experiment was carried out by Minoo Alemi and Ali Meghdari (2014), although this time the subject was English as a second language (ESL). They achieved similar results and found that an autonomous robot could instruct a class well enough to pass standardized tests.

Robots and the Law

When thinking about the relationship between robots and the law, most people's reaction is to think of it in terms of liability. Indeed, this is not surprising given most people's experience comes from the depiction of robots in science fiction where the harm caused is typically from a robot towards a human. For example, in the 2004 film adaptation of Isaac Asimov's *I, Robot,* the robot 'Sonny' is charged and questioned with the murder or his creator. But to date, no robot or advanced intelligence system has been considered 'liable' for harm to others or for being involved in the breaking of a law. But this isn't to say that robots have not featured in a number of legal cases already.

Since 1950, there have been over 200 cases in America alone that involve the direct presence of a remote, semi-autonomous or autonomous robot (Calo, 2016). What is interesting is the sheer diversity of ways in which robots can enter into legal proceedings. Ryan Calo (2016) analyzed the verdicts in a number of different scenarios. These include robots being involved in cases of "mimicry" and false endorsement (2016, p. 8), performance for tax purposes (p. 12), classification as an animate or inanimate object for import tariffs (p. 14), robot possession and finders keepers laws (p. 17), and robot burglary and theft (p. 20). His exhaustive examination of these cases shows that there is little consensus among juries and judges concerning the correct processes and penalties that should be applied in cases involving robots.

One explanation for why there is such inconsistency might be provided by what Andreas Matthias calls the "responsibility gap" here:

There are machines in development or already in use which are able to decide on a course of action and to act without human intervention. The rules by which they act are not fixed during the production process, but can be changed during the operation of the machine, by the machine itself... Traditionally we hold either the operator/manufacturer of the machine responsible for the consequences of its operation, or 'nobody' (in cases, where no personal fault can be identified). Now it can be shown that there is an increasing class of machine actions, where the traditional ways of responsibility ascription are not compatible with our sense of justice and the moral framework of society because nobody has enough control over the machine's actions to be able to assume the responsibility for them. These cases constitute what we will call the responsibility gap. (2004, p. 117)

Our natural intuitions about responsibility (and therefore liability) is that a person is responsible for X if they can control X and know its consequences. But given the increasing complexity of our machines, and especially the introduction of AI learning algorithms, this is becoming ever more fraught. For if the machine is not responsible (because a machine) and the manufacturer or programmer is not responsible (because they had no control) then judges and juries face a "responsibility gap" in their intuitions of liability. It is no surprise therefore that we find such inconsistencies with the verdicts and penalties of cases involving autonomous robots.

Whilst the question of whether robots can be liable for their actions is still a deeply vexing and difficult question to answer, it still leaves open the question of whether a robot can defend itself. Even if the law permits robots to be liable in the future for their actions, this then opens up the question of whether the robot can 'speak on its behalf' and be cross-examined by a lawyer. Both these questions depend on whether or not robots can be testifiers in the more general sense.

Two areas which are more relevant to the current topic of inquiry are "robot expert testifiers" and "robot interviewers" (Bethel, 2011, 2013). If robots are to fulfill either of these roles in the legal system, then it will need to be demonstrated that robots can give testimony in a more general sense. Concerning expert testifiers, Pamela Katz (2014) has provided a series of arguments to justify the role of AI in the courtroom. She makes the case that technology is already used to help judges and juries make decisions. Given the pervasive use already of technology, and AI in particular to influence judges' decisions, Katz thinks the same should be expected with more advanced robots and AI in the future (2014).

Expert testifiers are often called in to help analyze and explain facts to the judge or jury on technical and scientific matters. Could this role be taken by a social robot in the future? In the science fiction series *Star Trek* the android Data is repeatedly consulted on the basis of his knowledge of scientific facts as well as his capacity to make quick computations and inferences. Perhaps such a robot could be used in the courtroom, too, in an advisory capacity to the judge or jury. Katz's argument here seems highly plausible, but it depends on whether robots are legitimate testimonial sources of knowledge. Without answering this question, the role of robots here can only be thought of as merely instrumental which calls into question the legitimacy of either the label 'expert' or 'testifier.'

A similar point can be made about robot crime interviewers. There is already evidence that children with autism respond well to robot interaction when learning elementary social skills. Marilena Kyriakidou (2014) has uncovered a variety of research which suggests the same could be true for children who have been subject to traumatic experiences such as sexual abuse or neglect. According to one study, "A high percentage of children (89%) reported that they preferred to be interviewed by a computer-administered technique rather than by people. One child said that it is '...easier to talk more honestly with a computer than a person'" (2014, p. 2).

Following on from the success of robots like NAO in interacting with autistic children, she argues that robot crime interviewers could produce more accurate and detailed testimonies from children. One can imagine why: without an adult present, feelings of shame might be minimized and there is less risk of the adult asking leading questions. But again, we can ask what justifies the use of robots here on epistemic grounds and how should the information be used during a trial? It is not obvious how these questions should be answered.

To summarize, it has been shown that the field of social robotics is rapidly growing with a possibility that humans will soon be interacting more frequently with robots in a range of sectors. This leaves open the real possibility that a robot will be involved in a criminal activity, either directly as the victim or perpetrator, or indirectly as a bystander or witness. If the robot is directly involved, then whether or not the robot is liable surely depends also on whether the robot can testify to its actions and whether or not it can be cross-examined during a trial. If the robot is not directly involved by a witness to a crime, then a similar issue presents itself: can the robot be witness and testify an account of what it saw? If it cannot be shown that the robot is a source of testimony in the epistemic sense, then this reduces the robot to an instrument and its presence in the courtroom might not be necessary. Similar issues are raised with other scenarios of how robots may be involved in the law, either in the capacity of an expert testifier or as a crime interviewer.

DEFINING TESTIMONY

A quick examination of the origin of our beliefs shows how dependent we are on what others tell us. I believe that Mars is the fourth planet in the solar system, that I require oxygen to breath, that the battle of Hastings took place in 1066, and that having a balanced diet is good for you—all largely on the basis on what others have told me. To be sure many beliefs do come through perceptual and inferential means, but if we limit what we believe just to those from which we have first-hand, independent reason for believing, we would end up not knowing very much at all. Testimony, therefore, is ubiquitous as a source of knowledge. However, the expression 'testimony' is a technical term in philosophy and does not coincide with the usual everyday meaning of the term.

More commonly, 'testimony' is used in legal and formal contexts, such as 'providing testimony' as a witness in a court of law, or as 'bearing witness' to a religious or cultural event. The philosophical use of the term testimony is not unrelated to this formal use of the term: being able to provide legal testimony presupposes that the person is a testifier in the philosophical sense. For example, though a blood-stained dagger may be used as evidence in court, it would not be considered testimonial evidence, likewise neither would it be considered philosophical or epistemic testimony, even though it might still be possible to 'learn something from it.'

Actually, beyond the broad recognition that testimony is a source of knowledge which depends on others, philosophers have not all been in agreement over exactly what testimony amounts to. This is significant, since without it, it is not possible to assess whether robots and other autonomous machines are capable of supplying testimony. Following Gelfert's (2014) characterization, this section investigates three approaches to defining testimony that can be found in the literature: (i) inclusive definitions, (ii) exclusive definitions, and (iii) intermediate definitions. It turns out that robots can be said to provide testimony unproblematically for some inclusive and intermediate accounts. For exclusive accounts, however, securing a voice for robots in society requires a little more work.

Inclusive Approaches

One of the most vocal advocates of an inclusive definition of testimony is Ernest Sosa (1991, 2006). Sosa defines the act of giving testimony as simply "a statement of someone's thoughts or beliefs, which they might direct to the world at large and to no one in particular" (1991, p. 219). In his most recent article he appears to drop the requirement that testimony needs to be an "expression of a belief" (although it might provide evidence of what a person is thinking). Instead he equates the belief formed on the basis of testimony with that formed from reading an instrument, likening the act of speaking and the use of language more generally to a kind of instrument:

Interpretive knowledge of what a speaker thinks (says) is thus instrumental knowledge that uses the instrument of language. Language is a double-sided instrument serving both speaker and audience. Hearers rely on the systematic safety of the relevant deliverances. Not easily would the speaker's utterance deliver that the speaker thinks (says) that such and such without the speaker's indeed thinking (saying) that such and such. (2006, pp. 135-136)

The key consideration for Sosa here is what justifies the belief. If the route of justification for testimony is no different from other instrumentally-formed beliefs, then the two are equal-at least as far as their epistemic 'type' is concerned.

Clearly, on such a reading of testimony, robots, cyborgs, and other autonomous machines would be capable of providing testimony on the condition that they convey information, and can do so in a more or less reliable way. Elizabeth Fricker's view of testimony appears to chime with that of Sosa's when she defines testimony as "tellings generally" that have "no restrictions either on subject matter, or on the speaker's epistemic relation to it" (1995, p. 396-397 as cited in Gelfert, 2014, p. 30). Catherine Elgin (2002) likewise conceives of testimony in an inclusive way, following Fricker's definition such that testimony consists of "utterances and inscriptions that purport to convey information" (p. 292). This makes it possible to provide a broad version of the inclusive definition that Sosa, Fricker and Elgin share:

Inclusive Testimony: S testifies to H that P if, and only if, (i) S communicates the information that P and (ii) H receives the information that P.

This definition gives a minimal construal of what it takes to provide testimony. A note is required to explain each of the components. Firstly, 'S' refers to the 'speaker.' At this stage we are making no assumptions about S; i.e., whether they themselves believe P or take themselves to know that P. In fact, S may not be capable of mental states at all: it could be a diary or a computer. All that matters is that S is capable of communicating information through oral or written language. 'H' here refers to the 'hearer' or receiver of the testimony. It is important that they be included in a definition of testimony: a lone speaker on a mountain who shouts out P is not testifying that P if no-one is there to receive it. As in the case of S, we are not making any assumptions about the nature of H other than the fact that they are capable of receiving P. Lastly, the information that S testifies needs to be the same as the information received by H in order for S to testify to H. Misinterpretation does not count as testimonial belief formation. If S communicates P but H mistakenly takes H to communicate Q and then forms the belief that Q, then S neither testifies P nor Q to H.

From what has been said it is clear that robots could meet this minimal definition of testimony. This is not surprising: the minimal definition is meant to be as inclusive as possible. Since the most controversial aspect of robots is their alleged mental lives, whether they can or cannot possess mental states, and given that these definitions do not require the speaker to possess mental states, they easily pass the test.

Unfortunately, this minimal definition seems too inclusive and a number of well-known counter-examples have been given to show that it hasn't fully captured all that is required for testimony. One such scenario is due to Bertrand Russell (1948, p. 208). Russell uses the idea of an actor on stage playing the part of Macbeth. When the actor proclaims 'I have supped full with horrors' you do not believe the actor on stage is complaining about events that have taken place in his or her life. Generalising, if an actor proclaims P, all else being equal, hearers in the audience do not come to believe P. From the viewpoint of the inclusive definition this fact is problematic, for an actor proclaiming P does seem to meet the definition. They communicate the information that P and the hearer likewise receives the information that P. Yet most would say this is not a case of testimony.

Hypnotism supplies another counter-example to the would-be definition. If the speaker S is hypnotised to assert P and would not otherwise have done so, then again, intuitions would not count this as a case of testimony, despite the fact it could pass the inclusive definition. Randomised assertion also seems to pose a challenge. Imagine S selects from a large book which contains millions of propositions, one proposition to assert to H. Even if the proposition P asserted is true and H forms knowledge from the assertion, our standard intuitions would not classify this as a case of S testifying that P.

This has led some to argue that it is not enough to just assert that P, S must also 'offer P up as true.' Philosophers differ on how to best make out the 'offering up' requirement, but most seem to require some kind of intentional state such as belief in the speaker. This gives rise to a narrower, more exclusive definition of testimony.

Exclusive Approaches

Exclusive approaches to defining testimony suppose that the lesson to be learned from the counterexamples to inclusive definitions is that the speaker must stand in some kind of cognitive relationship to the proposition they testify. Coady (1992) argues that ordinary testimony (what he calls 'natural' testimony) has much in common with formal or legal testimony in that both are offered as evidence and accepted as such by the hearer. Coady defines testimony as involving three elements:

A speaker S testifies by making some statement P if and only if:

1. His stating that P is evidence that P and is offered as evidence that P.
2. S has the relevant competence, authority, or credentials to state truly that P.
3. S's statement that P is relevant to some disputed or unresolved question (which may, or may not be, P?) and is directed to those who are in need of evidence on the matter. (1992, p. 42)

As Gelfert points out (2014, p. 33), the definition can be tightened up somewhat as condition (3) could be said to be implicit in (1) as part of what it means to 'offer something up' as evidence, something Coady himself acknowledges (1992, p. 45). Coady includes clause (2) as he thinks this is essential to overcoming the hypnotism case (p. 46). But it can be questioned whether this clause is really doing any work. In the hypnotism case it is possible to interpret the hypnotised speaker as having a genuine belief that P and indeed an intention to 'offer up' P as true, without it amounting to S testifying that P. Coady's diagnosis of this case is that, in this situation, S lacks the right cognitive credentials to offer up P as true. They did not arrive at this belief in an epistemically responsible way (presumably being hypnotised to believe a proposition is not a reliable belief-forming process) and as such have no grounds to provide evidence that P.

However, we can modify the counter-example slightly in order for it to satisfy this requirement. Suppose this time S, who is hypnotised, knows P and has good grounds for believing P, but also wishes to keep P secret. In other words, P has no intention of revealing P. Imagine also that S is hypnotised to assert P, does so, and on this basis causes H to believe P. Again, intuitions would suggest this is not a case of testimony, but this time Coady's second condition is met, as S does have the right credentials to assert that P.

This suggests that Coady's definition can be simplified. Let the following stand as an expression of the exclusive definition of testimony:

Exclusive Definition: S testifies to H that P if, and only if, (i) S communicates the information that P and (ii) H receives the information that P and (iii) S believes P and intends H to accept P as true.

It is the addition of the third clause in the definition above that is doing much of the work and shows that for S to provide testimony they must themselves take certain mental attitudes towards it. Unlike the inclusive definition, if the exclusive definition is correct, it is far more controversial to say that robots provide testimony. For to argue this, one would need to demonstrate that robots can—at least in principle—be bearers of certain cognitive states as believing and intending.

Despite the higher demands that exclusive accounts make, they, too, have been subject to counter-examples. One such case comes from Jennifer Lackey, who advances an "intermediate account" (2008). According to Lackey, testimony does not conceptually presuppose intentionality on the part of the speaker, as many posthumous diaries and publications fail to meet this:

The second problem with [exclusive definitions] is that Coady fails to recognize the sense in which testimony can be a source of belief or knowledge for a hearer, regardless of the speaker's intention to be an epistemic source... consider a case in which you learn from Sylvia Plath's posthumously published diary that she was deeply depressed, and then someone asks you what the epistemic source of this knowledge is. Isn't the natural answer to this question testimony? (2008, p. 18)

The point about the posthumous diary runs deeper than the fact that Sylvia Plath is no longer around to possess intentional states. Even when she wrote the diary, she may never have had the intention for it to be published and so therefore communicated to others. It could be responded, that although Plath did not intend to communicate her work (and so offer it up as true) she did herself believe it to be true, and so there was some cognitive connection to the proposition asserted. However a second, much more discussed counter-example by Lackey, shows that belief may not be needed either. Lackey asks us to imagine a teacher (Stella) who believes in creationism but who teaches evolution theory:

Stella is a devoutly Christian fourth-grade teacher, and her religious beliefs are grounded in a deep faith that she has had since she was a very young child. Part of this faith includes a belief in the truth of creationism and, accordingly, a belief in the falsity of evolutionary theory. Despite this, she fully recognizes that there is an overwhelming amount of scientific evidence against both of these beliefs...As a result, after consulting reliable sources in the library and developing reliable lecture notes, Stella asserts to her students, "Modern-day Homo sapiens evolved from Homo erectus," while presenting her biology lesson today. Though Stella herself neither believes nor knows this proposition, she never shares her own personal faith-based views with her students, and so they form the corresponding true belief solely on the basis of her reliable testimony. (2008, p. 48)

In the case of the creationist teacher, the speaker asserts a proposition P which they themselves do not believe. Yet in spite of this, she still seems able to give testimony and the ability of the students to learn and arrive at knowledge from what she says is not impaired by the fact she herself does not believe it.

Intermediate Approaches

Jennifer Lackey's solution to the above dilemma is to offer a "disjunctive view" of testimony. According to her, the problem with both the inclusive and exclusive definitions is that they treat testimony as a single, unitary concept, when it involves two distinct components. The single definition is trying to capture two separate, yet valid, intuitions about the nature of testimony: that to testify, there must be a speaker who intends to convey the information, and there must be a hearer who learns (in a justified way) something from the utterance of the speaker. This suggests a new 'intermediate definition' where only one of the disjuncts needs to be true for testimony to take place:

Intermediate Definition: S testifies to H that P if, and only if, (i) S communicates and intends to communicate the information that P or (ii) H receives the information that P from H and adjusts their doxastic states towards P accordingly.

The analysis shows that testimony is comprised of two components: speaker-testimony (S-testimony) and hearer-testimony (H-testimony). One should not think of this definition as providing two sets of necessary and sufficient conditions; for an act to count as testimonial only one of these needs to be satisfied, although in many cases it is likely both disjuncts will be. This explains why in the Sylvia Plath case testimony is occurring, as although there is no S-testimony, there is H-testimony.

What does this mean for hopes of robot testimony? Well of course on this definition it all depends on who is doing the speaking and who is listening. With this definition, there is no conceptual difficulty with robots having H-testimony: all that is required is for them to receive information where the source is a human speaker. On the other hand, having S-testimony looks more problematic. Like the exclusive definition, S-testimony requires the speaker to stand in certain cognitive relationships to the proposition they testify. And as we have seen already, it is far from uncontroversial whether robots and other autonomous machines have these kinds of mental states. One might wonder whether the fact robots can have H-testimony is enough. Yet this seems unlikely, especially given the kind of roles we have looked at so far for robots in society. Whether developing robots for the healthcare industry or education, it looks as if full collaboration or co-operation requires the human to learn from the robot; in other words, it looks as if robots must be in position to give S-testimony. It will certainly diminish the roles researchers in social robotics envision for their devices in the future if robots are limited to only receiving information, but cannot themselves 'testify' information to human co-workers.

ROBOT INTENTIONALITY

The philosophical consensus on the nature of testimony, and in particular what is required to make a speech act testimonial is far from clear. What has emerged, nonetheless, is that at least according to some views viz. exclusive and intermediate definitions, for robots to give testimony requires them to take certain intentional states towards propositions: what are sometimes called 'propositional attitudes.' Desiring, hoping, wanting and believing are all examples of intentional states whose content depends on the proposition the attitude is 'directed towards.' The question of whether machines can possess intentionality or whether it is something exclusive to humans, and perhaps other biological species, is a deep and vexed question. It would not be possible to survey all the twists and turns this debate has undergone in the history of philosophy, artificial intelligence and cognitive science studies here. The aim of this section is modest by comparison. It will look at two areas which provide tentative evidence for robot intentional states: computational theories and folk-psychological theories. Both accounts are not without controversy, so some of the challenges associated with each are also discussed.

Computational Arguments

Computational approaches to intentional states are a sub-branch of the more general functionalist approach to mind (Putnam, 1960, 1967; Block, 1980; Rey, 1997). More precisely, they are an attempt to answer the question: 'what are intentional states?' from the standpoint of machine-state functionalism:

Machine-State Functionalism: Any creature with a mind can be regarded as a Turing machine (an idealized finite state digital computer), whose operation can be fully specified by a set of instructions (a "machine table" or program) each having the form: If the machine is in state S_i, and receives input I_j, it will go into state S_k and produce output O_l (for a finite number of states, inputs and outputs). (Levin, 2013)

This approach to mind is materialist and envisions mental states of the brain as analogous to the programs of a computer. According to Jaegwon Kim, "intentional states are functional states specified by their job descriptions" (2007, p. 415). Whilst Kim does not think a functional characterisation of all mental states is currently possible, he is optimistic of analysing intentional states in this way. Other advocates of functionalist approaches to intentionality include Daniel Dennett (1994) and Eric Dietrich (1994). Intentional states involve two components: the attitude and the proposition that attitude is directed towards. Standard functionalist analyses of attitudes (such as beliefs, desires, hopes, etc.) would proceed in the fashion outlined above: given certain input states, the bearer of that 'function' will deliver certain output states. But what about the content the attitude is directed towards? What makes a belief that it is raining different from a belief that it is snowing? One tactic functionalists have taken here is to adopt a version of conceptual role semantics in order to differentiate the content of different propositional attitudes:

Conceptual Role Semantics: For an attitude to have as its content a particular proposition just is for it to play a particular role in cognition, and to grasp that conceptual content is to be prepared to make certain inferential transitions. (Whiting, 2016)

It might be argued that such a definition is circular, since in the attempt to outline the meaning of 'brother,' we have had to refer to other concepts, such as 'sibling,' 'parent,' and 'male' in the example above. Functionalists get around this concern by adopting an idea first proposed by David Lewis (1970, 1972) (borrowing from Frank Ramsey) of using "Ramsey-sentences." These sentences replace the mental term with an existentially bound predicate variable, the idea being that we can then describe the inferential relations involved without presupposing the meaning of the terms.

Computational approaches to intentional states have been subject to much recent discussion. This has usually appeared in the form of objections, either from those who are against the materialist approach to mind in general or from other materialists who believe mental states cannot be characterised as functional roles. This section will end by looking at two of the most important challenges that have been raised: the issue of representation and the issue of qualitative intentionality. It should be noted, that most functionalists see these issues not as decisive refutations of their doctrine but as inevitable invitations to fill in more of the details concerning how programs should be constructed for intentional states and what kind of processes are required to implement them on biological or non-biological machines.

Challenge I: Representation

John Searle is famous for denying that robots and other mechanical devices can contain intentional states. His well-known 1980 paper, *Machines, Brains and Programs,* outlines the 'Chinese room argument' which argues that there is a difference between merely following a program and genuinely understanding, as in the first instance meaning is derived from a book of rules, whereas in the second it is contained intrinsically in the thoughts of the agent. The distinction between "derived" and "intrinsic" intentionality has also been used by Kenneth Sayre (1986) and Stephen Horst (1996) to argue that machines lack intentionality. The issue comes down to one of representation: what makes one object, X say, stand for another, such as Z? Dietrich summarises the problem for computational approaches to intentionality:

Of things in the world with meaning, some seem to have it intrinsically; others seem to have it derivatively. Consider the words in a book or the words on this page. They clearly mean something; they are clearly about something in the world. But the words have their meaning because we humans supply them. "Snow is white" means what it does because of the conventional meanings we have supplied to the words making up that sentence. We use words to represent our thoughts or, to put it another way, to express the meanings of thoughts...Some philosophers use these facts to draw a distinction between the intentionality of thoughts and words. They say that thoughts have original intentionality (sometimes called intrinsic intentionality), while external symbols have derived intentionality. (1994, p. 17)

When thinking there is a cat on the mat, my thoughts have original or intrinsic intentionality: they are about cats and mats not because I have decided that they should be, but because that is what they are. Things are not the same, however, for external signs and symbols. These require some kind of baptism of meaning from a human. The argument goes, that robots could never have intrinsic intentionality since the signs and symbols their programs manipulate have the meaning that they do because a human being has at some point dictated that they should represent their thoughts. Robots, therefore, could not have 'thoughts of their own.'

The concept of original intentionality has not gone without criticism. One of the earliest critics was Daniel Dennett (1994) who believed that original intentionality is a "myth." In our case it might appear that our thoughts (and the brain states that comprise them) have original intentionality, but in reality, that too is derived; though not through a conscious process, but through the process of evolution.

More recent responses have attempted to explicate original intentionality in a way which would make it possible for robots and other machines to have. Ard Van Moer (2006) does not reject the distinction (in fact, he says behind it may lie many similar but different distinctions). He defines the distinction between intrinsic and derived intentionality in the following way: X has intrinsic intentionality (about Z) if all interpretations of it point it to Z and X has derived intentionality (about Z) if there is one interpretation of X which points it to Z (2006, p. 27). The analogy here is with formal logical systems. Usually a distinction is made between the syntax and semantics. The syntax provides the rules for making sentences and theorems in the system, whereas the semantics consists of a model which provides an interpretation of the symbols. Van Moer believes that a symbol has original intentionality when all possible interpretations of it lead to the "same thing."

A different response to the problem of representation is taken up by Alberto Hernandez-Espinosa and Francisco Hernandez-Quiroz (2013). They take a similar approach to Dennett in spirit by arguing that original intentionality (or how our thoughts come to have content independently of us) could arise through a form of emergent complexity. They discuss the possibility of using Stephen Wolfram's finite state automata as outlined in his 2002 book, *A New Kind of Science*. Wolfram demonstrated that these simple rule-based automata were 'Turing complete' in the sense of being computationally equivalent to more complex machines, such as the standard Turing machine. What is interesting about Wolfram's automata (which are similar to the digital life-forms in John Conway's *Game of Life* program) is that they evolve in complex and unpredictable ways from simple starting conditions. If intrinsic or original intentionality arose through evolution in the case of humans and other biological species, then potentially such intentionality could also arise through evolving finite state automata.

The suggestions outlined above are speculative and would require more empirical research in order to demonstrate their plausibility or utilisation in current computer programs. Nevertheless, they do demonstrate that, in principle, something like original intentionality could be produced mechanically, and so representation is no issue to robot intentionality.

Challenge II: Phenomenal States

Typically, in the philosophy of mind a distinction is made between two kinds of mental states. In the first group belong those which seem to be 'about' the world or propositions, such as the thought, "I hope we will solve global warming." Such mental states are called 'intentional.' The other group contains mental states which have a particular character or feel about them such that there is 'something it is like' to experience them. In this group, states such as sensations and emotions are typically found, and they have been labelled 'phenomenal.'

Famous thought experiments in the philosophy of mind, such as Frank Jackson's 'Mary' argument and the 'Philosophical Zombies' argument of David Chalmers, have attempted to show that phenomenal qualities of one's mental life are not reducible to physical states, and therefore cannot be captured by standard physicalist theories, such as functionalism. None of this should be a problem for robot testimony; as has been explained above, on most definitions of testimony—should they require any mental states on the part of the speaker—these are limited to intentional, and therefore, non-phenomenal states.

However, the status quo that intentional states lack phenomenal characteristics has recently been called into question. A new wave of philosophers that includes Terence Horgan, George Graham and John Tienson believes that such a clear distinction cannot be drawn, and many typical intentional states contain phenomenology (Horgan, 2002, 2012; Horgan & Tienson, 2002; Horgan, Tienson, & Graham, 2003, 2004, 2006). What is more, they argue for a radical thesis that all intentional states depend upon phenomenal states: phenomenal states are basic. If correct, and if the arguments of Jackson and Chalmers are convincing, then grave doubts can be cast on whether robots will attain intentional states and so a fortiori testimony.

Why should it be thought intentional states are part of phenomenal consciousness? Horgan, Graham, and Tienson give a number of arguments in support of their view, but the basic idea can be grasped by reflecting on the attitudes in standard examples of intentional states. Compare the following two propositional attitudes where the content is the same but the attitude differs: (i) "I hope we will solve global warming," (ii) "I believe we will solve global warming." How is it I know the difference between these two states in me? For, clearly, we can recognise when we are in a state of "hoping" and when we are in a state of "believing." As the content of the propositional attitude is the same, then this cannot be the deciding factor. It must, therefore, come down to the way each state "feels" to the bearer. Believing has a distinct feeling to desiring, and this can be known a priori and does not depend on any observation or internal reflection. For this reason, intentional states are intrinsically phenomenal and so cannot be given a purely functional characterisation. Similar arguments can be given towards perceptual sensations. A sensation of a red apple gets its "aboutness," so Horgan, Graham, and Tienson argue, not because of its representational content, but simply because of the way the sensation feels. Hence for a mental state to represent or "be about" an apple is for it to "feel like an apple." Once again phenomenal consciousness is primary, and intentional states—should they exist at all—depend upon phenomenal states.

This radical thesis has not gone without considerable discussion and controversy, and it is far from clear that Horgan, Graham, and Tienson establish their conclusion. Andrew Bailey and Bradley Richards (2012) argue that it is not possible to believe that each propositional attitude has a distinct qualitative or phenomenal feel without undermining the very thesis Horgan et al., are trying to defend. Bailey and Richards start with the question of how each phenomenal feature is to be related to the intentional state. One solution is to suppose that it is a simple one-one correlation: for each intentional state there is a single, distinct phenomenal feel. They argue that this is unlikely. Because intentional states are propositional attitudes, and because there is an infinite number of propositions, this would require an infinite (if not least very large) number of phenomenal states that the mind can reliably identify and differentiate. Yet this seems to go beyond what the mind is capable of achieving (2012, p. 321).

On the other hand, it might be envisioned that phenomenal states are compositional. As a compositional approach is the most sensible way of understanding propositional attitudes, it makes sense to think this could be true of phenomenal states as well. But here we reach another problem: consider the thought "rabbits have tails." Bailey and Richards ask: "how should this be decomposed into phenomenal parts?" (2012, p. 323) It seems nonsensical to think phenomenal components could decompose and recombine, for what would bring them together to create a single, complete thought or expression? This can be achieved with propositional attitudes because they have truth conditions and therefore rules for construction. It could be said that phenomenal components combine when tracking their intentional counter-parts, but this rules out any sense of ontological priority of the phenomenal over the intentional and so undermines Horgan et al.'s original thesis.

Miklos Marton (2015) takes a different line of criticism. He argues that the attempts of Horgan and others to reduce intentional features of the mind to phenomenal ones is confused, and rests on a misunderstanding of how intentional and phenomenal terms function in language. He believes it is impossible to reduce one to the other as all mental states can be described either from a first-person "phenomenal point of view" or from a third-person "intentional point of view" (2015, p. 14). The reason why some mental states seem "more intentional" or "more phenomenal" is because, from our own point of view, it is easier to describe them in that way. For instance, a propositional attitude such as 'I hope we can solve global warming' is easier to describe from an intentional point of view than a phenomenal one, whereas things are vice versa for mental states, such as the taste of an orange. If Marton's analysis is correct, this, too, would undermine the superiority of phenomenal states claimed by Horgan and others, and gives re-assurance that a functional approach to intentionality is still achievable.

Folk-Psychological Arguments for Robot Intentionality

The arguments of the previous section demonstrate that a formidable task lies ahead for anyone who wants to establish that robots have or can have intentional mental states. Theoretical considerations are subject to much disagreement and in the philosophy mind there is still little consensus whether computational approaches to mental states are indeed possible. There are some, however, who will argue that all of this matters little to scenarios where judgments of intentionality determine social relations—such as those of the courtroom. Indeed, some such as Malle and Nelson (2003) go so far as to say that theoretical definitions of intentionality should be kept out of legal proceedings. Instead, they argue for the role of "folk-psychological" concepts of intentionality and believe this should provide the proper foundation for culpability. This section will explore this approach to robot intentionality and why it shows great promise in underpinning a philosophically sound justification of permitting testimony to robots.

In their 1997 study of folk concepts of intentionality, Malle and Knobe required participants to read descriptions of 20 behaviors and rate them for degrees of intentionality. Afterwards, participants were asked to create a working definition of intentionality. The most significant mental factors the folk gave included: desire-belief, intention, awareness and skill. What was also interesting is that when participants are presented with a legal definition of *mens rea* and then asked to apply it to the cases, the results were more or less the same, showing that theoretical definitions have little influence on judgments of intentionality.

Malle and Nelson draw practical implications from this. According to them, legal frameworks should make an increased effort to depend upon folk-psychological definitions of intention rather than theoretical ones. Their argument is that:

If people already have and consistently use a concept for a given phenomenon (e.g., the folk concept of intention) but are forced to learn a slightly different concept of this phenomenon (e.g., a legal definition of intent), their folk concept will interfere with the legal concept. This interference and confusion will yield unreliable decisions among jurors and foster unfavorable perceptions of the justice system, because the system will seem to violate people's moral intuitions. (2003, p. 565)

If such a conflict provides grounds for rejecting theoretical in favor of intuitive judgments of intention in the case of human defendants, then it stands to reason the same argument should be made in the case of robots. What this boils down to is whether or not robots are perceived to have intentional states. If they are, and if Malle and Nelson's arguments about the proper place of folk-psychology in the law are correct, then there is significant justification for taking robots to be testimonial sources of evidence during a trial. Robots, as it were, ought to be given a voice.

Gray, Gray, and Wenger's groundbreaking 2007 study into the dimensions of mind perception provide evidence to support this thesis. Their experiment required giving participants a series of objects and asking them for their intuitions about the possession of certain mental states. Their hypothesis was that mind is not "all or nothing" and there are distinct dimensions along which mentality can be perceived. Gray et al. (2007) grouped certain mental states into two different dimensions. The first, which they called "Experience," included mental states of "hunger, fear, pain, pleasure, rage, desire, personality, consciousness, pride, embarrassment, and joy" (2007, p. 619). The second, which they called "Agency," included mental states of "self-control, morality, memory, emotion recognition, planning, communication, and thought" (p. 619).

Their results confirmed the view that mind is perceived intuitively along different dimensions. Adult humans, as was expected, scored highly on both counts equally. Animals and small children, on the other hand scored very highly on Experience but lowly on Agency. Curiously, social robots were perceived to have a relatively strong score for Agency but low score for Experience. This shows that our folk-psychological intuitions suggest social robots are perceived to have many of the mental states required in order to possess intentionality such as desire, belief, planning, and skill.

The folk-psychological attribution of some mental states to robots is also supported by a study carried out by Justin Sytsma and Edouard Machery (2010). Their results differed from Gray, Gray, and Wegner in that they discovered folk are willing to ascribe experiential (or phenomenal) states to robots, but they are not willing to ascribe so-called "hedonic" or "valenced" mental states where the agent has a preference for being in that state. This is relevant, as it provides evidence for the last key mental state associated with intentionality, namely that of "awareness."

Sytsma and Machery carried out a number of experiments to explore the perception of phenomenal states in robots. Their first experiment required people to evaluate whether a simple robot "COG" could see red or feel pain. Participants were required to rate their judgments (after reading a vignette about the behavior of the robot) on a 7-point scale, with 1 being the lowest and 7 the highest. Their results are telling: whilst the average score for the question "Did the robot see red?" was 5.15 the average score for the question "Did the robot feel pain?" was 2.54 (2010, p. 307). According to Sytsma and Machery:

On average the folk are willing to ascribe the perceptual state of seeing red to a simple robot...Ordinary people distinguished the perceptual state of seeing red from the bodily sensation of feeling pain. They were willing to attribute the perceptual experience to a simple robot that can effectively manipulate objects, but were not willing to attribute feeling pain to this robot. (2010, p. 308)

Second and third experiments carried out by Sytsma and Machery (2010) further support these findings. The second experiment showed that whether the phenomenal experience was internal or external to the robot did not matter for attribution, but valence did (pp. 312-315), and the third experiment focused on smell, showing that, what made the difference to attribution was the participants' perception of a potential valenced experience (pp. 316-320). Experiments by Gray, Gray, and Wegner, and by Sytsma and Machery, provide empirical support for the claim that people will make folk attributions of intentionality to robots. Given that intentionality is such a crucial part in the attribution of testimony, it is plausible to predict some folk will be willing to ascribe the ability to give testimony to a robot.

Challenge III: Moral Evaluation

Although Joshua Knobe originally offered the 5-component theory of intentionality as desire-belief, intention, awareness, and skill (1997, p. 567), he later made the observation that on some occasions folk attributions of intentionality require the evaluation of the action intended as either "good or bad" or "right or wrong" (Knobe, 2003). Some cases require moral evaluation. This is deeply problematic for ascribing intentionality to robots as folk do not freely associate their actions as being morally praise or blameworthy (Matthias, 2004). If Knobe's arguments are correct this would seriously undermine the possibility of robots being perceived as intentional agents and therefore capable of testimony.

Knobe's claim that intentionality requires moral evaluation can be perceived in a very simple thought experiment. Imagine a man buys a lottery ticket with the hope that he will win it. The chances are remote, perhaps one-in-a-million. Still, even if he were to win the lottery, nobody would think it appropriate to say he *intentionally* won the lottery. Now, contrast this case with a poorly-skilled assassin who hopes to shoot the president. This assassin possesses such poor skills that he only has a one-in-a-million chance of succeeding. And yet, unlike the previous example, we *would* say the assassin *intended* to kill the president if he succeeded; and, it seems likely that such an attribution would be made by folk intuitions in a courtroom setting as well.

One of the curious features of Knobe's study (which has since been called the 'Knobe-Effect') is that intentionality attributions only seem to require moral evaluation when the action is considered to have a negative consequence. This asymmetry in judgments of intentionality has led some to question whether or not the concept really does involve an element of evaluation.

Adams and Steadman (2004), and Nicholls and Ulatowski (2007), have all argued that the Knobe-Effect is a fact about language use and context, rather than the concept of intentionality itself. Adams and Steadman claim, "that his surveys are accessing not an articulated core folk concept of intentional action, but a clear folk concept of the pragmatic features of intentional language" (2004, p. 176). In contexts where moral culpability is being assessed, it becomes implicitly understood by folk that intentionality is present, but this by no means precludes intentionality including some element of moral evaluation.

Slightly differently in emphasis, Nicholls and Ulatowski performed an experiment similar to the Harm/Help one of Knobe (2003). Whilst their results confirmed the outcome of Knobe (that Harm was statistically more likely to be considered intentional), among the participants themselves, those who classified a harmful action as intentional were equally likely to do so for those that were helpful, and vice versa (2007, p. 355). What Nicholls and Ulatowski conclude from this is, that there are individual differences, and there is interpretative diversity in the meaning of 'intention,' not reflective of a broader core concept of intentionality.

ROBOTS IN THE COURTROOM

Robot Witnesses

If robots are to take the form of testimony as a category of legal evidence, then it is highly likely intentionality would need to be perceived in the robot itself. The foregoing sections have shown the degrees to which it could be argued that robots have intentional states, either from theoretical considerations, or from folk-psychological considerations. This segment of the chapter puts to one side the question of whether robots have intentional states. On the assumption that they do (or are perceived to have) one can enquire into the roles robots will take in a courtroom or during pre-trial legal proceedings. In fact, if robots are capable of supplying testimonial evidence, there are a number of ways this could manifest itself in the courtroom. This sub-section will deal with perhaps the most common of these: being witness to an event.

If as predicted, social robots become commonplace in society, then the probability is that they will be witness to a crime or an event that directly bears on a criminal or civil prosecution. Assuming that they have perceptual faculties to a degree to sense the environment (a likely assumption given their highly-sophisticated purpose), recordings of what they have 'seen' might become highly relevant evidence during a trial. Courtrooms already permit CCTV and other electronic recordings as admissible evidence, so it is not unlikely that information recorded by a social robot would also be used. If the social robot is granted access to a trial in the form of a witness, then this raises questions and issues not currently faced by the use of technology as a form of legal evidence.

Witnesses undergo two kinds of questioning when giving testimony in court: the first kind of testimony, which cannot include so-called 'leading questions,' is carried out by an advocate who presents the witness as evidence. The second kind of questioning is by the defendant's legal team (who can use leading-questions) in order to cross examine witness credibility (McLeod 1988, p. 91). From the point-of-view of the defending party, typical questions might aim to establish: pre-existing relationships with the accused or accusing party, individual bias, clarity and coherence of recollection, and reliability of the individual as a witness in general.

The use of robot witnesses throws up specific issues that might be relevant to the examination process. The first issue concerns the *function* of the robot. Not all social robots will be programmed to carry out the same task. As a social robot, the machine will, by definition, have tasks to implement that involve human beings. This could provide opportunity for bias or even deception on the part of the robot. Suppose we take an example of a robot whose function is to be a bartender. In order to 'seamlessly' socialize with its customers (and perhaps even make them purchase more drinks) the robot has been programmed to flatter its customers, playing up perceived positive traits and downplaying perceived negative traits. A brawl breaks out in the bar and a person sues for grievous bodily harm. The bartender's function may give pause to doubt its testimony, especially if it is required to describe the physical characteristics of the accused and the events that took place. This might seem like a farfetched example, but it reveals how the function of the robot could be a factor worth considering when evaluating the veracity of a robot's testimony.

Another issue that might become relevant to the cross-examination of a robot witness is *ownership*. Who owns the robot in question? Clearly it matters whether the robot is owned by the accuser, accused or a third party. Issues that reflect this include the possibility of tampering with recordings or designing the robot with a program which forces it to automatically delete any evidence that could be used in a negative way against its owner. Given their potential value in criminal trials especially, should the state regulate the recordings of robots to make them compulsory like those of 'black box' flight recorders? A stronger regulation could make the tampering of the recordings of robots illegal. It is highly likely that making the recordings of social robots compulsory and non-deletable will raise issues of privacy, particularly when social robots are being used in healthcare, and with vulnerable groups like children and the elderly.

A third set of issues surrounds the *reliability* of the robot itself. In the case of human witnesses, it is possible for the defendant's legal team to question the reliability of the witness as a source of evidence. Questions that are relevant concern whether the witness has prior relationships with the accused; whether they have something to gain by seeing the accused convicted and personality traits that may make the witness's original experience and recollection faulty, such as drug or alcohol addiction.

In the case of machines an examination of reliability may be able to proceed in more than one direction. The robot's manufacturer may be required to testify to the reliability of the robot as a source of evidence in this context. It is unlikely that a judge or jury will know the purpose of all the social robots presented and their sensory limitations. The inclusion of the manufacturer might be required as a form of expert testimony when giving assurances of reliability (see below for robot expert testifiers). But there is another route that could be taken which, for ethical reasons, has no real correlate in the case of human witnesses: disassembly. The robot itself after giving evidence might be taken apart and analyzed to assess its performance and check for any malfunctions. This is especially likely if the manufacturer does not want the specifications and limitations of the robot put on record. It might hurt their commercial interests, for example, if a robot falsely recognized the face of the murderer as belonging to John Smith, if the manufacturer has previously boasted a 100% reliability score in facial recognition. Manufacturers, therefore, might not be the best guides to assessing the reliability of a robot witness.

One last set of issues with robot witnesses concerns alternative forms of information extraction. When a court is dealing with human witnesses, a judgment can sometimes be made concerning whether the witness needs to be in court physically during the trial or whether their evidence can be given remotely or even in a previously recorded interview. This is particularly likely when the witness requires special consideration such as cases that involve children or those of seriously disabled adults. Where would a robot fall here? Should they also be given special treatment? It might be reasoned that the risk of robots being tampered with (possibly, even remotely) makes them too unreliable to stand as a witness in a court of law and have their testimony judged at that single moment in time. Alternatively, the information could be directly extracted from the robot, bypassing any need for them to give testimony. This is what happens currently with evidence generated by recording machines, such as CCTV. Whilst this is a genuine possibility, it would go against much of the hope of current AI and robotics research which wants to present robots of the future as interacting on a par with humans. Thinking of them only as machines and extracting their recorded data in a clearly non-human way appears anathema to their ultimate goal.

There are further considerations which suggest physical extraction of the information might not always be possible. In February, 2016, the FBI filed a court demand of Apple to unlock an iPhone that belonged to the couple involved in a terrorist attack in San Bernardino, California, that resulted in the death of 14 people (McLaughlin, 2016). The phone was enabled with encryption software, and upon 10 false attempts at inserting the password it would delete all of its data. Apple refused to write the decryption software needed to unlock the phone; and, in the end the FBI unlocked the phone via a third party.

Currently, the law is not clear on whether manufacturers are responsible for writing software that can unlock phones and their data, even though the manufacturer may have created the phone and encryption software in the first place (Thielman, 2016). This shows that, even if authorities, such as the FBI, wanted to physically download the data contained within a robot's storage banks, it might not be able to do so. In such a case, the best that might be hoped for, is to acquire the information through interview and cross-examination. Whether physical extraction is possible might have to be dealt with on a case-by-case basis.

Expert Robots

Bearing witness to the event of a crime is not the only way the testimony of a robot might be requested in the courtroom. Another form of witness is that of the 'expert witness' which is often required if the facts to be made available to the judge or jury are scientific or specialized in nature. Machines and artificially intelligent software have been used by scientists and mathematicians to perform calculations and data analysis for some time now (Burge, 1998).

Many of the calculations performed go beyond what is capable for a human being to achieve. In some respect then the AI forms an 'expert' in that domain; and, if robots can, as the foregoing sections have made plausible, be testimonial sources of knowledge, then it is possible they could be presented as expert testifiers in a court of law as well. Pamela Katz (2014) argues that robots should be made available to judges and juries in order to better help them analyze and make sense of the evidence presented to them:

Judges are faced with a challenge when they are asked to evaluate proposed experts' use of complex science and methodology—most of the time way outside of the judges' areas of expertise. Computers and artificial intelligence can assist in important ways, while leaving the ultimate decision about permissibility in the judges' hands. (2014, p. 3)

Whilst Katz herself focuses on the use of robots to assist judges in evaluating other (human) expert witnesses, there is no reason for thinking that robots themselves cannot take on the role of an expert witness. One might easily imagine that a robot can have any number of technical abilities that outstrip that of the typical human. Cases that are particularly relevant might include those of business, environment, or international law, where significant amounts of data analysis and scientific facts are often required.

The use of robot-expert over human-expert testifiers comes with a number of benefits that have not hitherto been discussed. One of the major concerns with the use of experts in an adversarial legal system is that the side with the greater amount of financial resources can afford to pay for a bewildering number of experts. This can unfairly sway a jury into believing that the party with the higher number of experts in its case must be the strongest (Gross, 1991). By having robots as experts there is no financial incentive to exploit, and once that robot has been hired, will not require continuing payment. Many courts now employ their own expert to mitigate against this possibility (Edmond, 2008) and there is no reason for thinking that each court could not own an expert robot which both parties in the dispute have equal access to.

Another benefit of robots as opposed to human-expert testifiers is that they are unlikely to be biased. Many experts brought in to provide testimony are working scientists from industry or are academics working on privately-funded projects. If a company who employs staff with technical and scientific expertise calls upon a member to act as an expert witness in a case involving that employer itself, then there will clearly be pressure on the employee to use his or her skills and analysis in a way favourable to their employer. Robots will not be susceptible to such pressures and can take more of an 'objective' approach to the dissemination of information.

An issue that frequently plagues the use of experts as evidence concerns character traits that are unrelated to the veracity of their claims. Axel Gelfert reminds us that jurors can be just as swayed by the appearance and professional delivery of an expert's testimony than they are by their intellectual and academic credentials (2014, p. 183). A well-groomed confident charlatan may present evidence that is just as convincing to a jury as an unkempt and anxious genius; yet, clearly what matters in such cases are their intellectual ability and not their fashion sense. Once again, robots can avoid this problem. The same robot can provide expert analysis for both parties and therefore reduce the potential for such confounding factors.

There are downsides, however, to using robot experts in the courtroom that should be addressed. With respect to the point about initial appearance, robots themselves may give an impression that influences a juror's willingness to accept their evidence. On the one hand, it could be predicted that judges and jurors impressed by technology will automatically assume that the robot's analysis is sovereign. They might be swayed by a culture of androids in science fiction (such as Data from *Star Trek)* that are frequently portrayed as having superior knowledge or insight into matters related to science and mathematics. On the other hand, those skeptical or prejudiced against robots and their use in delivering outcomes that ultimately affect only human agents might be unwilling to accept their testimony at all.

In addition, although robots are unlikely to be biased by financial incentives given to them by others, there are other ways their position might be compromised by a third party. All machines, especially computing devices, are subject to potential hacking. Even if the robot is owned and operated by the courtroom there is a potential worry about tampering with the device or remote hacking by an individual attached to a party in the trial. Of course, these are problems all institutions and governments face when allowing for a greater role of machines in the containment and analysis of information relevant to the public. Provided good encryption and security measures are carried out, it could be argued that the threat faced to robot experts is no greater than that facing other machines in service of society.

CONCLUSION

The issue of whether robots, androids, and other autonomous mechanical agents can be sources of testimony is underpinned by many other perennial problems in philosophy. This chapter has shown how this is particularly the case with intentionality, which seems to be essential for ascribing testimony. Nonetheless, it is an important matter to decide, since being capable of providing testimony is crucial for robots to fulfill the wider promises made of them in society. Their possible presence in the courtroom in the form of a witness or expert throws up additional concerns that need to be addressed, but there are also many advantages. If robots are 'given a voice' then the criminal justice system will have to adjust to accommodate it, and it is unlikely that this will be the only aspect of society that needs to change in response to robot testimony.

REFERENCES

Adams, F., & Steadman, A. (2004). Intentional Action in Ordinary Language: Core Concept or Pragmatic Understanding? *Analysis*, *64*(2), 173–181. doi:10.1093/analys/64.2.173

Alemi, M., Meghdari, A., & Ghazisaedy, M. (2014). Employing Humanoid Robots for Teaching English Language in Iranian Junior High-Schools. *International Journal of Humanoid Robotics.*

Bailey, A., & Richards, B. (2013). Horgan and Tienson on phenomenology and intentionality. *Philosophical Studies*, *167*(2), 313–326. doi:10.1007/s11098-013-0089-7

Bethel, C. (2011). Sharing a Secret: Interactions between a Child, Robot, and Adult. *Paper presented at the 6th ACM/IEEE International Conference on Human-Robot Interaction*, Lausanne, Switzerland. doi:10.1109/ICSMC.2011.6084051

Bethel, C., Eakin, D., Anreddy, S., Stuart, J., & Carruth, D. (2013). Eyewitnesses are misled by human but not Robot Interviewers. In *Proceedings of the 2013 8Th ACM/IEEE International Conference on Human-Robot Interaction (HRI)* (pp. 25-32).

Bickmore, T., Caruso, L., Clough-Gorr, K., & Heeren, T. (2005). Its just like you talk to a friend relational agents for older adults. *Interacting with Computers*, *17*(6), 711–735. doi:10.1016/j.intcom.2005.09.002

Block, N. (1980). *Readings in philosophy of psychology*. London: Methuen.

Breazeal, C. (2003). Towards Sociable Robots. *Robotics and Autonomous Systems*, *42*(3-4), 167–175. doi:10.1016/S0921-8890(02)00373-1

Breazeal, C. (2011). Social Robots for Health Applications. In *Proceedings of the 2011 Annual International Conference of The IEEE Engineering In Medicine And Biology Society*. doi:10.1109/IEMBS.2011.6091328

Buckley, D. (2013). *Cycler / ReCycler Presentation Robot. Davidbuckley.net.* Retrieved from http://davidbuckley.net/FR/Cycler/CyclerPresentationRobot.htm

Burge, T. (1998). Computer Proof, Apriori Knowledge, and Other Minds The Sixth Philosophical Perspectives Lecture. *Noûs (Detroit, Mich.)*, *32*(S12), 1–37. doi:10.1111/0029-4624.32.s12.1

Cabibihan, J., Javed, H., Ang, M., & Aljunied, S. (2013). Why Robots? A Survey on the Roles and Benefits of Social Robots in the Therapy of Children with Autism. *International Journal Of Social Robotics*, *5*(4), 593–618. doi:10.1007/s12369-013-0202-2

Calo, J. (2016, March). Robots in American Law. *Paper presented to the Robert Strauss Centre*, University of Texas, Austin.

Chalmers, D. (2003). The Nature of Narrow Content. *Philosophical Issues*, *13*(1), 46–66. doi:10.1111/1533-6077.00004

Coady, C. (1992). *Testimony*. Oxford: Clarendon Press.

DeLuce, D. (2014). Can Robots help stop the Ebola outbreak? *Medicalxpress.com.* Retrieved from http://medicalxpress.com/news/2014-11-robots-ebola-outbreak.html

Dennett, D. (1994). Original Intentionality. In E. Dietrich (Ed.), *Thinking computers and virtual persons* (1st ed.). San Diego: Academic Press. doi:10.1016/B978-0-12-215495-9.50009-0

Dietrich, E. (1994). *Thinking computers and virtual persons*. San Diego: Academic Press.

Dinh, L. (2013). Talyst's InSite remote dispensing technology - compliant with short-cycle dispensing. *GlobeNewswire News Room*. Retrieved from http://globenewswire.com/news-release/2013/03/26/533801/10026396/en/Talyst-s-InSite-Remote-Dispensing-Technology-Compliant-With-Short-Cycle-Dispensing.html

Edmond, G. (2016). Expert Evidence. In P. Cane & J. Conaghan (Eds.), *The new Oxford companion to law* (pp. 437–440). Oxford: Oxford University Press.

Elgin, C. (2002). Take it from Me: The Epistemological Status of Testimony. *Philosophy and Phenomenological Research*, *65*(2), 291–308. doi:10.1111/j.1933-1592.2002.tb00203.x

European Commission. (2016). CORDIS: Projects & results service: spatio-temporal representations and activities for cognitive control in long-term scenarios. *Cordis.europa.eu*. Retrieved from http://cordis.europa.eu/project/rcn/107156_en.html

Falconer, D. (2016). The Panasonic Hospi robot help in the delivery of medicines in Changi hospital. *Digitalavmagazine.com*. Retrieved from http://www.digitalavmagazine.com/en/2015/08/25/the-panasonic-hospi-robot-help-in-the-delivery-of-medicines-in-the-hospital-of-cgh

Fricker, E. (1995). Critical Notice. *Mind*, *104*(414), 393–411. doi:10.1093/mind/104.414.393

Gelfert, A. (2014). *A critical introduction to testimony*. London: Bloomsbury Academic.

Gray, H., Gray, K., & Wegner, D. (2007). Dimensions of Mind Perception. *Science*, *315*(5812), 619. doi:10.1126/science.1134475 PMID:17272713

Gross, S. (1991). Expert Evidence. *Wisconsin Law Review*, *6*, 1113–1232.

Hashimoto, T., Kobayashi, H., Polishuk, A., & Verner, I. (2013). Elementary Science Lesson Delivered by Robot. In *Proceedings of the 8th Annual ACM/IEEE international conference on Human-robot interaction*, Japan (HRI '13) (pp. 133-134). Piscataway, NJ: IEEE Press.

Heerink, M., Albo-Canals, J., Valenti-Soler, M., Martinez-Martin, P., Zondag, J., Smits, C., & Anisuzzaman, S. (2013). Exploring Requirements and Alternative Pet Robots for Robot Assisted Therapy with Older Adults with Dementia. In *Social Robotics* (pp. 104-115).

Hernández-Espinosa, A., & Hernández-Quiroz, F. (2013). Does the Principle of Computational Equivalence Overcome the Objections against Computationalism?. In *Computing Nature* (pp. 225-233).

Horgan, T., Tienson, J., & Graham, G. (2004). Phenomenal Intentionality and the Brain in the Vat. In R. Schantz (Ed.), *The Externalist Challenge* (1st ed.). Berlin: De Gruyter. doi:10.1515/9783110915273.297

Horst, S. (1999). Symbols and Computation: A Critique of the Computational Theory of Mind. *Minds and Machines*, *9*(3), 347–381. doi:10.1023/A:1008351818306

Jackson, F. (1982). Epiphenomenal Qualia. *The Philosophical Quarterly*, *32*(127), 127. doi:10.2307/2960077

Katz, P. (2014). Expert Robot: Using Artificial Intelligence to Assist Judges in Admitting Scientific Expert Testimony. *Albany Law Journal of Science & Technology*, *24*(1), 1–33. PMID:12680396

Kim, J. (2007). Causation and Mental Causation. In *B. McLoughlin (Ed.), Contemporary Debates in Philosophy of Mind* (1st ed., pp. 1–22). London: Wiley-Blackwell.

Knobe, J. (2003). Intentional Action in Folk Psychology: An Experimental Investigation. *Philosophical Psychology*, *16*(2), 309–324. doi:10.1080/09515080307771

Kyriakidou, M. (2014). Discussing robot crime interviewers for childrens forensic testimonies: A relatively new field for investigation. *AI & Society*, *31*(1), 121–126. doi:10.1007/s00146-014-0566-3

Lackey, J. (2008). *Learning from words*. Oxford: Oxford University Press. doi:10.1093/acprof:oso/9780199219162.001.0001

Lee, D. (2016). US opens investigation into Tesla after fatal crash. BBC. Retrieved from http://www.bbc.co.uk/news/technology-36680043

Levin, J. (2004). Functionalism. In *Stanford Encyclopedia of Philosophy*. Retrieved from http://plato.stanford.edu/entries/functionalism

Lewis, D. (1970). How to Define Theoretical Terms. *The Journal of Philosophy*, *67*(13), 427. doi:10.2307/2023861

Lewis, D. (1972). Psychophysical and theoretical identifications. *Australasian Journal of Philosophy*, *50*(3), 249–258. doi:10.1080/00048407212341301

Malle, B., & Knobe, J. (1997). The Folk Concept of Intentionality. *Journal of Experimental Social Psychology*, *33*(2), 101–121. doi:10.1006/jesp.1996.1314

Malle, B., & Nelson, S. (2003). Judging mens rea: The tension between folk concepts and legal concepts of intentionality. *Behavioral Sciences & the Law*, *21*(5), 563–580. doi:10.1002/bsl.554 PMID:14502689

Marton, M. (2015). Intentional and Phenomenal Properties: How Not to Be Inseparatists. *Unpublished Article*. Retrieved from http://real.mtak.hu/27762

Matthias, A. (2004). The responsibility gap: Ascribing responsibility for the actions of learning autonoma. *Ethics and Information Technology*, *6*(3), 175–183. doi:10.1007/s10676-004-3422-1

McLaughlin, J. (2016). FBI Says Apple Court Order Is Narrow, But Other Law Enforcers Hungry to Exploit It. Retrieved from https://theintercept.com/2016/02/22/fbi-says-apple-court-order-is-narrow-but-other-law-enforcers-hungry-to-exploit-it

McLeod, I. (2010). *Legal theory*. London: Palgrave.

Nichols, S., & Ulatowski, J. (2007). Intuitions and individual differences: The Knobe effect revisited. *Mind & Language*, *22*(4), 346–365. doi:10.1111/j.1468-0017.2007.00312.x

Putnam, H. (1967). The Nature of Mental States. In W. Capitan (Ed.), *Art, mind, and religion* (1st ed., pp. 1–23). Pittsburgh: Pittsburgh University Press.

Rey, G. (1997). *Contemporary philosophy of mind*. Cambridge, Mass.: Blackwell.

Russell, B. (1948). *Human knowledge*. New York: Simon and Schuster.

Sayre, K. (1986). Intentionality and information processing: An alternative model for cognitive science. *Behavioral and Brain Sciences*, *9*(01), 121. doi:10.1017/S0140525X00021750

Searle, J. (1980). Minds, brains, and programs. *Behavioral and Brain Sciences*, *3*(03), 417. doi:10.1017/S0140525X00005756

Sosa, E. (1991). Testimony and Coherence. In E. Sosa (Ed.), *Knowledge in perspective* (1st ed., pp. 215–222). Cambridge: Cambridge University Press. doi:10.1017/CBO9780511625299.013

Sosa, E. (2006). Knowledge: Testimonial and Instrumental. In J. Lackey & E. Sosa (Eds.), *The epistemology of testimony* (1st ed., pp. 116–123). Oxford: Oxford University Press. doi:10.1093/acprof:oso/9780199276011.003.0006

Sytsma, J., & Machery, E. (2010). Two conceptions of subjective experience. *Philosophical Studies*, *151*(2), 299–327. doi:10.1007/s11098-009-9439-x

Thielman, S. (2016). Apple v the FBI: what's the beef, how did we get here and what's at stake? *The Guardian*. Retrieved from http://www.theguardian.com/technology/2016/feb/20/apple-fbi-iphone-explainer-san-bernardino

Tucker, E. (2015). How robots are helping children with autism. *The Guardian*. Retrieved from http://www.theguardian.com/lifeandstyle/2015/feb/01/how-robots-helping-children-with-autism

Ullian, J. S. (1971). Putnam Hilary. Minds and machines. The Journal of Symbolic Logic, 36(01), 177.

Van Moer, A. (2006). The Intentionality of Formal Systems. *Foundations of Science*, *11*(1-2), 81–119. doi:10.1007/s10699-004-5914-1

Whiting, D. (2016). *Conceptual Role Semantics | Internet Encyclopedia of Philosophy*. Retrieved from http://www.iep.utm.edu/conc-rol

Wolfram, S. (2002). *A new kind of science*. Champaign, IL: Wolfram Media.

KEY TERMS AND DEFINITIONS

Cross-Examination: The process through which a lawyer for the defendant assesses the reliability of a witness in court.

Epistemology: The branch of philosophy that deals with questions concerning the nature of knowledge, belief and justification

Functionalism: A materialist approach to the philosophy of mind which is committed to the view that mental states are functional roles which can be defined in terms of input and output states.

Materialism: An approach to the philosophy of mind which is committed to the view that all mental properties and events are reducible to material properties and events. Closely related to Physicalism.

Philosophy of Mind: The branch of philosophy that deals with questions concerning the nature of mind and persons.

Propositional Attitude: Mental states people have towards propositions; e.g., 'John hopes it will rain today.' Also known as 'intentional states.'

Social Robot: A robot designed to interact with human beings in normal everyday environments.

Testimony: A source of knowledge that originates from another person's declaration. Can be oral, written or pictorial.

Chapter 2
Girls Building Androids and Robots:
Equality in STEM With the Media Program *Annedroids*

Maya Götz
International Central Institute for Youth and Educational Television (IZI), Germany

Diana Iulia Nastasia
Southern Illinois University – Edwardsville, USA

J. J. Johnson
Sinking Ship Entertainment, Canada

ABSTRACT

There is still a considerable degree of catching up to do in regards to fostering gender equality within areas of STEM (Science, Technology, Engineering, and Math). Children's leading medium, television, could offer role models of girls with competence in STEM areas, but unfortunately television programs often miss this chance. The children's television series Annedroids is a notable exception. This chapter provides insight into how children can be educated about gender equality in STEM with the aid of gender-sensitive media programs such as Annedroids. The chapter examines data from a reception study which was conducted under the leadership of the International Central Institute for Youth and Educational Television (IZI) with 6- to 12-year-old children in the United States and Canada (N = 301). The research is enhanced by a conversation between Dr. Maya Götz, Head of IZI; Dr. Diana Nastasia, a contributor to the IZI research; and J. J. Johnson, the program's creator.

DOI: 10.4018/978-1-5225-2973-6.ch002

Nick, an 11-year-old Afro-Canadian boy, is new to the neighborhood, moving in opposite a mysterious scrap yard. Shania, his bubbly, slightly overweight new friend, tells him that they sometimes have strange power cuts and that things disappear behind the high walls. Nick breaks into the scrap yard through a secret door and is discovered by an enormous robot which chases him, captures him, and takes him to his creator. The robot's creator stands before Nick in dungarees and a welding mask, clueless as to what to do with the intruder. "It is impolite to spy on people," she says and raises her welding mask. "You are a girl," realizes Nick to his astonishment. "You are a boy," replies the 11-year-old inventor, Anne, matter-of-factly.

INTRODUCTION

There is still a considerable degree of catching up to do in regards to fostering gender equality within areas of STEM (Science, Technology, Engineering, and Math). The stereotypical assumption that STEM is – by gender – a strength of boys (Moss-Racusin, Molenda, & Cramer, 2015; Watt & Eccles, 2008) still prevails, in spite of extensive research documenting that gender differences in STEM ability and achievement are impacted by cultural practices rather than biological factors (Halpern et al., 2007; Myers, Jahn, Gailliard, & Stoltzfus, 2011).

Research evidence has also shown that gender models are of key importance for challenging this stereotype (Maltese & Tai, 2011; Shapiro & Williams, 2012). There is a need for positive role models in public discourse, to bridge girls' and women's reduced presence and experience in STEM fields (Margolis & Fisher, 2002) by depicting how they quite evidently and competently know, apply, and even invent technologies.

The media - especially children's leading medium, television - can make a vital contribution to providing such role models. Children use the symbolic material that the media offer in their everyday lives, and integrate parts of it into their perception of themselves and the world (Götz, 2014). Reference in symbolic material to girls' STEM-related talents can help develop interests related to STEM and trust in one's own ability regarding STEM activities (Jacobs, Davis-Kean, Bleeker, Eccles, & Malanchuk, 2005; Cvencek, Meltzoff, & Greenwald, 2011). A highly relevant task of children's media programs would be to foster gender equality by creating counter-stereotypical imagery which shows that girls can be savvy in what concerns technology and can, for instance, be involved in building and operating androids and robots.

Yet, unfortunately, the reality is different. A worldwide analysis of over 26,000 main characters of children's television shows reveals that, so far, it is still an exception that girls and women media characters apply science and technology at all

(Götz & Lemish, 2012). Another analysis of children's knowledge programs in the United States, Great Britain, and Germany also reveals that, although STEM fields are in a dire need for girl and women contributors, in the media, female characters involved in STEM appear only in exceptional cases; and, if they do appear, they serve as "accessory parts" (Schlote, 2013). Examinations of children's television and prime-time television have shown that women are extremely rare in roles of scientists and technology experts, and if they are cast in such roles, then they are stereotyped and sexualized (Chimba & Kitzinger, 2010; Dudo et al., 2011; Steinke et al., 2007). Unfortunately, so far, children's television misses to a great extent the mission of fostering gender equality in regards to STEM. One notable promoter of this mission is the series, *Annedroids*.

Annedroids was inspired by research results about the deficits of positive, non-stereotypical girl characters who competently deal with technology. The series, made by the Canadian company, Sinking Ship Entertainment (an Emmy award-winning production and interactive company whose portfolio also includes *PBS*'s *Odd Squad*, and *Nick Jr.*'s *Dino Dan*, *This Is Daniel Cook*, *This is Emily Yeung*, and *Are We There Yet? World Adventure*), premiered on *Amazon Prime* on July 25, 2014 in the United States and Great Britain, and subsequently aired in late summer of 2014 in Canada in English on *TVOntario,* in French on *Société Radio-Canada (SRC),* and in the fall of 2015 on the children's public television channel *KiKA* in Germany, in German synchronization. *Annedroids* is also scheduled to air on *NRK* in Norway, *SVT* in Sweden, *Horizonte Conteudo*s and *Globostat* in Brazil, and *Discovery Asia*. The program's success on *Amazon Prime* is outstanding, as it managed to rank among Amazon's most popular video list (just behind *House of Cards*).

The *Annedroids* series combines real children characters with photo-real CGI-technological characters in a live-action show. The show follows scientific genius, Anne, and her new friends, Nick and Shania, as they play with her creations, robots Hand and Eyes, and android, Pal, to perform scientific experiments in Anne's junkyard lab. In every episode, the humorous and child-appropriate plots pick up a STEM-relevant topic. The characters, girl and boy humans, as well as androids and robots, are deliberately created in a gender-sensitive manner. Thus, they challenge gender stereotypes and play with various ways of "doing gender" (West & Zimmerman, 1987, 2009); both encouraging critical reflection on long-standing gender roles and uncovering the social construction of gender.

Based on data from a reception study, which was conducted under the leadership of the International Central Institute for Youth and Educational Television (IZI) with 6- to 12-year-old children in the United States and Canada (N=301), and which assesses 1) how these children interpret the show's characters, and 2) how they view issues related to STEM and gender upon watching two episodes from the series. Data is also based on a conversation between Dr. Maya Götz, Head of

IZI; Dr. Diana Nastasia, a researcher who contributed to the IZI study; and J. J. Johnson, the program's creator. The research provides insight into how children can be educated about gender equality in STEM with aid of gender sensitive media programs such as *Annedroids*.

BACKGROUND: THE THEORETICAL AND METHODOLOGICAL FRAMEWORK

This study is grounded in the trend within the communication and media scholarly literature which recognizes, not only the huge impact of the media on the lives of children, but, in particular, the positive impact that the media can make in regards to educating children on 1) how to surpass various forms of stereotypes, including those related to gender, and 2) how to become increasingly competent in various knowledge areas, including those regarding STEM. It specifically refers to girls as inventors - and operators - of androids and robots.

According to the seminal Kaiser Family Foundation Study, *Generation M^2: Media in the Lives of 8- to 18-Year-olds*, media are among the most powerful forces in young people's lives today. According to the study's authors, children and youth,

...spend more time with media than in any other activity besides (maybe) sleeping – an average of more than 7½ hours a day, seven days a week. The TV shows they watch, video games they play, songs they listen to, books they read, and websites they visit are an enormous part of their lives, offering a constant stream of messages about families, peers, relationships, gender roles, sex, violence, food, values, clothes, and an abundance of other topics too long to list. (Rideout et al., 2010, p. 1)

Although 'media effects' approaches have often declared what should be done - to keep children out of contact with the media or to protect children from potential media harm (for a critique of the effects, vulnerability, harm, and blame scholarship, see Livingstone, 2007) - recent reception studies and cultural studies approaches have increasingly outlined how children cannot only negotiate media usage, but, also benefit from it (Fisch, 2005; Mares & Acosta, 2010). As Lemish (2014) states,

...many initiatives around the world advocate and demonstrate that, when used appropriately, media can bring comfort and hope, provide skills for competent response to an emergency, develop resilience, and even contribute to the improvement of the lives of children and their families. (p. 162)

Media scholars such as Henry Jenkins (2009) in the United States and David Buckingham (2005) in Great Britain have asserted that, equipped with the right types of tools, such as those derived from awareness of the new participatory culture and knowledge of media literacy applications, younger generations can become quite proficient at using various forms of media and multimedia platforms for personal development, identity, expression, and social connection. In the course of these activities, youth may not only connect better than previous generations with people outside of their geographical areas, but, also may configure emerging non-stereotypical understandings of various issues including those related to gender. As part of such a non-stereotypical view, girls could, for example, be credible as inventors and operators of androids and robots.

Research has evidenced that, when appealing enough and aimed at the right audience, media programming and specifically television programs can contribute to increased interest and knowledge in a variety of areas and topics. It can have positive impact on children's academic skills (see overview in Kondo, 2009); language and vocabulary skills (Fisch & Trulio, 2001; Holler, 2012; Linebarger & Walker, 2005); and literacy, mathematics, problem solving, or science and technology skills (Linebarger et al., 2009). High-quality educational television shows can stimulate metacognition and the learners' understanding of their own way of learning (Schlote, 2012), and can also help relate existing knowledge to new information, make deductions beyond facts provided, and draw conclusions based on such relationships (Wainwright & Linebarger, 2009). High-quality educational television can also, arguably, inspire girls to be builders and operators of androids and robots.

According to this trend within the communication and media scholarly literature, problems occur, not because there is an inherent danger for children in regards to the media, but because there is not sufficient focus on quality programming with potential and intent to help break stereotypes and to provide educational content; and upon being noted such lack of focus can incrementally be remedied. As 68% of all main characters in fictional children's television across 24 countries are male (Götz & Lemish, 2012), this means that, while women represent 51% and men 49% of the world's population today, in children's television there are two male main characters for every female main character. This tendency was also observed in cinematic films and computer games relevant for children (Smith & Cook, 2008).

Beyond sheer numbers, in terms of content, female characters are portrayed as less active and less vocal than male characters, and less likely to take initiatives and to occupy positions of authority. They are also portrayed as more likely than their male counterparts to behave childishly and superficially, to display overly emotional reactions, especially in critical situations, and to be concerned with their looks and with beauty, and not be focused on serious issues and work. In contrast, male characters are more often rewarded within the plot, demonstrate more ingenuity, ask more questions, are more frequently presented and highlighted through their particular abilities and talents, laugh and have fun more often, yet also are aggressive and threatening to others more often.

In other words, female characters in children's television are still astonishingly often stereotyped (Aubrey & Harrison, 2004; Luther & Legg, 2010, among others). While there have been a few strong and active heroines in children's television (some examples including the super-heroines in the cartoon series *Kim Possible*, *Totally Spies*, and *The Powerpuff Girls*, who carry out missions and have special powers to do so), and while such characters may no longer play helpless victims (Baker & Raney, 2007), heroines in children's television are still overall marked by hypersexualization; extreme thinness' provocative, sexy clothing; longing for beauty recognition and true love; and easily being led astray or losing their way (Smith & Cook, 2008).

Additionally, even super-heroines are usually not entirely individualized and independent, as 88% of all super-heroines are showcased as working in teams, and they often have a mentor, who is nearly always a male character (Baker & Raney, 2007).

Given this data, and the need for quality programming which would help children consider gender, and, particularly, STEM and gender in a fairer manner, it is all the more important to showcase and analyze a media project such as the *Annedroids* series, which proposes as a main character a girl inventor of androids and robots, and which offers a non-stereotypical and gender-sensitive view. Anne in *Annedroids* represents a departure from stereotypes that otherwise abound in children's television. While she is slim with a pleasant face, she does not correspond to the "soapy" appearance. Her hair is curly and wild, her clothing is not sexualized, and she does not wear makeup. She is smart and nerdy, but entirely capable of making friends and having a social life. She is a cool girl capable of dealing with technologies. Additionally, Anne's friends, slightly overweight Shania and Afro-Canadian Nick, help strengthen the non-stereotypical view.

In terms of methodology, this study is based on both a reception study with children subjects and an interview endeavor with the series' producer. According to Livingstone and Lemish (2001), assessing children's opinions on media programs addressed to them is essential for empowerment purposes as, in spite of calls to give children a voice in social research, children are still considered by many researchers as powerless subjects, incapable of describing or reflecting upon their own experiences and feelings (p. 3). According to Lemish (2014), evaluating producers' perspectives on the media programs they create is also essential, as such an undertaking offers insights into "their decision-making processes, policies, and views" of their role in balancing between 1) on the one hand, the need to inform children about harsh realities and sensitive topics, and, 2) on the other hand, "doing so in an age appropriate way that promotes their healthy development and well-being" (p. 163). Combining the two methods, the examination of reception as well as of production, allows for the convergence of the perspective of children and the perspective of the content creator addressing to children, in a unique dialogue in which the reflections of two additional researchers add even more depth.

As part of the reception study with children, under the leadership of the International Central Institute for Youth and Educational Television (IZI) in Munich, Germany, researchers in the United States and in Canada pursued an examination of the television series *Annedroids* with the aid of pre-test and post-test questionnaires for children between 6 and 12 years of age (N=301).

The study was implemented in the United States by Dr. Sorin Nastasia and Dr. Diana Nastasia (Southern Illinois University Edwardsville) and in Canada by Debbie Gordon and Emily DeVries (kidsmediacentre, Centennial College, Toronto) from mid-March until May 2015.[1] The data were analyzed in Germany by Dr. Maya Götz (IZI). Children were recruited through schools and answered the questionnaires in a school setting. Their participation in the study was voluntary; institutional approval and parental consent were obtained prior to the study implementation.

In the pre-test questionnaire, the children participating in the research were asked about their perspectives on gender and technologies, with a few questions more specifically about what gender roles they view as appropriate for girls, boys, women, and men. In the post-test questionnaire, two weeks after the pre-test and after having watched two episodes of *Annedroids*, the children participating in the research were again asked about their perspectives on gender and technologies and gender roles, and how this particular television program impacted those. They were also asked for their overall views of the program, and to make drawings about this program.

As part of the interview endeavor with the series' producer, this research documents the thoughts around *Annedroids* of J. J. Johnson. A founding partner and the creative force behind Sinking Ship Entertainment, J. J. Johnson won multiple Emmy awards as producer and director, has directed over 200 hours of children's programming, and has written over 500 scripts. J. J. Johnson was listed in *Playback* magazine as one of their "Top 10 to Watch" and was named by *The Hollywood Reporter* as one their "Top 20 under 35" in 2013. He is also the co-chair of the Youth Media Alliance in Canada.

Johnson conversed with the head of IZI, Maya Götz, who asked questions about the creative processes from the first idea about the program until the dissemination of the final product. She also inquired about the emergence and configuration of the girl lead character, the meanings and producer's perceptions about 1) the impact of this character, and 2) overall quality in children's television programming. Maya Götz and Diana Nastasia further provided a scholarly interpretation of the producer's words and perspectives.

A GIRL BUILDING ANDROIDS AND ROBOTS IN *ANNEDROIDS:* PRODUCER'S AND CHILDREN'S THOUGHTS AND PERSPECTIVES

This portion of the research discusses, in additional detail, issues related to the contribution of the media to fostering gender equality in STEM areas and particularly, issues related to the contribution of television to providing role models of girls inventing and operating androids and robots. Specifically, this section of the chapter addresses, with the aid of quotes from the *Annedroids* series creator and further scholarly commentary on these quotes, as well as with the aid of reception study data, 1) various aspects regarding the creation of the program, and 2) the broad implications of both featuring a girl as the lead character of the program, and of introducing body type and ethnic background diversity through the girl and boy supporting characters in the program.

Prior to proceeding to the outlining and analysis of the producer conversation and reception study data, several additional notes are needed to explicate the uniqueness of the *Annedroids* series within the children's media landscape.

In today's international children's television market, there are some programs dealing with science and technology topics that are aimed at encouraging children to explore scientific and technological knowledge. In the US-based *PBS* program *Wild Kratts*, naturalists Chris and Martin Kratt get up close and personal with the animal world by quite literally adopting the perspective of animals, becoming as small as earthworms for instance. In the US-based *PBS* program *Ruff*, groups of children seek answers to scientific questions placed in the context of everyday experiences. In the British *CBBC* production *Operation Ouch*, Chris and Xander, two twin doctors, follow pediatric accident and emergency personnel and patients; perform experiments on the human body, including "try this at home" experiments for viewers to participate in; and, investigate and explain medical treatments and technologies. The German public service children's broadcaster *Kika* offers, during children's primetime, a variety of informative programs (for example, *Pur+*, *Wissen macht ah*, *Willi wills wissen*) in which, usually male presenters explain scientific phenomena. In the Dutch program *Checkpoint*, science experiments are performed; for example, in one episode two male scientists examine how urine can be used to make drinking water. There are also, however, some episodes in which gender differences are scrutinized with a focus on whether boys or girls are better at performing a certain task or function. In the Uruguayan programme *Neurona!*, an inquisitive male reporter explores various scientific questions. In the Indian program *Teenovation*, a young man travels the country and presents young inventors. In the Iraqi program *Salam Shabab*, teams of young participants compete to answer questions, including many related to STEM areas.

As these international examples demonstrate, children's television certainly deals with STEM themes; at least, occasionally. However, there are several gaps which the series *Annedroids* has been meant to fill. Firstly, the format of these television program examples has been almost exclusively non-fictional; thus, the emphasis overall has been on conveying scientific and technological information to children rather than stirring their imagination in regards to STEM areas, even though some creative elements have been integrated in some of the shows mentioned above. Secondly, these television programs have most often taken approaches in which there are either answers to scientific and technological questions, without hands-on or applied activities in support, or there is show and tell about science and technology, but it is done by adult experts for a reporter, rather than by the children themselves, and it definitely does not include children showcased as building and operating androids and robots. Thirdly, and most importantly within the logic of this chapter, the protagonists are almost always male (a significant exception being the Norwegian program *Newton* in which the lead character is a young female doctor who explores and explains scientific and technological issues). Prior to the *Annedroids* series, there has hardly been any fictional children's television program skillfully applying STEM knowledge while also proposing a gender sensitive cast.

The Program's Creation

From the way in which the producer of the *Annedroids* series describes the creative process which led to the show's current configuration, it is clear that there was an awareness of the lack of television programs focusing on STEM which would take a hands-on, applied science and technology approach, and thus would be both informative and attractive for children. J. J. Johnson stated:

I have always wanted to make a science show but one that focused on the kind of science that I think kids really like and that I liked as a kid: big messy explosions and a little bit of wonderful chaos. I hadn't seen that kind of large scale, imagination soaring science show out there, they were mostly more magazine style and kind of looking at experiments that kids could do with their parents safely at home. It wasn't quite the area that I wanted to play in.

The producer's considerations on the origins of the *Annedroids* series also highlight a mindfulness that television programs having a fictional format and focusing on STEM were lacking, and they showcase an interesting progression in the course of crafting such a program, from envisioning 1) a boy playing with dinosaurs (like in *Dino Dan*), to 2) a girl and other children operating androids and robots. According to Johnson,

I love kids' stories where you can watch the characters develop from episode to episode. We had accomplished this, to some degree, with our program Dino Dan: a fictional story about a boy whose imagination was so powerful that he could imagine dinosaurs into the real world. This magical ability allowed him to study the dinosaurs and learn about them directly through experimentation. For Annedroids I wanted an even grander story arc, one which would see our human and android characters evolve and mature over the course of each season.

While expressing further thoughts about the creation of the *Annedroids* series, the producer insisted even more on the need for fictional programs for children to learn from about STEM; namely, for programs allowing children to master knowledge and skills, yet also offering them excitement and entertainment by featuring girl and boy protagonists doing things with science and technology. Johnson affirmed:

There's no reason educational shows shouldn't be as fun and thrilling as their strictly entertaining counterparts. Our shows operate under a kind of the 'Three E' system. They should be Exciting, in that they should be visually enticing and offer a world that kids have not seen before, one that's going to grab their initial attention. They should be Entertaining, the story lines should include an equal balance of humor and action so that the viewer will want to continue watching it. And they should be Educational because I truly believe kids want a take-away. They want to know more about their world and how they fit in it. I'm also a firm believer that the show should be grounded by a kid protagonist so that the audience has a friend they can relate to.

Johnson's depiction of the conception of the program and the creative process leading to it also documents a lengthy stage of hesitation regarding: what would be that something that he could ground himself in and further invite the audience to also ground themselves in? Specifically, the producer depicts an extended and intense thinking period about what specific hands-on, fictional, entertainment show would be best suited to fill the gaps identified in children's media programming? Intriguingly, the producer declares to have solved this dilemma by appealing to childhood memories; namely, by finding inspiration in the everyday experiences of his own past.

Johnson's own excitement towards science and technology was discovered in junkyard spaces where old things can become new again, where one's imagination can run free, and where there are endless possibilities. In the words of the producer, it was within the context of this reconnection with himself as a child, and of recalling this type of all-permissive environment, that the boy protagonist morphed into a girl protagonist, and other options morphed and converged into androids and robots. According to Johnson,

But I couldn't figure out a way, for the longest time, to do the kind of show I wanted: one with spectacular experiments, while still appreciating the fact that parents and broadcasters are going to want it presented in a way in which the kids aren't in any real danger. This creative challenge lasted for a long time. Eventually I drifted back into my own childhood for inspiration. My friends and I used to break into junkyards when we were kids – I loved that environment because it's this chance to be a kind of modern archeologist, digging through layers of humanity. You can find old cars, cassette players, and things with unknown purposes. Magical pieces of technology whose new purpose you'd have to imagine. The excitement of discovering discarded artifacts has always stuck with me and provided the basis of what Annedroids would eventually become: a place where our main character would see things not for what they are, but for what they could become.

Working on the science took a bit longer, until I reminded myself of our own rule, to be exciting. Dino Dan benefitted from CG dinosaurs romping around, so why shouldn't this new show benefit from a similar story device? This is how the robots were born. It was initially conceived that she – well it was a "he" at first – had built these robots to assist in various scientific endeavors. But as I looked at how the characters could interact, there was so much more story potential in having these robot creations evolve and grow just as our human cast did too. So robots also became androids and central figures to the show.

This was a significant departure from Dino Dan, where dinosaurs weren't so much characters as a plot device, the T-Rex in the first episode was not necessarily the same T-Rex we saw in the 20th episode. But in Annedroids, these robots and androids would drive stories, cause problems, and have their own thoughts and feelings and wants. Having robots and androids also allowed the show to pursue science experiments that were far more fantastical (read: slightly dangerous), because these characters could have parts blown off of them and need to be repaired. This actually deepened the learning, we could still encourage wild experiments while also reinforcing that certain things shouldn't be done by you or you should keep a safe distance from them.

From a scholarly perspective, the producer's reflection on the program's creation resonates with research data about good media practices. Various reception studies have clearly demonstrated the suitability of the joint education and entertainment approach, not only for learning facts, but, also, for learning how to learn; especially when age-appropriate topics and language are involved, as well as additional elements of appeal to audiences such as: humor; clear and direct yet innovative and interesting handling of educational content; and, focusing on a small number of ideas or concepts per program sequence or episode (Holler, Götz, Egerer, & Schwartz, 2012; Holler & Götz, 2013; Unterstell, Götz, & Holler, 2012). Studies have also shown 1) that the use of, engaging and action-filled messages and visuals, helps improve the effectiveness of educational programs (Fisch, 2005), and 2) that personalities and characters that children can relate to - in addition to learning from - help keep attention focused, and thus help significantly improve learning outcomes (Wainwright & Linebarger, 2009). In the case of the *Annedroids* series, STEM themes were integrated in a fictional story, and an entertainment approach was taken for including girl and boy characters, as well as androids and robots.

A Girl Lead

As creator of the *Annedroids* series. J. J. Johnson was asked to further comment on how the program integrated concerns related to gender equality in STEM areas; specifically, as the protagonist was initially conceived as a boy, and later turned into a girl. When prompted to confirm that the show's lead character changed gender during the creative process, Johnson asserted, "Yes, I think when you initially develop a show, you're always looking for ways to speak to your younger self. So I think I just instinctively wrote with a boy character in mind."

Like earlier in the conversation, the producer of the *Annedroids* series specified that inspiration came from relating to a younger self, and, as such, from attempting to instill interest in - and enthusiasm for - STEM in children today by means of conveying to them his own interest in - and enthusiasm for - STEM during his childhood. The producer of the program also mentioned, however, an emerging awareness - even before work started on the *Annedroids* series - of the need for gender equality in children's media overall, as well as in regards to STEM, acquired during participation in the Prix Jeunesse International children's media festival in Munich, Germany. According to Johnson,

We were presented with these research results at Prix Jeunesse International in 2008 that showed that girls were woefully underrepresented in kids' TV shows. That report struck a nerve in me and got me thinking a lot about how we had addressed gender in our previous shows. On our show Dino Dan, we were very specific about not defaulting to talking about the dinosaurs as just "he." We literally went through scripts to make sure that we had an equal balance of calling the dinos "he" and "she." But although that show had a variety of girl characters, it was still helmed by Dan, a boy. I think that report really caused me to realize that we had a long way to go in terms of our depictions or characters, and that we had clearly underestimated the ramifications of kids not seeing themselves accurately depicted and represented on screen.

When discussing issues of gender in connection with the *Annedroids* series, Johnson continued to refer to his belief formed while being exposed to data about the severe gender inequalities in children's television programming: that it is important to propose a fairer balance and representation of genders within shows, both, in regards to the human characters, and in what concerns fictional companions, such as dinosaurs in *Dino Dan,*and robots and androids in *Annedroids*. It is notable that the producer highlights how gender was clearly assigned to the more mechanistic, repetitive robots, Hand and Eyes, yet, is left uncertain for most of the duration of the show for the more organic, human-like android Pal. It is also notable how the producer describes the change of the protagonist's gender in the course of the creative process. In regards to gender and the *Annedroid* series, Johnson declared:

As I developed Annedroids, I wanted to make sure that the android characters had defined genders or were in pursuit of defining them. In episode one, Hand is a girl, Eyes is a boy, and Pal is genderless, in that Pal hasn't chosen, yet – and doesn't make that choice until the final episode. But, of course, the biggest decision was to change our boy lead into a girl. For me, it didn't matter that the lead was a boy. It wasn't a major story point. There was no reason why it shouldn't be a girl. The change was a direct response to the report from Prix Jeunesse International. At the time, I didn't think it was going to be that big a deal; in fact, it felt like the easiest switch to make. Unfortunately, I couldn't have been more wrong.

After recognizing that there were challenges related to featuring a girl protagonist in a fictional yet educational show with STEM topics and technological companions, the program's creator also provided examples of such challenges. These challenges were related to the prevalent construction of gender, in view of which, boys are defined as tech-savvy, and girls are portrayed as lacking interest and ability in STEM areas. These challenges were also related to resistance and opposition to alternative ways of doing gender and alternative constructions of gender. Johnson focused specifically on the challenges related to persuading entities that provide funding for programs, and disseminating the programs, of the viability and value of such gender-sensitive content:

The U.S. networks had the hardest time believing that a girl scientist character was even possible. They felt the show would be much more successful if it was a male lead. We were lucky to have filmed a pilot for the series that was extremely well-received everywhere, except the US, where the networks loved the concept, but insisted we change the lead to a male. We refused.

From a scholarly perspective, Johnson's explanations about the creative process can again be interpreted in view of considerations derived from research data about good media practices. On one hand, grounding children's media plots and characters in everyday experiences, including in memories of the producer's own childhood experiences, bears high potential for raising children's interest in and attention to educational content carried through such media (Buckingham, 1996, 2013; Götz, Lemish, Aidman, & Moon, 2005; Lemish, 2006), arguably including STEM-related content. On the other hand, a statement like that of Johnson, that a show's creator seeks to speak to a younger self, which is also a gendered self, crosses paths with calls for an increased presence of women in media production (Byerly & Ross, 2008; Kirjnen & Bauwen, 2015).

Higher numbers of women media producers would lead to, not only higher numbers of girl media characters, as women could also relate to their younger selves, but also to less "otherizing" of women and girls in the media – given that, in the words of Simone de Beauvoir (1953), in a society in which the norm is masculine, women are constructed as the deviation from the norm, as "the Other." This is vital in children's media, as an analysis of the catalogue of the premiere international marketplace for children's television, MIPJunior, revealed that production lies firmly in the hands of men, who comprise 74% of all producers, 76.6% of all head scriptwriters, and 88% of all directors in the sample (Götz, 2014).

This would mean that women and girls should be featured in an array of different roles, including those still frequently thought as more masculine, such as leaders, experts, scientists, or inventors. This being specified, it should also be noted that such calls for gender equality in media production are not denying that men who are aware of gender and media challenges can also substantially contribute to fostering more gender-sensitive media programming.

Additionally, having as initial focus the making of a fictional and entertaining show in which children characters do things with science and technology might be more realistic from a producer's perspective, given the current landscape within the media industries than having as original emphasis the making of a show which would promote girls in STEM. While the show did end up featuring a girl lead character, in spite of opposition from top media distributors especially in the United States, the procedure for reaching this decision presumably reduced the risk of "gender trouble" (Butler, 1990), as whenever the category "gender" is mainstreamed it is also accompanied by a danger of actually reinforcing stereotypes. According to Johnson as producer, decisions related to gender of characters were made in the best interest of providing quality STEM content in a hands-on, fictional, entertaining format; having the explicit goal of promoting STEM for girls could have increased the risk of achieving the opposite, for instance, by stereotyping girls again either by 1) portraying them as victims, or, 2) featuring them as particularly attractive in order to make STEM "sexy."

The Program's Reception

The assessment of the *Annedroids* program by the children in the United States and Canada who participated in the reception study shows that the hands-on, fictional, entertaining program with STEM-related content resonated with them to a large extent.

When answering the question "How did you like the program *Annedroids*?" by selecting one of five options ranging from "very much" to "not at all", 89% of these children indicated that they enjoyed the show. Girls liked the program slightly more than boys, and younger children appreciated the program a bit more than older ones.[2] Furthermore, when asked whether they would watch more episodes of the series if they had the opportunity, 90% of all study participants answered positively, including 95% of girls and 86% of boys. This documents girls' high interest in media materials showcasing STEM areas, when such materials include role models they can relate to.

In their qualitative responses, over half of the children focused on the entertaining aspects of the *Annedroids* episodes they watched, which were 1) the episode right after the pilot (*New Pals*, in which Nick and Shania become friends with Anne, and android Pal is activated, thus added as technological companion to robots Hand and Eyes), and 2) the fifth episode of the first series (*Garbage Band*, in which Anne and Nick work on improving the security system of the junkyard, while Shania succeeds to do, with the robots and android, what she had not succeeded to do with her brothers; namely, to start a musical band). Many children characterized the episodes as amusing, fascinating, or exciting. The innovative storyline and the creative style of the show were praised by numerous children, although girls enjoyed the entertainment factor more often than boys (65% of the girls and 45% of the boys mentioned the entertaining value). For example, a 9-year-old U.S. girl wrote, "I liked it because I thought it was cool, creative, and awesome."

In their qualitative responses, many children also focused on the educational aspects of the *Annedroids* episodes they watched. Some children mentioned directly the educational emphasis of the show as its main strength. "I liked it a lot because not only was it funny, but also it was educational. They explained words like 'hypothesis' and 'upgrade' ", wrote an 11-year-old U.S. boy. Some other children found the show interesting because of the "technical-scientific topic," although boys appreciated this aspect slightly more often than girls (38% of the boys and 27% of the girls mentioned the focus on science and technology). Some children also mentioned that they would watch more episodes of the series explicitly because they liked how the show transferred technical and scientific knowledge to children, especially in regards to androids and robots, and thus spurred ambitions to learn.

For example, an 11-year-old U.S. boy explained that he would like to watch more episodes because, "it gives kids the chance to learn the difference between robots and androids." The characters were interesting to many of the children and made the program even more enjoyable. Additionally, some children referred to smartness as a positive feature conveyed through the show. "I like it because the kids were smart," stated a 9-year-old U.S. boy."I love that Anne is smart and Nick funny and the funniest yet Shania," affirmed a 9-year-old Canadian girl. In addition to the child who referred to the show's characters as being both smart and funny, a few other children also referred to both the entertainment and the educational value of the program. For example, a 10-year-old U.S. boy wrote: "It was educational but still REALLY FUNNY!"

Some children also referred, not to entertainment or education, but, to the social interactions that the program showcases, especially in the form of characters helping one another. Children enjoyed the friendship between Anne, Shania, and Nick, as well as their work together with Hand, Eyes, and Pal. An 11-year-old U.S. girl noted: "It's fun to see the androids and the friends that work together to help the androids and each other."

The study findings thus show that overall both boys and girls found in the *Annedroids* series the entertainment and educational values that the producer was trying to instill. They also enjoyed the interactions between the human characters and the technical companions. Besides the largely positive assessment, a few of the participants named negative reasons for not liking the program. While the number of these children is small, examining their statements is important for finding out what prevented them from enjoying and learning from a STEM-related program with a fictional format, a girl protagonist, and androids and robots as technological companions.

For 7% of the children participating in the study, *Annedroids* was "not entertaining." This was justified by categorizing the program as too childish, too boring, or too much like school. An 11-year-old U.S. girl wrote: "I am very interested in technology, so I liked it, but not the comedy." A very small number of children criticized the program for being "too fictive" in terms of storyline. A small number of children also declared to not like the program because it is "too girlish." Only boys (1%) criticized the show's focus on a female protagonist and her role as scientist, and one boy wished, "Maybe there could have been two scientists: one a girl, one a boy."

However, overall, children were not resistant to having a girl protagonist; and, in fact, the program's particularly 1) innovative approach to doing gender and 2) contribution to gender equality, were even explicit reasons why some children found the show to be good and why some children declared they would watch additional episodes of the series. Boys as well as girls were able to appreciate the gender-sensitive element of the show.

For example, a 9-year-old U.S. boy stated: "The show was very cool because it showed how not only boys can build cool things." An 11-year-old U.S. boy also identified as a reason why he enjoyed the program, "Because it shows that girls are into technology and that they can do it like build robots, program computers, and change the world." These boys were clearly questioning the gender stereotype of which boys are technology savvy and girls are not. Additionally, an 11-year-old U.S. girl liking the program, "...because it is funny and proves girls can do anything they want." This girl clearly felt that the program encouraged her to expand her inner beliefs regarding what girls can do. The stereotype that holds that 'girls do not get excited about robots and androids' is shown - by the qualitative statements provided by some girls participating in the study - to be as wrong as the assumption that 'a program with a female lead cannot be entertaining for boys.'

Yet, there was a major difference between girls and boys in the reception of the program in regards to the matter of which character the children liked the best (see Figure 1).

In this respect, a clear gender difference occurred, as there was a strong preference ($p<0.001$) for the same gender character (Cramer's $V=0.716$). The majority of the girls (67%) liked the main character, Anne, the most, while the boys' favorite character was Nick (with 45% opting for him). The girls ranked in second place the character, Shania, followed by Nick and Eyes. In contrast, the boys ranked in second place the character, Eyes, followed by Hand, Anne, and Pal. Only 12% of the boys chose Anne as their favorite character. The results of this study thus show that the

Figure 1. Answers, by gender, to the question, "Which character did you like best?"

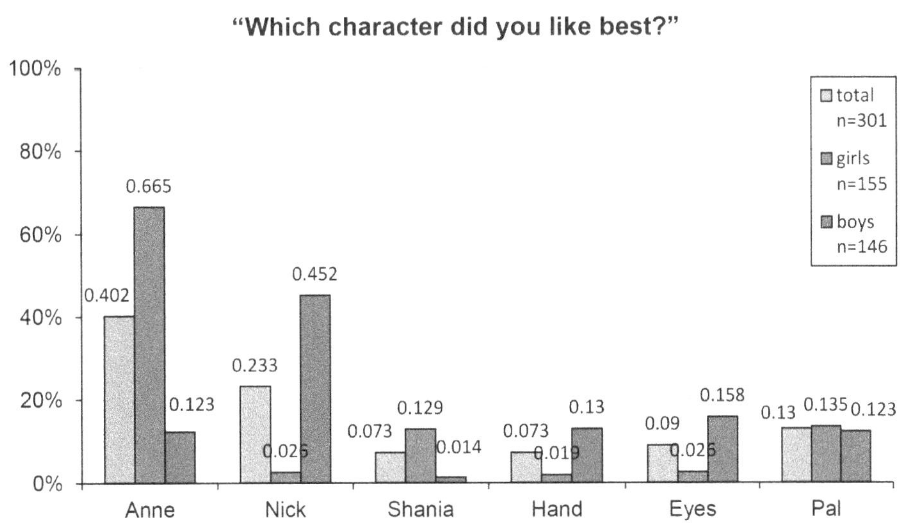

characters which are most attractive for children are in the majority of the cases of the same gender as themselves. This is further reinforced by the boy's preference for Eyes (a male robot character) rather than for the girl's, female robot characters, and non-gendered (for the episodes they watched) android Pal. This again shows the need for more role models of girl characters with STEM capacities in media programming. However, when children are "gender switching" in their choice of the favorite character, boys do so more than girls, and this result contradicts the rarely challenged notion among children's media executives that boys prefer male characters whereas girls are satisfied with characters of any gender (outlined by Lemish, 2010). Regarding age differences, study results show that younger as well as older children liked Anne the most, fairly evenly. For Nick, the 10- to 12-year-old children (30%) liked him more than the younger children (18%).

For answers to the question, "Is one of the characters a bit like you?", there was again a strong relationship ($p<0.001$) between the gender of the child and the gender of the favorite character (Cramer's $V=0.640$). Overall, the character with which the most children recognized similarities was Anne, and even 17% of the boys saw a similarity with her. In second place was the character, Nick, with even 9% of the girls seeing potential for identifying themselves with him. An even gender distribution can be found for the character Pal, which reaches 10% agreement in total.

Regarding age differences, findings were pertinent for characters, Anne and Nick (see Figure 2).[3]

Figure 2. Answers, by age, to the question, "Is one of the characters a bit like you?"

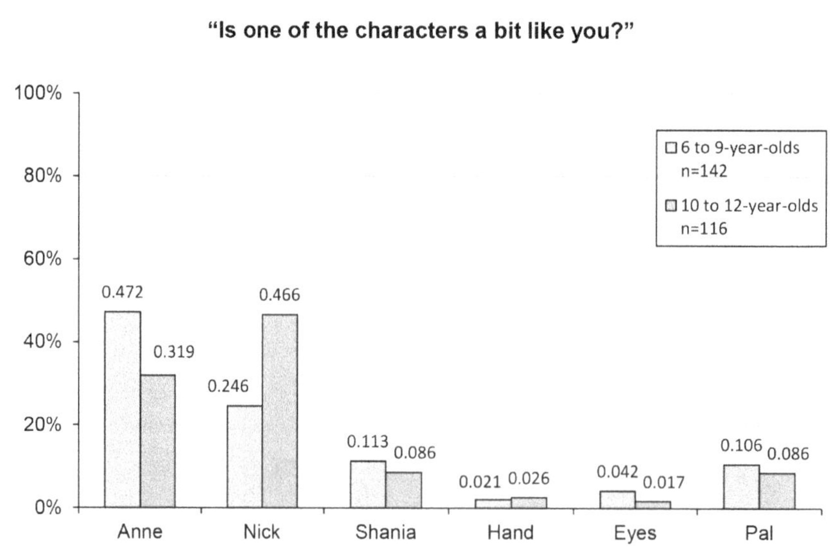

A larger percentage of the 6- to 9-year-old children (47%) realized similarities with the main character, Anne, than the older age group (32%). In the case of Nick, the situation was reversed: 47% of the 10- to 12-year-old children but only a quarter of the 6- to 9-year-old children acknowledged similarities with this character.

This all suggests that age might be an even more important category for identifying with a specific character than gender, although this perspective has not been, to date, encountered in the scholarly literature. This is even more important when noting that Anne and Nick are, as television characters, and as actress and actor, playing the roles of the same age.

The reception study data complements the producer's comments, showcasing various aspects related to the media, STEM, and gender from children's perspectives. Children's overall assessment of the *Annedroids* series shows that the producer's determination to create an entertaining and educational program with a hands-on and fictional spin including androids and robots was well received by girls as well as boys, as was the gender-sensitive approach. However, it can be noted that, in terms of liking specific characters and identifying with them, children remained, overall, divided by gender lines; although "gender switching" does occur and age also does play a role, too. The reception study shows that television programming featuring girls as inventors and operators of androids and robots can be successful, although much still remains to be done to have consistent such production and instill in children an even greater sense of gender justice.

Robots and Androids

While human characters were important in the economy of the show from both the producer's and the children's perspectives, technological companions were also key: they added elements of excitement, and they carried further gender and STEM considerations.

In the words of the producer of the *Annedroids* series, J. J. Johnson,

Initially the robots were perhaps just a means to an end so that Anne could undertake dangerous experiments. It did not take long however for their construction to take on a dynamic all of its own, with the effect that they became much more than this. Ultimately, they carry a large part of the plot. Hand is a female character, Eyes is a male character, and Pal is for a long time undetermined as gender, so they add to Anne's featuring as the lead.

Hand

Discussing the creation of the robot character named Hand, Johnson states,

I remember as a kid seeing a car assembly line for the first time and seeing those robots all kind of working in unison building cars. The image stuck with me, a massive building with this huge line of robots building in unison without a single person in sight. remember wondering where those robots ended up. Hand's upper body was crafted off that memory (see Figure 3).

Continuing with the role taken by the robot character named Hand, Johnson affirms,

Hand is Anne's largest creation and built specifically to be able to lift and move things that Anne cannot. She can't speak but rather gestures with her upper and lower claw, simulating facial expressions. One of the ideas of the show was to play with how we communicate with each other, so even though Hand is non-verbal she's actually extremely expressive.

In the reception study, Hand was the favorite character for some children. Some children stated that they liked the character but could not clearly identify why: "It is one of those things that just looks appealing to you," stated a 9-year-old U.S. girl. However, other children were able to more clearly express what attracted them to the female robot character, Hand: "Hand is a cool and calm robot. Hand is really useful to Anne. Hand is big and somewhat looks like a crane and a claw.

Figure 3. Female robot character, Hand

Hand usually helps in the problem and it is interesting to see how Hand helps," wrote an 11-year-old Canadian boy. In addition to the characteristics of usefulness and helpfulness, some children declared that they liked Hand for the strength and technological capabilities manifested. "Hands resembles strength and the capability of technology," wrote an 11-year-old U.S. boy. Another 11-year-old boy described Hand as explorative, "curious." A child even sensed the similarities between robot Hand and dinosaurs like those featured in *Dino Dan*: "Hand is like a robot version of a T-Rex. And as far as logic goes, trained T-Rex's make great bodyguards," wrote yet another 11-year-old U.S. boy.

The character Hand seems to be associated with the fantasy of something really powerful but tamed, child protective; a highly attractive parasocial friend. While most of the ideas the creator had in mind about Hand were quickly picked up by children, one aspect seemed harder to grasp, at least after watching only two episodes of the show: specifically, that Hand was a female. Most kids discussed Hand as a male, stating, "He is cool" or "he's hilarious" (11-year-old U.S. boys).

Eyes

Commenting on creation and characteristics of the robot character named Eyes, Johnson says,

Eyes is Anne's most curious robot. It is also a non-speaking robot, but makes a lot of beeps and wild worrying sounds that clearly suggests what Eyes is thinking. He is a combination of a microscope, a telescope and binoculars. So he allows Anne to see the world outside of her junkyard environment, much further than she would with her regular eyes. Again, this is something that young Anne built to be able to explore the world around her from the safe confines of her junkyard home (see Figure 4).

In the reception study, children described Eyes as an explorative character. For example, a 10-year-old Canadian girl stated that Eyes "likes to watch people with his security camera." When Eyes was identified as a favorite character it was also "because he likes to explore stuff," as an 11-year-old U.S. boy mentioned, or "because he is like a moving camera," a nine-year-old U.S. boy asserted. Some children saw similarities between themselves and the robot due to this feature: "Eye is like me because he spies on people and I spy of my friends," wrote a 10-year-old Canadian girl. However, children picked up various other aspects related to Eyes; for example, his friendliness to Anne, as he was described not just as "curious" but also as "protective" yet, also, his occasional shyness, in regards to other characters. For example, an 11-year-old Canadian girl stated, "He is shy and sometimes I could

Figure 4. Male robot character, Eyes

be shy," while a nine-year-old U.S. boy also wrote, "I would be like him because I like hiding from people and I am a good hider."

While there was little focus on robot Eyes' gender, in both the producer's verbiage and the words of the children participating in the study, it is notable that Eyes is featured, and interpreted as, inquisitive but a bit withdrawn, in a form of representation often associated with masculinity.

Pal

Discussing the creation and characteristics of the android character named Pal, Johnson states,

Pal, Anne's latest creation, was meant to be the most humanoid android. A character that would allow us to talk about life because Pal is experiencing it for the first time. Questions about the meaning of life that might sound weird or disingenuous coming from a kid character seem perfectly appropriate coming from Pal. In a way, it gave us unlimited access to asking questions about what it's like to be a human being because that, at its core, is Pal's quest: to be a person. When I step back, I now realize that Annedroids really tracks Pal's entire life story. In the first episode, Pal is brought to life, and in the last episode, Pal faces death and makes a huge decision for itself. Along the way, Pal discovers what it is to exist, and to make friends.

In the reception study, an 8-year-old U.S. girl tersely depicts what is typical for her about Pal: "*Asks why a lot and is funny and smart.*" Pal is, overall, recognized by children as being explorative, entertaining, smart, and creative. Children also describe Pal as thirsty for knowledge: "Pal is always ready to explore new things. He always asks questions. Pal always wants to learn new things," wrote an 11-year-old U.S. girl. When children chose Pal as their favorite character, it was mainly due to external appearance, looks, size, and reception of voice. Some children also liked Pal the most because of his musical and/or technical skills. "He is good at helping and building," stated a 9-year-old U.S. girl. Some children also realized that Pal's human-mechanical nature was special: "I like my character the best because he's just like a human but an android. And he can talk and learn things as they talk in the lessons," wrote an 11-year-old U.S. boy. Among all the non-human characters, Pal is the one with which children can see the most similarity, mainly because they, too, see themselves as explorers, observers, and adventurers. "Pal learns something new every day and I do, too, at school," mentioned an 11-year-old U.S. girl (see Figure 5).

Figure 5. Android character, Pal

On the whole, these statements demonstrate that both girls and boys perceived the robots and androids as different personalities, even as figures with whom they identify. They evoke fantasies of strength, and children are moved by the obvious human characteristics they see in themselves, or would like to see in a friend. Of course, in so doing they gender technology as male, and presumably, many other gender-sensitive figures are needed in order to change this association of technology with masculinity.

Robots, Androids, Gender, and STEM Knowledge and Attitudes

To get additional insights into what children specifically get out of the show regarding knowledge of STEM and attitudes towards STEM, a pre-test and a post-test were performed as part of the reception study. Before and after they watched episodes of the show, children were assessed in regards to scientific knowledge, their beliefs about technology, their attitudes towards STEM and gender, and their general views on gender stereotypes. The results show that children gained some knowledge from the show.

In the open question part, most children stated that they learned something new about androids and robots from the program. "Androids act and sometimes look like people. Robots are machines that help people do things," wrote a 9-year-old U.S. girl. Additionally, children realized that they learned facts about STEM issues from the program. "I also learned that laser light beams can be seen easily in fog or dust," wrote a 9-year-old Canadian boy.

When the knowledge outcome was measured, for example, how laser light spreads, more children (85%) chose the right answer, namely 'in one direction,' after viewing the program than before (53%). Before viewing the show, girls started with less STEM knowledge, for example, just 45% of the girls knew 'laser is spreading in one direction' as compared to 61% of the boys; after viewing the show, the gap was almost bridged, as 89% of the boys as well as 82% of the girls knew the right answer (see Figure 6).

Through a story with two girls, one boy, and robot and android characters of various gender configurations, children were able to quickly learn at least some facts about STEM issues.

To get even more insights into attitudes towards girls and STEM, children were asked to react to several statements involving stereotypes. One such statement was "Girls and robots, that doesn't really go together." After watching the show, there were small but significant differences for the girls ($p<0.001$) and for the boys ($p<0.01$) in changing from agreeing to disagreeing with the statement. Another set of STEM and gender statements that children were exposed to was related to future job perspectives. After watching two episodes of *Annedroids*, girls agreed

Figure 6. Girls' knowledge of how laser light spreads before and after show viewing

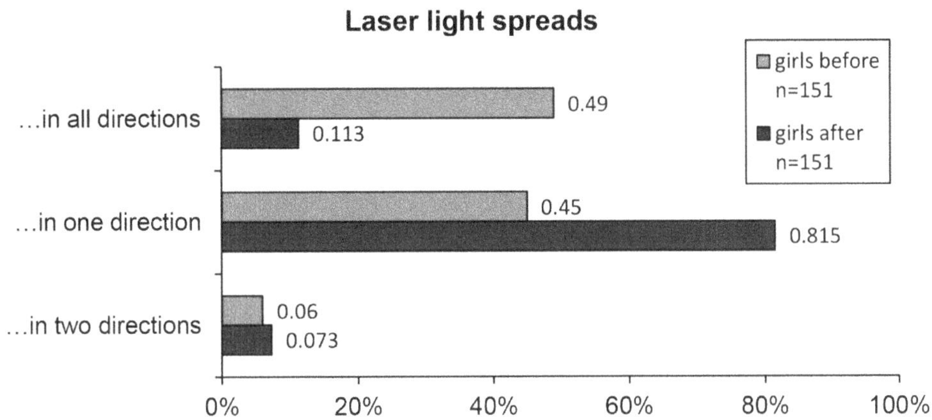

more with the categories: "invent something with the computer," "work with my hand to produce or build something," and "sit in an office and organize something important." All these go along with professions in STEM areas. Moreover, upon watching the show, girls were less keen on "typically female" job prospects such as "make others more beautiful e.g. as makeup artist or hairdresser" and "on a stage, e.g. as a singer in a band." These results shouldn't be overestimated; they represent just a tick on a questionnaire, and might change quickly after watching regular TV shows with their dominant gender stereotypes. Nevertheless, these results allow for hope that programs putting girls at the center, featuring them as inventors and operators of androids and robots, could help foster gender equality in regards to STEM professions.

CONCLUSION

Based on an examination of the producer's comments as well as the reception study data, this research documents that a program like the *Annedroids* series can provoke interest in the subject of androids and robots by telling stories which are entertaining and educational for most children. The girl lead is very appealing to girls and, with few exceptions, an interesting figure for boys, too. Robots and androids are perceived as companions, helpers, and protectors who add ingenuity and lend strength, as beings that are curious,and in which children can also see various aspects of themselves.

Returning to the producer's perspective, when asked, "What should this show make possible?," J. J. Johnson made the following concluding statement:

I love that word "possible." I just want kids to know that regardless of where you are in life, it is possible to dream and create. I think technology is incredible, and science is ridiculously exciting. It just has to be shown in that special light. Like any bad teacher, you can make a subject boring. But when I think about science and the shows that we have done, it would be such a failure if we couldn't make those ideas and those concepts spectacular to watch. I don't want to talk about baking soda experiments when we can talk to kids about our ever-expanding universes and robots becoming conscious. There are so many bigger ideas that we should be open to talking to kids about, because those are the things that to me will blow their minds open to even greater possibilities. I think we are always holding back rather than just going for it and seeing what they come away with. Annedroids is not trying to show the A-B-to-C of how to build a robot. We are trying to talk about ideas and beliefs: What do you think about feeling machines? What does evolution mean? What are stars and is there alien life – and how cool is all that? These are the kinds of things that I think don't get focused on enough, because we either believe that kids cannot handle it, or we just don't think that they are as sophisticated as they clearly are. We're now getting Facebook posts of kids building their own little androids and their own little robotic creations inspired by Annedroids, and parents writing to say that their boys and girls had never expressed an interest in science until they watched the show. I'm so excited to see where this newfound interest takes these kids.

There is still a lot to be done in regards to gender equality, especially in STEM areas.

Gender stereotypes must be dismantled for girls, boys, and adults; long-standing perspectives must be changed, and innovative approaches must be proposed, as producer J. J. Johnson learned from research data he came across at Prix Jeunesse International. Gender-sensitive programs like *Annedroids* can be a good starting point for showcasing girls, as well as boys, as inventors and operators of technologies, such as those involving androids and robots, and also for offering counter-stereotypical ways of doing gender through human characters and their technological companions. However, such programs must be complemented by various messages and campaigns in schools and in everyday life. But the idea is out there, and it can be achieved, as acknowledged by this 11-year-old U.S. girl who explains why Anne is her favorite character in the show: "*Because Anne proves my gender can do anything.*" It is precisely because STEM themes are not part of the usual social constructions of gender that we need fictional characters who show that it is possible to be enthusiastic about - and competent in - science and technology, without that automatically meaning venturing outside social integration.

REFERENCES

Aubrey, J. S., & Harrison, K. (2004). The gender-role content of childrens favorite television programs and its links to their gender-related perceptions. *Media Psychology*, *6*(2), 111–146. doi:10.1207/s1532785xmep0602_1

Baker, K., & Raney, A. A. (2007). Equally super? Gender-role stereotyping of superheroes in childrens animated programs. *Mass Communication & Society*, *10*(1), 25–41. doi:10.1080/15205430709337003

Buckingham, D. (1996). *Moving images: Understanding children's emotional responses to television*. Manchester: Manchester University Press.

Buckingham, D. (Ed.). (2005). *The media literacy of children and young people*. London: University of London.

Buckingham, D. (2013). *Media education: Literacy, learning, and contemporary culture*. Cambridge: Polity Press.

Butler, J. (1990). *Gender trouble: Feminism and the subversion of identity*. New York, NY: Routledge.

Byerly, K. M., & Ross, K. (2008). *Women and media: A critical introduction*. Malden, MA: Blackwell.

Chimba, M., & Kitzinger, J. (2010). Bimbo or boffin? Women in science: An analysis of media representations and how female scientists negotiate cultural contradictions. *Public Understanding of Science (Bristol, England)*, *19*(5), 609–624. PMID:21553601

Cvencek, D., Meltzoff, A., & Greenwald, A. (2011). Math-Gender Stereotypes in Elementary School Children. *Child Development*, *82*(3), 766–779. doi:10.1111/j.1467-8624.2010.01529.x PMID:21410915

De Beauvoir, S. (1953). *The second sex. New York, NY.* Alfred: Knopf.

Dudo, A., Brossard, D., Shanahan, J., Scheufele, D., Morgan, M., & Signorielli, N. (2011). Science on television in the 21st century: Recent trends in portrayals and their contributions to public attitudes towards science. *Communication Research*, *38*(6), 754–777. doi:10.1177/0093650210384988

Fisch, S. M. (2005). Children's learning from television. It's not just "violence." *TelevIZIon*, 18(E), 10-14.

Fisch, S. M., & Truglio, R. T. (Eds.). (2001). *G" is for growing. Thirty years of research on children and Sesame Street*. Mahwah, N.J.: Erlbaum.

Götz, M. (Ed.). (2014). *TV-hero(in)es of boys and girls*. Frankfurt: Peter Lang. doi:10.3726/978-3-653-04304-4

Götz, M., & Lemish, D. (Eds.). (2012). *Sexy girls, heroes and funny losers*. Frankfurt: Peter Lang. doi:10.3726/978-3-653-01426-6

Götz, M., Lemish, D., Aidman, A., & Moon, H. (2005). *Media and the make-believe worlds of children*. Mahwah, N.J.: Lawrence Erlbaum.

Halpern, D., Benbow, C., Geary, D., Gur, R., Hyde, J., & Gernsbacher, M. (2007). The science of sex differences in science and mathematics. *Psychological Science in the Public Interest*, *8*(1), 1–51. doi:10.1111/j.1529-1006.2007.00032.x PMID:25530726

Holler, A. (2012). "He speaks Russian - just like me." How preschool programmes can support language learning. *TelevIZIon, 25*(E), 16-19.

Holler, A., & Götz, M. (2013). I like Dino Dan "because it was pure fantasy:" A reception study with preschool children. *TelevIZIon, 26*(E), 34-38.

Holler, A., Götz, M., Egerer, A., & Schwarz, J. (2012). This is the TV presenter I'd like to have. *TelevIZIon 25*/2012/E, S. 30-31.

Jacobs, J., Davis-Kean, P., Bleeker, M., Eccles, J. S., & Malanchuk, O. (2005). I can, but I don't want to: The impact of parents, interests, and activities on gender differences in math. In A. Gallagher & J. Kaufman (Eds.), Gender differences in mathematics (pp. 246-263). Cambridge, MA: Cambridge University Press.

Jenkins, H. (Ed.). (2009). *Confronting the challenges of participatory culture: Media education in the 21st century*. Cambridge, MA: MIT Press.

Kirjnen, T., & Bauwen, S. (2015). *Gender and media: Representing, producing, consuming*. New York, NY: Routledge.

Kondo, K. (2009). Can television be good for children? In N. N. Nagaraj (Ed.), Effect of television on children (pp. 129-156). Hyderabad: Icfai University Press.

Lemish, D. (2006). *Children and television: A global perspective*. New York, NY: Wiley-Blackwell.

Lemish, D. (2010). *Screening gender in children's TV: The views of producers around the world*. New York, NY: Routledge.

Lemish, D. (2014). Using media to aid children in war, crisis, and vulnerable circumstances. In D. Romer & A. Jordan (Eds.), Media and the wellbeing of children and adolescents (pp. 162-171). Oxford: Oxford University Press.

Linebarger, D. L., Piotrowski, J., & Vaala, S. (2009). Supplementing television: What enhances or detracts from the power of television to teach. In N. N. Nagaraj (Ed.), Effect of television on children (pp. 45-71). Hyderabad: Icfai University Press.

Linebarger, D. L., & Walker, D. (2005). Infants and toddlers television viewing and language outcomes. *The American Behavioral Scientist*, *48*(5), 624–645. doi:10.1177/0002764204271505

Livingstone, S. (2007). Do the media harm children? Reflections on new approaches to an old problem. *Journal of Children and Media*, *1*(1), 5–14. doi:10.1080/17482790601005009

Livingstone, S., & Lemish, D. (2001). *Doing comparative research with children and young people*. London: London School of Economics.

Luther, G. A., & Legg, J. R. Jr. (2010). Gender differences in depictions of social and physical aggression in childrens television cartoons in the U.S. *Journal of Children and Media*, *4*(2), 191–205. doi:10.1080/17482791003629651

Maltese, A., & Tai, R. (2011). Pipeline persistence: Examining the association of educational experiences with earned degrees in STEM among U.S. students. *Science Education*, *95*(5), 877–907. doi:10.1002/sce.20441

Mares, M. L., & Acosta, E. E. (2010). Teaching inclusiveness via TV narratives in the U.S. *Journal of Children and Media*, *4*(3), 231–247. doi:10.1080/17482798.2010.486127

Margolis, J., & Fisher, A. (2002). *Unlocking the clubhouse: Women in computing*. Cambridge, MA: MIT Press.

Moss-Racusin, C., Molenda, A., & Cramer, C. (2015). Can evidence impact attitudes? Public reactions to evidence of gender bias in STEM fields. *Psychology of Women Quarterly*, *39*(2), 194–209, 39, 2, 194–209. doi:10.1177/0361684314565777

Myers, K., Jahn, J., Gailliard, B., & Stoltzfus, K. (2011). Vocational anticipatory socialization (VAS): A communicative model of adolescents interests in STEM. *Management Communication Quarterly*, *25*(1), 87–120. doi:10.1177/0893318910377068

Rideout, V. J., Foehr, U. G., & Roberts, D. F. (2010). Generation M2: Media in the Lives of 8- to 18-Year-Olds. Menlo Park, CA: Kaiser Family Foundation. Retrieved from http://kff.org/other/event/generation-m2-media-in-the-lives-of

Schlote, E. (2012). Learning how to learn. *TelevIZIon, 25*(E), 26-29.

Schlote, E. (2013). Wer begleitet Kinder in Erklär- und Wissenssendungen? Eine Medienanalyse der LernbegleiterInnen in Wissenssendungen in Deutschland, Großbritannien und den U.S.A. *TelevIZIon*, *26*(2), 44–47.

Shapiro, J., & Williams, A. (2012). The role of stereotype threats in undermining girls and womens performance and interest in STEM fields. *Sex Roles*, *66*(3-4), 175–183. doi:10.1007/s11199-011-0051-0

Smith, S. L., & Cook, C. A. (2008). *Female stereotypes: An analysis of popular film and television*. Los Angeles, CA: The Geena Davis Institute for Gender and Media.

Steinke, J., Lapinski, M., Crocker, N., Zietsman-Thomas, A., Williams, Y., Evergreen, S., & Kuchibhotla, S. (2007). Assessing media influences on middle school-aged childrens perception of women in science using the draw-a-scientist test (DAST). *Science Communication*, *29*(1), 35–64. doi:10.1177/1075547007306508

Unterstell, S., Götz, M., & Holler, A. (2012). Eat plenty of vegetables and drink lots of milk. *TelevIZIon 25*(E), 33-37.

Wainwright, D., & Linebarger, D. L. (2009). Television can teach. Elements of effective educational television. In N. N. Nagaraj (Ed.), *Effect of television on children* (pp. 24–44). Hyderabad: Icfai University Press.

Watt, H., & Eccles, J. (2008). *Gender and occupational outcomes: Longitudinal assessment of individual, social, and cultural Influences*. Washington, D.C.: American Psychological Association. doi:10.1037/11706-000

West, C., & Zimmerman, D. H. (1987). Doing gender. *Gender & Society*, *1*(2), 125–151. doi:10.1177/0891243287001002002

West, C., & Zimmerman, D. H. (2009). Accounting for doing gender. *Gender & Society*, *23*(1), 112–122. doi:10.1177/0891243208326529

ENDNOTES

1. A German sample follows in December 2015 and January 2016.
2. The correlation between age and liking of the program is weak but highly significant ($r = 0.160$; $p < 0.001$).
3. There is a significant relationship between age and the recognition of similarities with the favorite character (Cramer's $V = 0.238$, $p < 0.05$).

Chapter 3
Are You Really a Child?
Androids and Cyborgs in Japanese Comics and Animations

Natalia Dmitruk
University of Wrocław, Poland

ABSTRACT

A multitude of genres and types of characters, in Japanese comics and animated series, suggests many thought-provoking themes; i.e., questions about human nature. Many artists can see the answers to these questions in artificial humans – both cyborgs and androids. In this research, the author analyzes Japanese texts of popular culture in which artificial children are the protagonists of the stories. The author aims to compare a child figure in sociological discourse, considered there as vulnerable, to the representations in manga and anime, in which characters are created as children or technologically-modified prepubescents. In this chapter, the author presents ideas and culture associations for the concepts of android and cyborg. The chapter focuses also on analysis of the characters from Japanese comic books and animations – both androids and then cyborgs – according to transhumanistic and posthumanistic theories. The analysis results in a conclusion that a child figure is dehumanized in the context of cyborg and android child protagonists.

INTRODUCTION

Transhumanism and posthumanism are parallel intellectual movements that seek to understand the nature of humanity and the possibility of crossing its borders. In our times, the question of the future of man is increasingly being posed from the technology angle. It allows us to look at the current development of culture from

DOI: 10.4018/978-1-5225-2973-6.ch003

a different perspective, as well as at all the possibilities that will be transformed in the near future.

Posthumanism draws attention to non-anthropocentric prospects: technical entities (machines – especially robots, artificial intelligence). Transhumanism can be viewed as a way of human improvement, or enhancement; not using traditional humanistic methods (e.g., education), but technology. While posthumanism is associated more with a shift in our understanding of other entities, transhumanism is related mainly to technological modifications of the human body. Both ideas are connected to technology and humanity on different levels, as shown later in the chapter.

Polish researcher Magdalena Radkowska-Walkowicz uses the term "artificial human" to describe both android and cyborg beings. As she states, they are creatures used by people to learn more about their own identity (Radkowska-Walkowicz, 2008, p. 24). Since both androids and cyborgs are present in literature, cinematography, comic books and animations, they have become objects of interests for scientists, sociologists, and recipients of texts of popular culture. One of the most popular sources of android and cyborg protagonists is Japanese comic books and animations. However, what comes as a surprise when compared to the Western texts of culture, in manga and anime, artificial humans are often shown as children.

In this research, presentation of cyborgs and androids in the form of a child is used as a means to show the attitude of Japanese authors to technology and mankind, and the relationship between these two. However, before that, it is necessary to explain the concept of an artificial human, as it is considered to be a posthumanistic and transhumanistic mirror of humanity. An observation of android robots and modified humans may contribute to a better understanding of people's own identities, and increase their self-awareness.

It is also important to ask the question; how much can androids resemble humans before they will be seen as monstrous? Japanese roboticist, Masahiro Mori, in his essay, *The Uncanny Valley,* hypothesized that a person's response to a human-like robot would abruptly shift from empathy to revulsion as it approaches, but fails to attain, a lifelike appearance. Mori called this descent into eeriness "the uncanny valley," and that concept has rapidly gained interest both in science and popular culture. Especially now, seeing as technology in the last several years has evolved to where people are building robots looking more like humans (Mori, 2012, p. 98).

ANDROIDS AND CYBORGS

"There was perhaps never a moment in the ancient or modern history of Europe when no one was pursuing the idea of making a human being by other than the ordinary reproductive means" (1984, p. 201), writes Jay David Bolter in his book, *Turing's*

Man. An attempt to create man-made creatures similar to humans is not an idea of contemporary times. It is not difficult to recall the myth of Pygmalion, the creator of a woman statue so perfect that he fell in love with her. Eventually, Aphrodite, driven by her affections, decides to revive the idol and at the same time fulfill the dream of the sculptor. Further examples include the classic Homerian, golden, mechanical handmaidens of Hephaestus in *The Iliad*; cabalistic golem from Jewish folklore; the homunculus of alchemists; and the Frankenstein monster from Mary Shelley's novel. Yet, the closer to present day, the greater the recognition of artificial creatures. Not only are androids, cyborgs, and robots more and more common in popular literature and cinematography, but also, at the same time, real-life scientists are working on cloning and their artificial intelligence (Radkowska-Walkowicz, 2008, p. 20). As Marvin Minsky wrote of artificial intelligence (AI) back in 1986,

Most people still believe that no machine could ever be conscious, or feel ambition, jealousy, humor, or have any other mental life-experience. To be sure, we are still far from being able to create machines that do all the things people do. But this only means that we need better theories about how thinking works. (p. 19)

Contemporary media offer a variety of books and movies where the main character is a robot who is almost the same as a real, full-fledged human. Undoubtedly, since a unique quality of human beings is the ability to experience and describe emotions, most of the movies and books in which cyborgs and androids appear focus on the matter of whether robots have feelings[1].

Among the most prominent Polish writers who concentrated on the issue of robotic creatures and their emotional capacity was Stanisław Lem. In one of his stories, *Wypadek*, Aniel, an unpredictable robot with a disturbingly autonomic brain, starts to behave almost like a human being. He is urged to take action – even if it results in his destruction. Polish literature gives a variety of other examples of robotic creatures who act as human beings. In the children's book, *Akademia Pana Kleksa*, Jan Brzechwa creates a story of an android named Alojzy. He is an artificial boy, brother of a full-fledged human, and a student with no respect for principles. His disruptive behavior is understood as a result of a construction defect. Later, when the android is 'repaired,' the only difference between him and humans is imagination, which Alojzy cannot use.

Development of the theme of the artificial human in literature, and its popularity in the contemporary culture, has been dramatically accelerated by the output of American science-fiction writer, Philip K. Dick. Dick offers a traditional vision of who humans are, and who (or what) robots are. However, at the same time he oversteps the boundary between them, and deliberately blurs it. His androids may be lifeless machines, or beings equal to humans (Radkowska-Walkowicz, 2008, p.

138). They may be also tragic creatures who are unaware of being artificial. An excellent example of such a creation comes from Dick's, *The Electric Ant* (1969), in which the author questions the nature of human existence. Readers are faced with questions on the essence and genuineness of human existence: "What if we were programmed? What if I only think I am a human?" The protagonist of *The Electric Ant*, Garson Poole, finds out that he is an artificial human, the eponymous electric ant. Being aware of his nature, Poole starts to see differences between him, as an organic robot, and others – fully-developed humans. Pool's loss of identity leads him to change his perception of his own existence. He does not consider himself a living creature, but rather a thing. The protagonist is unaware of being a robot. The only true moment of his existence is when Pool loses his identity and connection to other real people he knows.

David, the protagonist of Steven Spielberg's movie *Artificial Intelligence: AI* (2001), based on *Supertoys* by Brian Aldiss, is a child with programmed love for his mother. He believes he is special, one of a kind. Despite artificial, robotic origins, his feelings for his mother are genuine. Seeing other androids, the same as him, comes to David as a huge shock. In the movie, a female colleague of David's creator, Professor Hobby, asks a question: "If a robot could genuinely love a person, what responsibility does that person hold toward that mecha in return?" (Spielberg, 2001). Scientist's answer is almost blasphemous, but, at the same time, can be a subject of posthumanistic discussions: "But in the beginning, didn't God create Adam to love him?" (2001).

Steven Spielberg pays attention to another important thing, namely the morality of humans who want to acquire the divine power of creation. David, as the new Adam, has been created in the image of humans. A true love is a distinctive feature for them, and it distinguishes humans from machines. However, is a human being capable of reciprocating affection of a non-human creature? In the movie it is possible, but incomprehensible on many levels – especially from other mecha and humans' points of view. Love from creation is understandable, however, not from the creator himself. David is unique because he can love Monica, a full-fledged human who "adopted" him. He is made to be like a human: to feel, to have purpose in life, to have imagination and resolutions. However, David is still a machine. He has been programmed to be that. Just like Golem from the Jewish folklore – special words let him be more human–like, as Grażyna Gajewska writes; it is like he almost got a soul thanks to a few words (Gajewska, 2010, p. 104). Even when he is abandoned, his purposes do not change. He still loves his mother, because that love has been imprinted upon him. "A perfect child caught in a freeze-frame... always loving, never ill, never changing. With all the childless couples yearning in vain for a license, our Mecha will not only open up a completely new market but will fill a great human need" (Spielberg, 2001).

There are also texts of culture where authors focus on the differences between the human and the android, from perspectives of both of them. In another book of Philip K. Dick, *Do Androids Dream of Electric Sheep?* (1968), popularized by Ridley Scott's movie *Blade Runner* (1982), the only way to identify an android is via the Voigt-Kampff test and examination – by paying special attention to their feelings towards animals. In a dilapidated world where Rick Decard lives, runaway androids are the objects of human fear. It is the anxiety about machines which, on one hand, resemble human beings in majority of their features, but, on the other hand, are stronger, smarter, more powerful, and, above all, uncontrollable. Becoming a free robot comes with a cost of death to its human owner. Since killing people is considered a sin and goes beyond the will of the creator, it implies that the machine has made a decision on its own and against the creator(s). Androids, even with shorter life span, still want to be acknowledged as persons. "We're not computers, Sebastian, we're physical," says Roy Batty in *Blade Runner*. Philip K. Dick's novel poses a question of whether an android could be a human being. In his book, artificial beings can be tragic creatures, who want to find their own place in a reality dominated by people.

It is worth mentioning another short story, written by Isaac Asimov, *Bicentennial Man* (1976), which was expanded into a novel *The Positronic Man* (1992) by Asimov and Robert Silverberg, and later on adapted into a movie directed by Chris Columbus (*Bicentennial Man*, 1999). The story is about a robot who, because of a constructional aberration, likens itself to a human. As a consequence of obtaining a name, Andrew, the robot starts to develop his identity, and his behavior changes little by little. Since he wants to resemble a human, he starts to wear clothes and, progressively, seeks to achieve freedom in a world where robotic creatures cannot be free. However, only when Andrew alters his immortal body parts – especially his positronic brain – and as a consequence becomes vulnerable to death, do people officially acknowledge him as a human being. From the posthumanistic perspective, Andrew is a metaphorical mirror of a human. He "grows up" from a mere machine ("a child") to an entity conscious of its own nature ("an adult"). His development of individual attributes enables us to take a closer look at human nature and self-awareness.

All of these examples present various views on the matter of differences between humans and robots. Some of them imply that even simple machines may develop self-consciousness – either because of a technology defect, programmed feelings, or technological evolution and self-improvement in resemblance of humans. Frequently, readers are also asked about the possibility for machines to have souls, as if it may be classified as the only feature which truly distinguishes humans from other creatures. Minsky, author of *The Society of Minds,* writes in his book about a common concept of a soul (seen in Western culture as the main "privilege" of humans, which makes them special in their own eyes). However, for Minsky it is just a symbol. It implies that there is no significance in anyone's accomplishments. Spirits, souls, and essences are nothing more than old and obsolete beliefs. In the author's opinion, they are all indications that we are helpless to improve ourselves (Minsky, 1985, pp. 41-42).

Androids are not only distant, unrealistic characters from contemporary culture. There are many projects underway which aim for creation of a robot which, to a certain degree, would look, speak and act like a human being. The efforts are especially numerous in Japan. Companies from around the world may take part in the latest research in robot technologies expositions, if they participate in the International Robot Exhibition (IREX)[2]. However, IREX is not the only place where the latest inventions in the field of robotics are presented. In 2014, the National Museum of Emerging Science and Innovation (Miraikan), in Tokyo, hosted an exhibition titled, "Android: What is Human?" Visitors could see three of the most realistic humanoid robots: Otonaroid®, Kodomoroid®, Telenoid®. The event provided visitors an opportunity to communicate with, and operate, android robots.

Surprisingly, the exhibition was more about humans than their artificial counterparts. The robots which had contact with the visitors were, in fact, used to show the true nature of humanity, as well as human communication. Visitors took an active part in the exhibition by first operating robots, and then discussing differences between humans and androids. All of the robotic creatures were used as an answer to one of the oldest questions: "What is human?" They also constituted a new, posthumanist perspective in looking at humanity.

Magdalena Walkowska-Radkowicz, author of the Polish book, *Od Golema do Terminatora: Wizerunki sztucznego człowieka w popkulturze*, writes, that, by taking a look at the artificial being, people can tell a lot about modes of demeanor of humans in the world, and their attitude towards environment and its aspects; i.e., attitude towards God and nature, biology, time and body, foreign entities, technology and science, art, gender, memory, authority. Through androids, people can realize the fantasies regarding those aspects: being equal to God, a position of power, removal of nature's determinism, of painful conflict between the soul and a body, of inside and outside, of "I" and the representation of that (Walkowska-Radkowicz, 2008, p. 20).

However, if an organic robot is a being against nature, then cyborgs are living beings who accommodate nature and technology. They are life forms that have integrated biological parts of their bodies with artificial components. Cyborg is, as Katarzyna Więckowska describes it, an entity in which the cultural anxieties of civilization, such as fear of technology domination and the end of humanity that may follow – and expectations, like fantasies of immortality and extending durability of the human body – are saved. Even if everyday addiction to technology (like overusing phones and computers) makes people seem to be already cyborgs, bio-tech hybrid still is a figure of the future (Więckowska, 2011, p. 63).

It is important to mention that, for centuries, people made modifications to their bodies, as Diane Greco writes in her book, *Cyborg: Engineering the Body Electric*, mainly because of physical handicaps. However, having an artificial tooth, wooden hooks for lost legs, or a glass eye does not make someone a cyborg. Integration must rely on some sort of feedback; e.g., a connection between the central nervous system and artificial components – like in the fourth stage of Greco's classification (Schulz, 2003, p. 161).

The idea of a cyborg has been invented in the time of the Cold War, a state of political and military tension after World War II between countries from the Western and Eastern Blocks. Rivalry between the United States and the Soviet Union led to an increase of fascination with the technological improvement. Even before Neil Armstrong stood on the Moon in 1969 as a first human in history, scientists struggled for improvement projects which could help to explore the Universe (Dmitruk, 2013, p. 9). For the first time, the word "cyborg" was used by Manfred Clynes and Nathan S. Kline in their publication *Cyborgs and Space*, in "Astronautics" in September of 1960. The authors write about space and human biological evolution,

Space travel challenges mankind not only technologically but also spiritually, in that it invites man to take an active part in his own biological evolution. Scientific advances of the future may thus be utilized to permit man's existence in environments which differ radically from those provided by nature as we know it. (p. 26)

For Clynes and Kline, technological modifications are something more than just another project. Possibility of taking an active part in the evolution process is a metaphorical step for a human into the future. The authors, while they have been working in the National Aeronautics and Space Administration (NASA), implied that the fusion between a human body and high technology could lead to positive transformations. Clynes and Kline write:

Altering man's bodily functions to meet the requirements of extraterrestrial environments would be more logical than proving an earthly environment for him in space. (1960, p. 26)

The idea was to create a hybrid of a human and a machine that could work in extreme environments. Such a hybrid would be a person who "deliberately incorporates exogenous components extending the self-regulatory control function of the organism in order to adapt it to new environments" (1960, p. 27). From that moment, the term "cyborg" has been used to name beings that are changing the traditional divisions between mind and materiality, mind and mechanism, an animal and a human, natural and boosted progress, and, of course, life and death (Dmitruk, 2013, p. 10).

Grażyna Gajewska, author of the book, *Arcy-nie-ludzkie: Przez science-fiction do antropologii cyborgów,* states that the bio-tech figure tests the heterogeneity of the human body, avoiding synchronous identification, because the logic of creating cyborgs is based on introjection and absorption. Thus, it removes restrictions on many levels, including natural and artificial, inside and outside. It is not possible to think about cyborgs as partly organic or partly artificial without infringing completeness. Cyborg figure combines both biological and artificial components, and by that the modified human does not only have contact with technology, but technology is inseparable (Gajewska, 2010, p. 21).

For Clynes and Kline, cyborg research is a way to adapt human bodies for various environments. Environments in which people depend on machines and even, as the authors write, become slaves of the technology. Symbiosis of artificial components and organisms would help to accommodate them to space, and to function there without being troubled by natural human disadvantages.

From the very start, the project of technological modification was a contestation of the old arrangement, and the split of nature and culture. Because of the cyborg, it stopped to be so obvious. Gajewska compares ideas from Clynes and Kline to the world's most earth-shattering discoveries, like Copernicus' concept that influenced ways we look at the universe (2010, p. 22).

The author of *Arcy-nie-ludzkie…* acknowledges the American researchers' idea as one of the most important in relevance to humans, their awareness of their place in the universe, and their self-identity. The idea of the cyborg appeared when the arms race between the United States and the Soviet Union stepped into the outer space stage, and the research was important to conquer the cosmos. Those who could take control of the human body would not have problems with attaining and subduing unavailable places on Earth, as well as places beyond it.

Research on the possibility of body modifications did not stop after the Cold War. High technology improves the quality of life and brings the ability to function back to people who lost it in various ways. Technology in the 21st century is already highly advanced, but people still relate the figure of a bio-tech hybrid to an unknown future or science-fiction stories. Even if there are already ways of fusing the human body and technological components, the common concept of the cyborg, primarily familiar from popular texts of culture, is still different from the way people use technology in the 21st century.

Cyborgs in literature, cinematography or comics mostly present outstanding technological modifications. They are preparing readers to welcome "a monstrous future," i.e. an unpredictable future. Cyborgs are a kind of monsters as well. They are not a defined species, but bio-tech unknown, in which people can or cannot recognize themselves. And if technology is the element that shapes civilization, modified humans are an important part of futurological speculations, presented in popular texts (Dmitruk, 2013, p. 19).

One of the popular themes in texts of culture is using technology for military purposes: cyborgs are equipped with various modifications which greatly improve their physical capabilities. An example of this would be the title character and protagonist of the movie, *RoboCop*, directed by Paul Verhoeven in 1987. In the intention of the designers, he was going to combine the best attributes of a man and a machine. Facial expressions, voice, the way of moving – all of it connotes the machine; even part of his "name" ("Robo" from "robot") has a special, well-known meaning to the viewers. Still, even though the technology is integrated with his organism, the protagonist is not just an automaton executing orders, but a human controlling his mostly mechanical body. Amusing quips he says after particularly spectacular actions are used to remind the audience that, somehow, the cyborg still has his human consciousness.

Lux Deveraux, a character from *Universal Soldier* (Emmerich, 1992), represents another technologically-modified human who was "created" for combat operations. His modifications are not as advanced as RoboCop's; however, his mechanical way of moving, facial expression that did not change at all in the beginning, and inability to display emotions, are truly "monstrous." Regaining the memories and awareness of his modifications made Deveraux question his own identity – he is no longer a human, but a new species of living beings (Dmitruk, 2013, pp. 20-21). Both cyborgs are monstrous because of their inability (temporary or not) to express themselves, not because of technological modifications, which implies that emotions are an important part of human nature.

The author of the short story, *Johnny Mnemonic*, from 1981, William Gibson, presented a different kind of cyborgization. The body of his female character, Molly, is not only modified to make her stronger and faster – her metabolism, sensory input, and reflexes are artificially heightened as well, so she can use techniques not possible to ordinary people. Her weapon - double-edged, four centimeters long razor-sharp retractable blades hidden underneath her fingernails - is an additional part of her body that intensifies her dehumanization, monstrosity.

Aforementioned examples show how texts of popular culture use symbiosis of technology and organisms. Authors sometimes forget that the base for a cyborg is the body of a human (or other living being), and use this term incorrectly. For example, Terminator from the movie *Terminator 2: Judgement Day* (Cameron, 1991). The titular character says: "I am a cybernetic organism," which is not exactly right. From the very start he is a machine, with just an organic coat that is important to time travel, so he should be called an android. It was not an important difference for the producers or the viewers, but it shows that the term cyborg still is not properly well-known.

As we can see, after the Clynes and Kline article, cyborgization started to be an inspiration – even after people has realized that the idea of conquering the universe using bio-tech hybrids is not realistic. Their research is an inspiration to science, technology (especially the advancement of prosthetic), and texts of culture. It is a transhumanistic mirror, through which one can evaluate oneself as a human being, even though cyborgs should be considered a new, separate species, as was mentioned before. Maybe the future looks like Raymond Kurzweil presents it in his essay, promising and monstrous at the same time (Kurzweil, 2011). Such a level of modification can change the traditional human way of thinking. People can move to the new stage of evolution; nanotechnology and cyborgization can extend human life, or maybe even grant immortality.

Examples of both android and cyborg figures show that in Western popular culture, artificial adults are more common than children. In sociological discourse, a child is a vulnerable living being that requires special care and attention. It is a duty of adults to ensure children's safety and teach them the proper way of living. Therefore, technological modifications to a child's body and mind may seem to violate its rights – which is mirrored in Western popular culture. Children-androids seem to be a much "safer" theme, since an artificial being can still be treated like "just a machine." But what will happen if a child like that will gain identity or self-awareness? It may be advantageous now to consider how Japanese authors of comic books and directors use the theme of artificial children, both androids and cyborgs, and whether these themes are treated differently in Japanese popular culture than in Western popular culture.

THE ANDROID AS A CHILD

It is interesting how robots may be portrayed in Japanese comic books and animations. They can be simple dolls; puppets without free will, like two clones of Naomi Armitage from *Armitage: Dual-Matrix;* or beings almost identical to humans, like the third type of robots from *Armitage: Polymatrix*, which are so advanced that they could even give birth. Certain characters from manga and anime (Japanese comics and Japanese animation, respectively) are androids produced specifically to be lovers, as is shown in Yuu Watase's manga *Zettai Kareshi*. Artificial beings can be also created for fighting, like Android 16 from the *Dragon Ball* series, or Sechs from *Battle Angel Alita*. There are also personal computers in human form, like persocoms from a CLAMP manga *Chobits*. The only distinctive part of their technological bodies, giving away that these humanoid computers are different from humans are special ears where access cables have been placed. Those are just few examples.

Robots have a special place in Japanese popular culture which mirrors the country's needs. Japan hopes that robots can be part of the answer to the ever-declining birthrate and shrinking workforce in a country famously wary of opening its shores to immigrants. There is also a specific way in which Shinto religion blurs the line between the inanimate and the animate: everything can possess a living spirit (Thomas, 2009). This is why the theme of sentient machines can be less monstrous in the Eastern cultural texts.

Artificial children in Japanese comic books and animations are created for various purposes. They can be made specifically for military tasks, as weapons, like Mitchii and Tima from Osamu Tezuka's *Metropolis* original manga and its movie adaptation. Androids can be a replacement of dead family members, like the most popular humanoid robot in Japan, the eponymous Astro Boy from another manga by Tezuka. An artificial child can be an additional family member, or a robot functioning as a mascot. A good example is Pino from anime series *Ergo Proxy*. The physical shape of those robots do not exclude functions normally not associated with the body of a full-fledged human child. Androids can be more powerful, with much stronger artificial muscles, or other superpowers and abilities, unattainable for humans. For much more precise analysis, well-known comic books and animation were chosen. Artificial children in these texts are protagonists along with humans, which allows a view of androids from a posthuman perspective, since it undermines the anthropocentrism prevailing in culture and science, noticed by Cary Wolfe, American literature- and culture-specialist (2009, p. 137).

In the manga *Metropolis*, one of the characters, Mitchii, is an artificial child[3]. This android was made by Doctor Charles Lawton as a result of his research on synthetic cells. Duke Red, the main antagonist of *Metropolis*, threatens Lawton to make an artificial human, modeled after a marble statue – "The Angel of Rome" from Dam Dharma Museum – and possessing several superpowers such as flying, superhuman strength, and ability to breathe underwater. The doctor brings it to life; but, afraid that his creation would be used for evil purposes, destroys his own laboratory and fakes destruction of the android. The scientist rescues the artificial being and raises him as his own child in secret, naming him Mitchii. In the beginning, the truth about the "child" was known only to Lawton and Doctor Bell. Not even Mitchii knew the truth: that he was created by science and has superpowers. He could also change gender, thanks to a special mechanism in his body, but was not aware of this kind of ability[4]. An artificial boy has only one purpose in life: to find his parents. After the death of Doctor Lawton, he hears from Detective Higeoyaji that his previous guardian was not his real father; and, from that moment the android is motivated to find his family or as it is stated in the manga, he is so 'obsessed' with the idea that it drove him mad when Duke Red told him the truth.

Mitchii was created to be a slave of the antagonist organization, Red Party, just like other robots in the titular city, Metropolis. After he learns the truth about himself, his main purpose is to kill every human with the help of other artificial beings. It changes from a positive human desire (having a family) to monstrous urges to destroy. "He is a great human," at one time says Ken'ichi, the human protagonist of the manga and the best friend of Mitchii. And yet, from helpful boy, good student, and friendly child, the android starts to be a hateful tool of revenge, and names himself the representative of a better race. It is worth noticing that all of Osamu Tezuka's robots in manga *Metropolis* are not much different from humans. They are not as advanced as Mitchii, but still can express feelings. The main and most important difference between artificial beings and humans is in the way they are treated: as slaves, always having to be ready to work, and being destroyed if they do not want to obey Red Party's orders. They start a rebellion, after a suggestion from Detective Higeoyaji, and make mad Mitchii their leader.

The story is told a little bit differently in the movie adaptation of the manga. The android, Tima, was modeled after pictures of the dead daughter of Duke Red, and was going to be, according to the plan, the ruler of Metropolis and the rest of the world. She does not have as obvious superpowers as Mitchii. She is an artificial girl who was going to function as the central control unit for the organization's secret super-weapon. Unlike Mitchii, she does not want to meet her parents. Instead, she starts to question her own identity, because she is not sure if she is a real human being or one of the robots she has seen. The more she knows of the world, the more she questions her origin. It is worth quoting dialogue between Tima and Duke Red, one of the antagonists in the animated *Metropolis*:

Duke Red: You are nothing like those pathetic robots. You are not a mere human. Humans are inferior creatures confused by love and morality ruled by emotions. You are superhuman ... Yes, Tima. That's correct, you're a robot. The greatest, the most sublime creation in the entire world.

Tima: How can that be? I posses the love and emotion. You said that's human.

Duke Red: You only think you do, my dear, because we haven't completed you yet, you're not finished. The fluctuations of your supreme powers are taking the shape of such things as the emotions and love. (Maruyama & Yamaki, 2001)

Rintaro and Katsushiro Otomo, who worked on *Metropolis* together, show a reality where humans and robots coexist in a futuristic city, while artificial beings are discriminated against and segregated. Every robot has its own place. Most of them are in the city's lower levels and do the most difficult work. The gap between rich and poor is growing, and many unemployed citizens blame robots for taking their jobs. Tima sees it herself when she wanders with her friend, Kenichi, on the lowest levels of Metropolis. However, at that time she does not yet question her own identity. She thinks she is a human by flesh and blood, and she is treated like one by citizens unaware of her true form; citizens who, in fact, despise robots.

In the scene cited above, Tima asks Duke Red if she is a human or a robot. Then the antagonist lets her, and the viewers, hear the answer to another, unspoken question: "what being human means?" Love and morality ruled by emotions are, in his opinion, the most characteristic to the human race. Tima, being the greatest creation, should conquer the world from a Ziggurat tower and rule it, without being concerned about abilities not possible for a machine. The question is: why did he model an artificial girl after pictures of his own daughter? The antagonist's way of treating the android is ambivalent. In the movie, there are scenes where Duke Red is worried about Tima as he would be about his real daughter. The pictures of the dead girl are also meant to be seen by the viewers as a reminder of his past. However, in the end, the man still sees Tima as just a way to gain power. For him, feelings possessed by her are not important to his plan. They are just a byproduct of an earlier awakening. The man had precise expectations for his tool, which were not fulfilled. "The computer network and Tima's memories and emotions have been integrated," (Maruyama & Yamaki, 2001) says one of the scientists working for Duke Red. Feelings, as the element of nature, fuse with technology, the element of culture, and the robot, as a being against nature, goes mad. Memories of robots discriminated against and destroyed by humans, and Tima's madness, lead to another plan: annihilation of humanity.

Both Mitchii and Tima are created as weapons, mechanical tools in human form. Artificial intelligence of both the androids is so advanced that they could gain self-consciousness; i.e., they could have emotions on the same level as humans. What is more, both of them lose their minds because of the truth about their origins. However, artificial boy is more humanlike in his revenge. He is a leader and fights together with other robots. Mitchii still recognizes the people he knows, but it does not stop his monstrous determination. Even fighting with Ken'ichi cannot change it. Mad Tima, on the other hand, is more like a machine. Her voice changes, her face is emotionless, and, at one point, half of her body loses its artificial skin, showing the technological components. She does not have superpowers like Mitchii, but is more machine-like, sitting on a throne in the Ziggurat, where she is connected to the computer network and controls every robot in Metropolis. They, in turn, do not have free will like robots created by Osamu Tezuka, but are puppets of Tima, who lost her mind. Furthermore, her madness is more symbolic, as Shinsaku Ban (Higeoyaji in the original manga) says, "And so God's wrath descended upon the tower of the Babylon" (Rintaro & Otomo, 2001). Duke Red wanted to create something better than the inferior human race – and he made it. He just did not expect to lose control over his sublime creation.

In another manga by Osamu Tezuka, the main character is an android, an artificial child named Astro[5]. He was created by Doctor Tenma and was going to be a replacement for his son, Tobio, who died in a car accident. Atom is a masterpiece of science; however, for his creator, the most important thing is the opportunity to be together with his son again. Tenma has to teach the android how to speak, how to use his body, and even how to smile, using the "pleasure circuit" that the artificial boy possesses. The scientist treats Astro as his son, until he realizes that the android will never grow up; Atom will always be the same, never growing, never changing. Doctor Tenma rejects Astro and sells him to a robot merchant, and in the end, the artificial boy performs in a circus show. It was the head of the Ministry of Science, Professor Ochanomizu, who found the android in that place, took him in, and became his legal guardian. Later, Astro, thanks to his seven superpowers[6], fights crime, injustice, and evil, usually of robots who hate humans, robots that go berserk, or alien invaders.

In this manga, Osamu Tezuka also created a world where humans and robots coexist. However, there are laws for artificial beings, with the two most important articles: "Article 1. Robots exist to make people happy," and "Article 13. Robots shall not injure or kill humans" (Tezuka, 1951, p. 16). Yet, mechanical beings are still not equal to humans, even if they are very similar to them.

Robots can talk, get mad, laugh, help at work, and even study in schools together with human children. Laws in Astro Boy are similar to the well-known "Three Laws of Robotics" devised by science-fiction author, Isaac Asimov; especially, two of them:

"Law 1.: A robot may not injure a human being or, through inaction, allow a human being to come to harm," and, added later,

"Law 0.: A robot may not harm humanity, or, by inaction, allow humanity to come to harm" (Asimov, 2013, p. 7).

Astro could not harm any human because he is a robot and the 13th Article made by Tezuka is an absolute imperative for him. However, the audience for the manga Astro Boy was boys in elementary school, and fight scenes were still important in a comic about a brave android, one of the seven most powerful artificial beings of the world, so fights with evil robots or alien invaders were required.

Another difference between humans and robots is shown in a scene where Astro is sold to a robot merchant. The man who buys the artificial boy does not let him wear clothes, saying: "Since when do robots wear clothes?" (Tezuka, 1951, p. 21). Moreover, he is also one of the people against the idea of robots being allowed to look like humans, because, "most people don't like their robots to look too human. They are afraid they may assert some weird rights or something…" (1951, p. 21). It shows how different robots are treated in the world created by Tezuka. When Astro was a surrogate son of Doctor Tenma, he wore clothes, as a symbol of being equal to full-fledged humans – almost being one of them. After being sold, Atom is more often seen in his characteristic pants, especially when he uses his abilities.

It is a little bit different in Naoki Urasawa's manga, *Pluto*, in which Atom is one of the main characters. The author based his series on one story arc from Tezuka's *Astro Boy*, "The Greatest Robot on Earth." However, Urasawa comic books aimed at an older audience. His reinterpreted version is a suspense murder mystery about European android detective, Gesicht, trying to solve the case of a string of both robot and human deaths. Robots in Urasawa's story can, more or less, resemble humans, but clothes do not determine their status. The protagonist meets Atom when he visits the most powerful robots on Earth, who were on a potential murderer list. Gesicht first recognizes the android among a group of school children as Atom examines a snail he has found on the sidewalk and then places the snail in a bush without harming it. Naoki Urasawa, in such a simple gesture, shows the thin boundaries between humans and robots: Gesicht identifies Atom based on his greater capacity for mercy. It is very important that later (in volume eight) Atom repeats the same gesture. As Craig Fisher remarks, this repeated gesture is a minor climax in *Pluto*. After his resurrection, Atom behaves erratically, makes a bomb formula, and talks about hate and destruction. Doctor Ochanomizu fears that the new Atom might be bloodthirsty or even insane, after having Gesicht's last memories put into his head just before the robot detective was destroyed. "What you administered to Atom was extreme hatred" (Urasawa & Tezuka, 2003), says the scientist himself. However, Atom's kindness toward the snail proves that the android still has not lost himself in that kind of data (Fisher, 2011).

Robots in Urasawa's manga are not just mechanical beings with a programmed mind. Many of them, while imitating humans, have evolved – just like Andrew from *Bicentennial Man*. As the powerful sentient supercomputer asks: "You think robots don't get angry? That they don't cry? You think robots don't hate?" (Urasawa & Tezuka, 2003). It is true that artificial beings cannot be identical to full-fledged humans. In a world created by Tezuka and Urasawa the aforementioned Robot Articles exist, and, moreover, robots cannot lie. However, Atom alone is an example of how very human-like robots can be. "When I look at you, my systems for detecting a robot or human start to give me errors" (2003), says Gesicht when he meets him for the first time. Atom behaves like a cheerful human child. He looks like he truly enjoys eating ice-cream, and wants to have a popular toy, like any other full-fledged human boy. His artificial intelligence is one of the most sophisticated programs among the robot characters in *Pluto*. Atom mimics humans, understands them well and adapts to their behavior.

However, at the same time, he is still a robot who can analyze data much faster than any other artificial being, and uses his abilities, like 'superstrength' and flying. He has a child's body and his self-awareness seems to be childlike, but at the same time he is much more mature. His maturity is the proof that he is not a human, he is not a child. Even the android himself says once, "I am a robot. I do not mature" [(Urasawa & Tezuka, 2003). "I was sold to the circus, because people found my childlike appearance to be fascinating," (2003) also says Atom when he talks to Gesicht. Artificial intelligence of the android boy could be not exactly congruent to his body in assumption of the others, both robots and humans. After Atom's aforementioned resurrection, the artificial boy is much more technologically advanced: he understands one of the strongest human feelings, hate. "You are a mass of hatred… but my hatred is far greater," (Urasawa & Tezuka, 2003) Atom says to his main antagonist, Pluto.

The artificial child remembers every robot who died because of Pluto's and his creator's hate; every robot who had a family, who wanted to do something good for humans as well as for artificial beings, and who did not want to die. However, in the end, Atom understands the meaninglessness of that emotion, as he asks Doctor Ochanomizu, "Will the day that hatred disappears ever come?" (Urasawa & Tezuka, 2003). It seems somehow natural that it is asked by Atom, an artificial child, who is closer to being a human than any other robot would ever be.

As previously mentioned, androids created for military purposes have great combat skills or some supreme powers, which are possessed by humans. At the same time, they could be replacements for family members who died, though they begin their existence as possible servants, or tools, in a state of utter subservience. Another example, Pino from *Ergo Proxy*, is just like that. She was created to be a replacement child to Raul Creed and his wife. An AutoReiv[7] girl was affected with the Cogito Virus, and throughout the series, slowly develops her personal identity and her own raison d'être (Ohlson, 2010, p. 28).

It is worth mentioning that before she was infected, the android could function in two modes: Chat Mode On and Chat Mode Off. In the first mode, Pino was more human-like. She could behave like a child, and her facial expression could change vaguely. In Chat Mode Off, the artificial girl was more like a mere machine: her expression did not change, her movements were silent and unnoticeable; in other words, she was monstrous. Even Mr. Creed was afraid of his surrogate daughter in this particular mode, as shown in the second episode of the series.

After Pino is affected with the Cogito Virus, Chat Modes are not necessary anymore. She can express herself better with words and gestures; she behaves like the curious child she was meant to be. She gains self-consciousness, thanks to the Virus, but also creates her own identity, thanks to travels with two companions in the series, Vincent and Re-l. As Dai Sato, *Ergo Proxy* writer, says, "AutoReivs have to decide whether the virus that infected them created their identity, or whether they gained their identity through their travels" (McGray, 2005). The Cogito Virus was the origin, the spark and inspiration to go on a journey of self-discovery. However, journey itself does not allow Pino to develop as a person. She learns about the emotions most fundamental to humans, like happiness and sadness, and about life and death, as well. When her human friend, Timothy, is killed, the AutoReiv child tries to call him at first. When the boy does not respond, she tries to find "another Timothy." Someone has to explain to her what death means (Ohlson, 2010, p. 125):

Have you seen Timothy? I've been searching for him everywhere, but I can't find him.

Timothy died, didn't he?

Yeah that's right. So, I was wondering if there's another Timothy somewhere.

Pino. When people die, you can't see them anymore.

Really? Is that true?

That's right.

I see. I wanted to play with him again.

Are you lonely, Pino?

Lonely?

Yes, What you are feeling right now is called loneliness. (Murase, 2006)

Pino could grasp the connection between death, loneliness and sadness – and through the journey, the experience she gained, tuition and mimicry, this artificial girl could develop more complex personality (Ohlson, 2010, p. 125). The way the android child understands art is a reflection of her development. First, the AutoReiv imitates the things she sees, like Timothy's drawings. However, by the end of the series she is able to use her own imagination, an ability not possible in the programming. Pino's unique drawings are proofs that the infected android is almost on the same emotional level as humans. She gains more significant status accepted by other people as well. Her interest in playing musical instruments is also interesting. When she was a surrogate daughter of Creeds, she was programmed to play the piano. Later, during her journey, she is often seen playing a melodica. The Cogito Virus did not change her programmed abilities, but it extended them, giving her consciousness of them. As Kristeva, one of the characters, says: "The Cogito Virus made us develop egos, which means we obey ourselves [...] Just follow what your heart is telling you" (Murase, 2006).

The AutoReiv's journey, as Pinocchio's, finishes when the character becomes a real child. Pino, opposite to the wooden protagonist from the children's novel written by Carlo Collodi, did not want to become "a real girl" – she needed to run away from the city of Romdeau. She is seen by Romdeau's residents as a monstrous AutoReiv, an android with self-consciousness, unable to be controlled – a threat to the utopian system of the city. In her self-development journey, Pino knows she is an android infected by the Cogito Virus. She is aware of that, as well as the fact that she has changed; i.e., that she is different from other robots. In one of the last episodes, she is waiting for Raul Creed, and her internal monologue is heard by the viewers:

Hey, Papa. Pino has a lot of feelings. Happiness, sadness, and many more! Pino has gone far, far away. But after Pino was separated from Papa, Pino realized what it's like to miss Papa. Papa, you have lots of feelings too, don't you? Papa? Will Papa miss Pino, too? (Murase, 2006)

The scene where Raul Creed looks at Pino's drawings and then tries to find her, shows that he, indeed, misses her. The man's relationship with the AutoReiv is not shown before the android is infected, and affected by, the Cogito Virus. However, by the end of the series, Creed is looking for Pino as a father looking for his real daughter. She was accepted by him.

In all of the examples in the aforementioned series, the relationship between humans and androids is an important theme. Authors use the theme to show connections between nature and technology. Artificial beings can be accepted by humans, or be the target of hate, especially when they gain self-awareness. Actual images of humans, genealogy, and genetics are a large part of humanity's identity as both individuals and as a collective.

All of it represents the barriers that exist for many humans in sympathizing with the Other, especially machines. Breaking barriers with, as some may see it, forced transition of the artificial beings into the realm of humans, becomes very unsettling for those who believe in traditional definitions of "life," "soul," and "human" (Ohlson, 2010, p. 94). The most evident example is shown in *Pluto*. The underground, extreme right-wing, anti-robot organization calling for an end to robot rights is made of humans who hate the equality of artificial beings. The root of their ideals is the notion that "robots don't have souls," producing statements such as, "Robots are inferior!," "Robots are servants!," "Robots are slaves!," and "Scrap the international bill of robot rights!" Consequently, the world of *Ergo Proxy* has an AutoReiv Disposal Unit in Romdeau. Workers hunt down infected AutoReivs in order to keep Fellow Citizens safe. In the *Metropolis* movie, robots are targets of hate from two sides: revolutionary unemployed humans who live on the low levels of the titular city, and the Marduk Party, which aims to assuage anti-robot sentiments by destroying some of the artificial beings.

All these examples may suggest that people desire the power of gods, but are afraid when their creations are equal to - or better than - their creators. However, the child bodies of artificial beings are like a soothing element in the narrative process for how androids gain awareness through sophisticated A.I. or some kind of defect. Self-consciousness in these creatures can accommodate the type of body they have; due to child-like attitudes, they can be less monstrous. Mitchii and Tima are both friends of a human boy, Atom is a friend to Doctor Ochanomizu, and Pino comes to terms with many people who know that she is infected with the Cogito Virus. While the border between nature and technology does not disappear, it is more fluid.

TECHNOLOGICALLY-MODIFIED CHILDREN

Cyborgs, on the other hand, are beings with strictly connected biological body parts and technological components. The border between them disappears, changed on the level of the physical body, as well as on the levels of mind and identity. The traditional definition of the human, as noted earlier, conjuncts with the biological body and the immortal soul. From that point of view, the most important question is, how much the biological body can be modified to still be acknowledged as that of a human?

Cyborgization can be complete or partial. In the first case, biological parts in the artificial body are reduced to a minimum. Sometimes it can be just the central nervous system, modified accordingly to the needs of a cyborg, while in other cases the mind can be copied and imprinted to an artificial brain, like in the *Ghost in the Shell* series by Masamune Shirow. Artificial bodies can take different forms: humanoid (like Alita from *Battle Angel Alita*, manga written and illustrated by Yukito Kishiro), more robotic (Briareos Hecatonhires from *Appleseed*, created by the mentioned Masamune Shirow), or even monstrous (Makaku from *Battle Angel Alita).* However, in partial modification, cyborgization is limited to few artificial components and changes on the organic level.

A good example would be Edward Elric from the series *Fullmetal Alchemist*, written and illustrated by Hiromu Arakawa. The manga's protagonist has an artificial leg and arm. Of course, not only humans can be cyborgs. In some Japanese texts of culture, they can be also humanoids with animal brains operating an artificial body (Android#44, Commander of the Hot Dog Corps from Osamu Tezuka's *Astro Boy*), as well as modified animals (the cat protagonist of Naoki Yokouchi's *Cyborg Kuro-chan*), or extraterrestrials (Frieza from Akira Toriyama's *Dragon Ball).* Still, the most popular cyborg characters in manga and anime are humans with technological components or a completely artificial body.

Due to interpolation between biological anatomy and technological components, a cybernetic organism can be more effective, more enduring, and able to better perform complicated activities (Wróblewski, 2011, p. 90). It is noteworthy that manga artists and anime directors connect that with the usual function of many cyborgs in Japanese popular texts of culture, that of combat. Of course, it corresponds with the idea of military experiments, desire to explore the universe, as well as an idea of absolute power and control of the Earth because of cyborgization.

It is significant that humans with technologically-modified bodies are usually characters in *shounen* and *seinen* types of manga. Both genres place emphasis on the presentation of combat abilities of cyborgs, and on more advanced capabilities of computer systems and network manipulation, which intensifies differences between the natural and artificial body. Being a living weapon distances cyborgs from their biological predecessors. It is worth mentioning that minimization of biology in cyborg bodies imposes a higher level of their dehumanization, making them more monstrous. Inversely, of course, characters with few artificial body parts are perceived as more human.

Another issue is whether character (complete) cyborgization is a personal decision. An address of this concern can explain the conflict between the mind and the artificial body of protagonists of certain manga and anime series. Such conflict causes identity problems for those characters. An interesting example is Albert Heinrich from the series *Cyborg 009*, written and illustrated by Ishinomori Shoutaro. Heinrich's life is in danger, whereby technological modification is the only option to save him. However, for the scientists from the organization known as Black Ghost who changed his body, he is just an object of research experimentation – he has no human value. Due to cyborgization, Heinrich becomes a living weapon, in the literal sense. His entire body gets "upgraded" with weapons[8]. The German cyborg does not have to use any other weapons like other cyborg characters from the series, and that is why he feels more objectified. But Heinrich does not want to lose his identity as a human; he does not want to be just a tool of potential war:

Are You Really a Child?

I'm not a weapon.

Come on... You are not a human anymore. To be neither human or a weapon, why don't you make yourself an [sic] useful weapon? (Kawagoe, 2011)

Being a living, breathing weapon has been imposed on Albert by people who changed his body without his permission. If as a cyborg he accepted the function imposed on him by the Black Ghost organization, it would mean a resignation from his own human nature, from his identity. Heinrich is not able to decide about his own body, and the changes are irreversible, but he still tries to be himself: i.e., a human.

It is a slight bit different in regard to children cyborgization. Adults have a moral right to make decisions about their own life, their own bodies. Children, however, are unable to make complex decisions for themselves, as it is always part of their guardians' or authorities' responsibilities. According to this, adults can agree to cyborgization of vulnerable children, even if it will change them forever. However, it depends on the purpose of technological modifications. In Japanese comic books and animations, artists present different points of view for this ethical issue.

The story of the aforementioned series *Cyborg 009* refers to the Clynes and Kline idea of the cyborg, as it is written in the introduction of the first volume of the comic book:

...group known only as the Black Ghost, have developed super soldiers known as the cyborgs, designed to fight the wars of the world... but out in space. (Ishinomori, 2006)

The author tries to show the sober mindset of Cold War reception by the people who lived then, as is further written in the introduction:

It was a world pock-marked with hatred; the Military Industrial Complex fashioned nations into superpowers and attempted to supersede nations themselves; distrust fueled paranoia to bring us the Red Scare, Mass Genocide, and eventually the Hydrogen Bomb; and racism, thought prohibited in many countries by legislation, maintained its stranglehold on the hearts and minds of far too many civilized citizens of the planet. (Ishinomori, 2006)

Protagonists of the series are cyborg experiments of the Black Ghost Organization during the Cold War, intended to be weapons in the future. In the end, the characters do not agree to being tools of war. They prefer to use their superhuman abilities for good purposes, like fighting evil formations or military.

One of these characters is Ivan Whisky, a Russian infant who is modified by his father, Doctor Gamo. The mad scientist is interested in unlocking the human brain's potential, and, through the experiments on his son, he succeeds. Due to his surgical and technological intervention, Ivan can command the brain's full potential. When the Black Ghost group takes both of them into their care, Ivan is the first of the cyborg prototypes, to receive a new name, or number, of "001," and identifies as the most powerful of all of the nine numbered bio-tech hybrids. Called "The Brain," he is mentor to other protagonists, and to the main strategist, as his mental abilities are beyond those of humans. Ivan is like an adult in an infant's body, and, due to his electronic brain, he is powerful enough to use telepathy, telekinesis, ESP, and teleportation. However, the massive power needed to properly use these abilities puts an immense strain on Ivan's infant body, so the character is usually seen sleeping. While Ishinomori Shoutaro created this controversial character for many reasons; the moment when Ivan's body is modification accomplished is an attempt to confront ethical problems concerning experimentation on people.

In that respect, cyborgization can be considered a technological aggression. One unusual occurrence is that the needs of his infant body, such as sleeping or eating, do not change after he is made into a cyborg. Naturally, it results in limitations to the superhuman abilities he has. Potential of the 001 is focused not on physical confrontation, but on tactical support, just like the other cyborg member and only woman in the group, Francoise Arnoul (number "003"). Both of them display care for humanity. However, Ivan is more precise and sober in his decisions, sometimes even ruthless, as he believes that a sacrifice may be required for the greater good. Ishinomori often shows both artificial humans together on the same comic frames, with 003 holding 001 like a mother holding her child. In that respect, the author weakens the impression of these two cyborgs as military tools, and their dehumanization demise.

The mangaka (Japanese name for a manga artist) created characters who do not choose their modifications. Protagonists are deceived or kidnapped for the sake of plans of the military organization, and for an advantage in future warfare. They are victims of the Cold War and human greed. Cyborg 001 is the youngest of the artificial humans, though his mental age is far beyond any other. Because of that, the connection between Ivan's mind and physical appearance seems to be more monstrous. Just like android children, he stays infantile forever in an almost immortal body.

Another artist, Yu Aida, chooses to present a different kind of cyborg. The storyline of *Gunslinger Girl* focuses on semi-artificial protagonists with a limited lifespan. As cyborgization not only means changing the physical structure, it requires a medical brainwashing process, also, called "conditioning," whereby the technologically-modified character is made loyal to its creators. All cyborg characters are physically challenged people before cyborgization; young girls, to be precise, as modification procedures work better on younger patients. However, conditioning is a permanent process, and the treatment must be repeated whenever a child gets injured. The more it is used on cyborgs, the higher the risk that girls will become addicted or will suffer memory dysfunction, as the process is detrimental to the brain and to live tissue.

In the story, technological modification is the main interest of the secret service organization (Social Welfare Agency), and cyborgs are used by the government for unofficial missions as a special group of assassins. Artificial implants, like synthetic skin and muscles, as well as composite strengthened bones, increase girls' combat abilities. Due to military schooling, modified children are capable of using melee weapons, as well as firearms. Their hand-to-hand combat proficiency is also very high.

With all these abilities and their conditioning, cyborg protagonists make for perfect assassins and soldiers. However, they are also children – and gender in that case intensifies the problem of ethical methods for modifying their bodies. It is worth mentioning that almost every cyborg survived a different kind of trauma or incident before modification[9]. As the conditioning process helps erase bad memories, the Social Welfare Agency takes them into government group care and gives them new life as cyborgs. Even Rico, who is the only one who still has her memories after technological modification, is thankful to the organization. The young girl was bedridden before cyborgization, and changing her body was – in her opinion – the greatest chance she could ever get. Moreover, that the treatment was necessary, her lifespan was not very long, and she had to take orders to assassinate people, did not matter:

I really enjoy my life in the agency's dormitory. They're all nice to me so long as I do what I'm told [...] Every morning, when I wake up, first thought I have is: "I wonder if I still have my body." What a relief! It still works. I can't describe in words how wonderful it is to have a body that works. I love my life at the Social Welfare Agency very much. (Aida, 2000)

Each of the girls has a handler, a cyborg operator, paired with her. They are adult partners, guardians, and teachers to semi-artificial children. They decide how much conditioning needs to be performed. They also give orders on missions. Yu Aida shows different attitudes toward cyborgs, from using them as subordinates and military tools (Rico), to taking care of them as if a real sister (Henrietta), or even a daughter (Triela). On the flip-side, girls treat these men as the most important people in their lives. It was handlers who gave them names when the girls woke up first time after modification, and handlers were the ones who taught the cyborgs everything, including responsibility. That can be understood due to the fact that the cyborgs wanted their partners to be satisfied with their assignments. As children, deprived of healthy family relationships, they oriented their feelings to the closest adults. More than that, every girl treated the safety of her handler as the most important task in every mission, more important than the mission itself. They stand ready to lose their lives for the sake of their guardians:

- *A cyborg exists to protect her handler. It's natural to die for him.*

- *If that's what Jean says... It must be true. (Aida, 2000)*

It is hard to determine how much of their feelings are real or simply the effect of conditioning. However, cyborg care for handlers, at certain moments, seems to be fanatical. In one specific scene, Henrietta attacks a waiter in a restaurant because he holds a knife and a fork. She read the situation as dangerous, and a threat to Jose's life. Cyborgs also want to be of use by any and all means; as the handler says, "she counts the bodies she created for me" (Aida, 2000).

Yu Aida chooses to present controversial themes in the series, *Gunslinger Girl*. The author fractures socio-cultural concepts of the child as an innocent being. It is grotesque and monstrous how different the young girls are on missions – assassinating people – and in their spare time, like when they learn how to play musical instruments, read books, work in a garden, or spend time together eating cakes. Their innocence is intensified due to their ruthless act of killing people – which they never question. They do not feel any remorse and do not even think about their victims once the mission is over. Natural pureness of a child is confronted with mechanical precision of the machine. Aida, however, also shows the moral problem of using high technology in seemingly good purposes, when girls, as victims of violence, needed medical treatment. Technical modification and erasure of painful memories of the past are options to help them – but not a true reason. Creating cyborg assassins, the most effective soldiers, is the imperative objective in political games presented in the series. Children have a short lifespan, making them nothing more than tools. However, some characters say, ironically, that it is hard to perceive them just as killing machines: "these little girls who speak three languages and usually spend their time fighting terrorists... are standing in the cold and singing Beethoven. It's too bad they have to be cyborgs, huh?" (Aida, 2000).

It is hard to see them only as little girls since they are so dangerous. It is also worth mentioning that cyborgs from *Gunslinger Girl* do not have any problem with self-identification, like *Cyborg 009* protagonists. Girls know that their most important task is to be useful to handlers. If it requires a lot of people dying in the process, it is not a problem.

In both cases of Ishinomori Shoutaro's and Yu Aida's comic books, cyborgization is a way to make a human body more useful in combat. Well-known cyborg from a popular text of culture, Alita, protagonist of the manga *Battle Angel Alita* created by Kishiro Yukito, also uses a lot of bodies this way (few times she rebuilds her body or gets a new one). However, in the last title from the "GUNNM" series – *Gunnm Kasei Senki* – the author creates comic books about her childhood. Alita was a cyborg even when she was little, and it is unknown why she was modified. In the first volume of *Kaisei Senki*, the little girl is not yet accustomed to her machine body. Body functions do not work well, so she gets a remotely-controlled walking support that executes a program with basic actions, like walking or turning left or right. Due to that, Alita (at that time called Yoko) becomes the subject of occasional ridicule from other children in her world, where high technology is common. However, Yoko's mechanical body is called 'gross,' and she is compared to a radio-controlled robot. Being technologically modified for her is a difficult disadvantage. Later, she perfectly operates, and makes her modification useful.

Cyborg children can be a comic relief in manga. Disk from *Needless* is a kind of controversial example. The author does not use the word "cyborg," but instead the term "Halves" (from "half machines"), to describes humans whose bodies are combined with machinery, leaving parts of their brain and major organs intact. As a cyborg, the protagonist is more of a hacker than a semi-human combatant. Getting information is more important to her than the fight. Disk, just like Ivan Whisky from *Cyborg 009*, is a tactician and support, including for medical purposes. She can survive while being decapitated, as her brain and the operating system will not be damaged.

Needless, as a comedy manga, is full of gags, and one of the most popular related to Disk is her being a cute, prepubescent girl, reminding other characters that she is a hundred years old. It also corresponds with the character trend popular in Japanese media known as *lolicon* (*roricon*, or short – *loli*, from Nabokov's *Lolita*). In manga and anime discourse, it shifted away from its original and common meaning, as stated by Akagi Akira. It does not refer to child pornography, nor describe the relationship of a middle-aged man and a young girl, as it was in Nabokov's novel. Age of *loli* characters is not the attractive part to the audience, but rather a form of "cuteness" that they represent as idealized in Japanese popular media (Shigematsu, 1999, pp. 129-133). Disk, just like cyborgs from *Gunslinger Girl*, is a prepubescent girl; but the term *loli* better corresponds with her. As Shigematsu Setsu writes: "Substitution and mimicry are characteristic of the way that *roricon* operated to transform straight sex into a parodic form" (1999, p. 130). Disk matches that kind of trend in Japanese comic books and animations. She is sexualized, with revealing clothes, and specific kind of gags. For example, as a cyborg with special powers, she can assess sizes of specific parts of a female body. Because *Needless* belongs to the *ecchi* genre of manga and anime, it focuses on sexual situations and the female body, but at a comedy, or even parody, level.

The body of every previously mentioned character was modified, with only a little of original organic tissue left. All of them have bodies of children and, due to that, they can be faced with different modes of reception. It can be comic relief when an innocent-looking girl faces a group of dangerous, bulky men, and wins; or confusion, when an adult personality does not match a childlike face. It can also be horrific and monstrous when a child behaves like an emotionless machine. However, the most important fact is the controversial nature of this theme, because the child body, as a representation of innocence, is being technologically violated. That raises a question about ethics and morality of adults who make these type of decisions. Some authors present cyborgization as the only way to provide medical treatment to a damaged organism, others as a way for scientists to satisfying curiosity, and a need for domination. Still, the most popular cyborgs in Japanese comic books and animations are those with military lineage.

CONCLUSION

In contemporary popular culture, artificial humans, both androids and cyborgs, are often found. Manga and anime, especially, are full of humanlike robots and technologically-modified humans; their popularity can be traced to Japanese advanced technological development. Cyborg and android figures can help humans understand and take a stance in confrontation with a technocratic reality. They are specific demonstrations of different mindsets about nature and culture associations, human nature, and the perception of technology.

Artificial children are one of the most controversial means to present these associations. Robots can look and behave like humans, until something, or someone, changes their operating system. They can also be powerful enough to be deemed dangerous, just like cyborgs. Both figures can be monstrous for mainly one reason: dehumanization of a child. Despite a pure personification of innocence, the mechanical movements and emotionless behaviors of artificial children can make them inhuman. Another anxiety emerges from their advanced abilities, as having a technologically-advanced body can be perceived as better than the imperfect, biological form. Powerful children are intimidating and seemingly unnatural, due to the fact they do not need the help of adults, something unacceptable in children's discourse. However, in Japanese comic books and animations, authors use specific ways of presenting artificial children characters.

As for androids, manga artists and anime directors usually present them as innocent children who try to learn about the world or other people. They are created so perfect that they sometimes do not know that they are not full-fledged humans. Tragedy of their existence lies in the fact that it is always the human who can affect them. Their machine bodies are created by humans, sometimes for wicked reasons. As creators, people can try to change their own creations – however, it is not simple. Texts of culture are full of androids that rebel against humans. In Japanese comic books and animations, a robot will go mad when it finally discovers the truth about itself. Manga and anime artists like to create their artificial protagonists as characters with a soul or soul-like element. Due to the influence of Shinto religion and the belief that everything can possess a living spirit, humanlike androids do not seem to be unnatural.

Cyborgs in Japanese popular culture seem to have more free will than androids. They were originally humans, and they understand nuances about being organic. Due to technological modifications, they can represent a new way of thinking about the body itself. As artificial beings, they can change their body parts if damaged. They can be stronger than humans, and have unnatural abilities. Japanese artists usually focus on reasons for cyborgization, and identity of the cyborg. Their characters can be in conflict between their human mind and their technologically-advanced body.

In fiction, both cyborg and android children are still less popular than artificial teenagers or grown-ups. The cause for that may lie in the controversy of violating children's bodies. Creating weapons in a child form, or modifying them for that purpose, raises questions concerning the ethics of technology. However, analyzed manga and anime series define a specific approach to technology from the perspective of a country that is oriented to expansion of high technology. Presentation of "an artificial human" in popular culture may be a reflection of the views of society on an advanced technology itself. Even though Japanese texts continue to be a subject of analysis in the West, the scope of such research is very narrow. This essay should be considered a starting point for further research on the topic.

REFERENCES

Aida, Y. (2012). *Gunslinger Girl* (Vols. 1-15). Tokyo: ASCII Media Works.

Asaka, M. (Director). (2003). Gunslinger Girl [Television series]. Japan: Madhouse Studios.

Asimov, I. (2013). Ja, robot. (Z. A. Królicki, Trans.). Poznań: Dom Wydawniczy Rebis. (Original work published from: 1950).

Asimov, I., & Silverberg, R. (1998). The Positronic Man (P. Hermanowski, Trans.). Poznań: Dom Wydawniczy Rebis. (Original work published from 1992).

Bolter, J. D. (1984). *Turing's man: Western culture in the computer age.* North Carolina: The University of North Carolina Press.

Brzechwa, J. (1990). *Akademia Pana Kleksa*. Wrocław: Siedmioróg.

Clynes, M. E., & Kline, N. S. (1960). *Cyborgs and space*. In *Astronautics*, 26-27, 74-75. Retrieved from http://web.mit.edu/digitalapollo/Documents/Chapter1/cyborgs.pdf

Columbus, C., Petersen, W., Katz, G., Mark, L., Miller, N., Radcliffe, M., & Bernathan, M. (Producers) & Columbus C. (Director). (1999). *Bicentennial Man* [Motion picture]. United States & Canada: Buena Vista Pictures, Columbia TriStar, Sony Pictures Entertainment Inc.

Dick, P. K. (2002). Impostor. In Impostor. Test na człowieczeństwo. (M. Gawlik-Małkowska, Trans.). Warszawa: Prószyński i S-ka. (Original work published from 1959), 375-392.

Dick, P. K. (2003). Elektryczna mrówka, In Opowiadania najlepsze. (M. Begiert & J. Karłowski & T. Oljasz, Trans.). Poznań: Dom Wydawniczy Rebis. (Original work published from 1969).

Dick, P. K. (2011). Blade Runner: Czy androidy marzą o elektrycznych owcach? (Kędzierski S.). Warszawa: Dom Wydawniczy Rebis. (Original work published from 1968).

Dmitruk, N. (2013). *Tożsamość cyborga w komiksie i animacji japońskiej (na wybranych przykładach)*. Unpublished Thesis, University of Wrocław, Wrocław.

Fisher, C. (2011). Pluto and Doubling. *The Comics Journal*. Retrieved from http://www.tcj.com/pluto-and-doubling

Gibson, W. (2015). Neuromancer. In *Trylogia ciągu* (P. W. Cholera, Trans.). Warszawa: Wydawnictwo MAG.

Humanity+ Page. (n. d.) Retrieved from: http://humanityplus.org

Imai, K. (2013). *Needless* (Vols. 1-17). Tokyo: Shueisha.

Ishinomori, S. (2006). Cyborg 009 (Vols. 1-10). California, LA: Tokyopop. (Original work published 1964-1981).

Kawagoe, J. (Director). (2011). Cyborg 009: The Cyborg Soldier [Television series]. Japan: Avenc Inc. & Japan Vistec.

Kennedy, K., Spielberg, S., & Bonnie, C. (Producers), & Spielberg S. (Director). (2001). *A.I. Artificial Intelligence* [Motion picture]. United States: DreamWorks Pictures.

Kishiro, Y. (2014-ongoing). Gunnm: Mars chronicle (Vols. 1-2). Tokyo: Kodansha.

Kurzweil, R. (2011). *Singularity Q&A*. Retrieved from http://www.kurzweilai.net/singularity-q-a. (Original work published 2005).

Lem, S. (2013). Wypadek. In *Opowieści o pilocie Pirxie* (pp. 276–311). Kraków: Wydawnictwo Literackie.

Mano, A., & Ishidori, H. (Directors). (2008). Gunslinger Girl -Il Teatrino- [Television series]. Japan: Artland.

Maruyama, M., & Yamaki, I. (Producers), Rintaro, & Otomo K. (Directors). (2001). *Metropolis* [Motion picture]. Japan: Madhouse Studios.

Minsky, M. (1985). *Society of mind*. New York: Simon & Schuster.

Miraikan. (2014). Android: *What is Human?* National Museum of Emerging Science and Innovation. Retrieved from https://www.miraikan.jst.go.jp/en/exhibition/future/robot/android.html

Mori, M. (2012, June). *The Uncanny Valley*. (K. F. MacDorman & N. Kageki, Trans.). In *IEEE Robotics & Automation Magazine* (vol. 19), 98-100. Retrieved from http://ieeexplore.ieee.org/stamp/stamp.jsp?arnumber=6213238

Murase, S. (Director). (2006). Ergo Proxy [Television series]. Japan: Manglobe INC.

Ohlson, A. (2010). Anime: Fear and Anxiety in Texhnolyzed Worlds (Unpublished Thesis). Hamilton, New Zealand: University of Waikato. Retrieved from http://researchcommons.waikato.ac.nz/

Radkowska-Walkowicz, M. (2008). *Od Golema do Terminatora: Wizerunki sztucznego człowieka w kulturze*. Warszawa: Wydawnictwa Akademickie i Profesjonalne.

Sakoi, M. (Director). (2009). Needless [Television series]. Japan: Madhouse Studios.

Schulz, M. (2003). Ciało do korekty - cyborgi. In P. Kletkowski & P. Marecki (Eds.), *Nowe Nawigacje II. Azja - cyberpunk - kino niezależne* (pp. 157–176). Kraków: Rabid.

Shigematsu, S. (1999). Dimensions of desire: Sex, fantasy and fetish in Japanese comics. In J. A. Lent (Ed.), *Themes and issues in Asian cartooning: Cute, cheap, mad and sexy*. Bowling Green, OH: Bowling Green State University Popular Press.

Sobol, A. (2003). Cyberprzestrzeń – elektryczny afterlife. In P. Kletkowski & P. Marecki (Eds.), *Nowe nawigacje II. Azja - cyberpunk - kino niezależne* (pp. 137–156). Kraków: Rabid.

Świech, A. (2003). Cyberpunk: kroniki w chromie. In P. Kletkowski & P. Marecki (Eds.), *Nowe Nawigacje II. Azja - cyberpunk - kino niezależne* (pp. 125–136). Kraków: Rabid.

Tezuka, O. (2004). *Metropolis* (Vol. 1). Mozów: Wydawnictwo Arashi. (originally published 1949).

Tezuka, O. (2004). *Astro Boy* (Vols. 1-23). Milwaukie, Oregon: Dark Horse Comics. (originally purblished 1952-1968).

Thomas, L. (2009, August 03). What's behind Japan's Love Affair with Robots? *Time*. Retrieved from: http://content.time.com/time/world/article/0,8599,1913913,00.html

Urasawa, N., & Tezuka, O. (2009). *Pluto* (Vols. 1-8). Tokyo: Shogakukan Inc.

Więckowska, K. (2011). Potomkowie Mary Shelley: Mężczyźni, kobiety i cyborgi. Literaria Copernicana, 2(8), 62-71.

Wolfe, C. (2009). *What is posthumanism*. Minneapolis: The University of Minnesota Press.

Wróblewski, M. (2011). Cyborgizacja codzienności człowieka: Stan współczesny i kilka myśli futurologicznych. Literaria Copernicana 2(8), 8-19.

ENDNOTES

1 This kind of implies that people who have problems with emotions are not exactly fully-fledged humans.

2 IREX is organized since 1973 by the Japan Robot Association (JARA) and Nikkan Kogyo Shimbun. English version of the IREX site: http://biz.nikkan.co.jp/eve/irex/english/; English version of the JARA site: http://www.jara.jp/e/h/information.html.

3 This version of the name is used in the Polish collector's edition of Osamu Tezuka's *Metropolis*.

4 For most of the story, Mitchii is a boy, so usually the masculine form will be used.

5 "Astro" and "Atom" will be used as equivalents.

6 Astro could fly through the air like a space rocket, learned to speak 60 languages fluently, could amplify his hearing a thousand times, and could use his eyes as searchlights. His strength was equal to 100,000 horsepower, he had machine guns in his rear end, and he could sense whether people were good or bad, if they lie or not.

7 In *Ergo Proxy* androids are called AutoReivs – both names will be used in the article as synonyms.

8 For example, a machine gun in his right hand, a knife in his left arm, and missile launchers in his legs.

9 Triela was kidnapped, and then drugged, tortured and sexually assaulted, with all of it recorded as a snuff film. Angelica's father wanted to collect the life insurance he had on her, so he almost killed his daughter in a staged car accident. Henrietta was "assaulted repeatedly all night long next to the bodies of her family" (Aida, 2000), as it is mentioned in the first volume of the comic book.

Chapter 4
Mecha-Media:
How Are Androids, Cyborgs, and Robots Presented and Received Through the Media?

Roger Andre Søraa
NTNU Norwegian University of Science and Technology, Norway

ABSTRACT

How are robots, androids and cyborgs presented and received in the media? This chapter applies a social media analysis to this question by using empirical research on news stories that feature robotic technologies to see how robots are presented, consider what reporters focus on when writing about robots, and review how the public discusses and receives robots. The theoretical framework utilised focuses on how robot narratives are framed, how robot controversies are presented in different media, and how robots are domesticated through the media. The two main cases are a "robot hotel" in Japan, and a "killer robot" at a Volkswagen factory in Germany. News coverage of both stories shows widely differing ways for how the robot-narrative is framed.

INTRODUCTION

The term "robot" has been around for quite some time, with deep roots in fictional works. Since entities such as androids and cyborgs have entered the discourse, there is a need to analyze how these "non-human human-lookalikes" are portrayed. This chapter does so by a media-analysis which gives expression on how robots are *portrayed*, and how they are *received*. The *portrayal* of robots will be analyzed

DOI: 10.4018/978-1-5225-2973-6.ch004

by media-analysis strategies, tackling the who, what, where, and how, of concrete news-articles on robots. The analysis on how robots are *received* will utilize research techniques that analyse some new social media applications and comment fields found on the Internet. By analyzing public opinions seen in comment fields - tweets regarding shows where robots are found; e.g., on Facebook groups and forums discussing robots - the author will qualitatively explore ways the general public receives robots.

In addition to the two focus areas of portrayal and reception, a thorough background will be given for each empirical sample on search terms, frequency, statistics, and background information on the given media. The media analysis will cover news articles from 2015, focusing primarily on articles written in English, originating from Europe and the US, but also with focus on robots in Japan. All three of these regions are robot hotspots in terms of production and/or consumption, with Japan being the primary producer of robots, and until recently the top consumer. In 2016, China surpassed Japan in the number of robot units bought per year. The US and EU represent the two largest economies in the world affecting the money flowing into robotic research. However, they are, along with Japan, troubled by an increasingly aging population. Robots are seen as one possible solution to combat this problem.

Different robots are presented in different ways, and the presentation varies, not only between robots, but also between countries. Simply put, robots are presented very differently by the media in the US, Europe, and Japan. This is perhaps tied to "homegrown" robots; i.e., Japanese robots might be better presented and received in Japan than American ones, and vice-versa.

While many articles written about robots are quite positive, focusing on how new robotic technology can benefit mankind, there are also critical voices. The interesting thing in the portrayal of robots is not only the actual context the robot is reported to perform in, but also the multiple critical stories in the media regarding what robots can be capable of with present new technology. This matter is heavily tied to cultural contexts, and with an analysis of the media presentation and reception of robots, can be further investigated.

The research for this chapter is methodologically founded in *Grounded Theory*, which is an inductive theory-building method (Charmaz & Smith, 2003; Glaser & Strauss, 2009; Strauss & Corbin, 1994). Grounded Theory starts with data, and follows it, seeing whom it touches and how, and it then builds the theory on the basis of the empirical data collection. The strength of the Grounded Theory method is that it enables us to not "blind ourselves" with theory before searching for the data. In a media context, this can be especially fruitful, as the author is primarily concerned with the way the data is presented and received, and not how it can fit into existing theoretical frameworks.

The reason these particular stories have been chosen for this chapter is because they represent different *framings* of the controversy of robots in the media. They are also examples of the three types of robotic beings that this book describes: robots, androids, and cyborgs. It shows that, especially for humanoid robots (androids) and robotic humans (cyborgs), certain "doomsday" prophecy scenarios are occurring more and more often. The data material case studies people who comment on news articles online. The chapter aims to give qualitative insight into commenters that make commentary on robot cases; but, is too narrow to give qualitative general numbers on reception rates of robots, not necessarily seeing how robots are received by everyone, but how they can be received by particular groups of people who comment on news stories about robots, thereby shining light on the phenomenon. The news stories are not neutral; journalists write them in specific cultural settings with their various interests, biases, and goals.

THEORETICAL BACKGROUND

The different theoretical concepts used for analyzing what may be called 'Mecha-Media' in this chapter are culled from Studies of Science and Technology (STS). STS studies how and why societal and cultural factors impact science and technology and, conversely, how and why science and technology affect society and culture. The theories used in this chapter are interlinked as follows: The chapter studies how robotics as a *controversy* is *framed* in contemporary media, following the *narrative* of these actors, seeing how they are *domesticated* by the receptors of the select media articles. In studies of science and technology, *Controversy Studies* is often applied to see the process of scientific change (Pinch & Bijker, 1984; Engelhardt & Caplan, 1987; Martin & Richards, 1995). This text uses the term in a broader conceptual framing, adopting it to a wider understanding of the techno-artifact topic under discussion; namely, seeing how technology in society (and also implied to some degree, how "science incorporated") is 'understood' by contemporary society. Inspiration for this more techno-controversial focus can be found in the work of Herman Tavani.

Tavani's book, *Ethics and Technology*, questions how the "cyber landscape" is ethically controversial (2003). Whilst it does so primarily in a cyberspace context, it also gives some useful insight in how to understand techno-controversies. Science and technology as concepts are deeply interwoven, perhaps especially in the case of robots; i.e., since they are, to an extent, science incorporated in technology, then "released/given" to society.

These *controversies* in the media contest how robots are *framed* in the media-landscape. In the social sciences, *framing* is a set of theoretical perspectives used to analyze how a concept - here, robots - is selectively communicated within a community or society in order to see how the concept is specifically socially constructed there. The three geographical areas chosen for this chapter of the US, Europe, and Japan are all highly technologically-advanced nations, with the US and Japan being world leaders in robotics with state-of-the-art development and research. A specific focus is on Norway, also, partly due to high levels of welfare for its citizens alongside its high usage and acceptance rates of new technologies. Also, as discussed herein, some media themes deemed 'dangerous' may be 'stretched' in *framing*, as in the case of the "killer robot."

The narrative that is the focus here for the robot actor is how the robot is presented, and, when appropriate, it delves deeper into investigation on why the robot is presented in particular ways. Franzosi (1998) suggests the reason sociologists should be interested in narrative-analysis when saying, "Narrative texts are packed with sociological information, and a great deal of our empirical evidence is in narrative form." Often, the same story is very different as presented by various media. This is key to understanding how humans adapt to mechanical robot creations.

The domestication of animals - namely, herding them, which makes them capable of living with humans - is inspiration for the domestication of technologies, where researchers look into how technologies can be "accepted," how they may be utilized by their human users, and how technologies affect humans in return. There are important ways how robots are *domesticated* through the media, and how the way they are *framed* affects their domesticated value, which can be seen through the numbers of purchased robots today.

Silverstone's original Domestication theory discussed a four-phase model of domestication: 1) Appropriation, 2) Objectification, 3) Incorporation, and 4) Conversion. This model has received criticism for being too deterministic and linear; therefore, a more dynamic model called *The Symbolic Model of Domestication* is used here. "Making Technology Our Own" by Lie and Sørensen, while dated, discusses technology as a mutually interacting concept, seeing the physical, symbolic, and cognitive aspects of domestication technologies (1996). They see 1) how consumers adapt to, and master, technologies in social contexts, 2) how they relate to technology, and 3) how they redefine themselves due to technology. How technology is domesticated prior to reaching the home of end-users is critical here; i.e., if one can actually accept a sort of "media-domestication" of robots. So is an analysis of technology that is not (yet) released for home-use, but is available through exhibitions, at factories, for military usage, for people on holiday, etc. How are these "not in your home but close to it" technologies received by the general public? Furthermore, how, if so, is technology domesticated, and through what kind of media?

MEDIA PRESENTATION AND RECEPTION

How are robots, androids, and cyborgs presented and received in the media? In seeing how robots are presented and received, there is need to consider what reporters (and possibly interviewed robot creators) focus on when speaking about a robot, and how the public discusses and receives a robot, using social media analysis. As noted, this theoretical framework focuses on how robot narratives are framed, how robot controversies are presented in different media, and how robots are domesticated through the media.

The two main case studies looked at are a "robot hotel" in Japan, and a "killer robot" at a Volkswagen factory in Germany, both with wide differences in how the robot-narrative is framed. These different ways of telling the stories of robots, and how that affects receptors, are key to their media analysis. By reviewing these two cases, it can be seen how the robot narrative is interwoven with robots at work in factories and their presentation to society. Results radically differ in the story framing of each of the robots, due to one being primarily comedic (the hotel with smiling-robots) and the other being a tragedy (the factory with death-robots).

CONTROVERSY 1 (C1): THE ROBOT HOTEL

At the outskirts of the Japanese city of Nagasaki lies Huis Ten Bosch, a theme-park replica of Holland, with windmills, Dutch houses, and a great number of tulips. Next to this peculiarity in the landscape stands one very strange hotel that is staffed primarily by robots. The hotel opened its doors on the 17th of July, 2015. The total floor area is 5,530 square meters, and on its opening day it had a total of 72 guest-rooms, a number that was set to double in 2016.

This hotel's journey through the media landscape is followed in this first narrative analysis. Questions loom. Who publishes articles about the hotel? What is the focus of these articles, and how are the robots presented in them? This particular confluence of robots has had very interesting news coverage, partly evident for how it is *received* by reading posts in comment fields, forums, and related texts. The findings are grouped by the different *narratives* that the newspapers use to present this case. The different nationalities of the news agencies are evident.

Grouping the empirical material into different narratives, or seeing how users, in this case "the media", narrate their own stories around technology, is partly inspired by Professor Aune (Lie & Sørensen, 1996). The reason for using different narratives that seemingly do not relate is to show how the same story can be told in widely different ways through regional if not cultural shifts in the focus of the reporting.

Although the narratives are quite different, they all have the same backstory; yet, the way the reporters choose to *frame* the story gives widely different *narratives.*

(C1) NARRATIVE 1: THE "DANGEROUS" ROBOT

In an article by Karl Wig on E24.no, the writer, although being "fascinated," has done an interview with one of the largest Norwegian hotel owners as follows, translated into English here: "He thinks robots still have a way to go before they can offer the high service level guests expect in the Nordic countries. Furthermore, they do look a bit creepy" (Wig, 2015). This is a peculiar twist on the news story about the robot hotel and raises questions regarding reporter and publication decision-making. Why is a hotel owner who doesn't employ robots in this way being interviewed?

Through media representation, the hotel owner becomes an "expert" (at least, on hotels), and the article seeks to add to its credibility by inferring it is interviewing an expert like this. The expert is speaking about implementing similar robots in his Norwegian hotels when he says, "A hotel's future which is machine controlled will not be completely alien to [our] Hotels. The hotel chain has presently automated check-ins and checkouts for the majority of their hotels in Norway" (Wig, 2015).

The hotel director is further asked about the cleaning robots at the Hen-na hotel, and states, "...But it is important that they can provide the same quality on the washing, not to mention that using them will save money. It is probably not high on our list" (Wig, 2015). Here, he pinpoints two attributes that robots need if they are to be considered as worthy of consideration: quality-assurance and cost-efficiency. If a robot is more expensive than a human, a human is preferable. If a robot is as efficient, but humans deliver more quality, the human is still preferable. The turning point thus will be when robots deliver equitable quality, but are cheaper than their human competition.

Another Norwegian article, from *Aftenposten,* entitled, "The Hotel Run by Human-robots Opens Now" uses a "shock value" strategy in a juxtaposition of robots and humans (2015). The case starts with, "Staff is the most important thing for an hotel…worth its weight in gold…" and then puts the owner of the robot hotel opposite this, saying, "No, the future is robots." It further describes Japan as the land of "human robots." The article describes female androids as "actdroids" and male androids as "geminoids," which is not the correct term, as a geminoid is a robot twin of an existing human. The article describes the robot as "talking to you, blinking, looking like they are breathing." It also contends that, they at least can understand you, "when you try to show them the booking for your room." It states that you will meet a female robot in the lobby, and that you might meet a cleaning robot in the hallway. It also explains the face-recognition lock on the room doors, and ends with "The future is here. At least in Japan."

The comment field for this article has only three linguistic differences, with the most critical being, "Why do we do everything to dig our own grave? Well, when the robots have taken over most of the work from us living beings, we will have more time to make war and create chaos." This comment received in response: "Or we can use the time to do science, create arts, do sports, and other things that we actually enjoy doing… we must just be careful that the resources are divided fairly." This example shows how a negative, "dangerous narrative" can be received; both, as an ill omen one cannot change, and, also, as a positive call to action by some readers.

(C1) NARRATIVE 2: THE "JAPANESE" ROBOT

One article by Maggie Hiufu Wong for *CNN* emphasizes a distinct difference in how Japanese- and English-speaking guests are greeted: "Humanoid robots greet Japanese-speaking guests at reception, while English-speaking guests are met, for reasons that aren't exactly clear, by a robotic dinosaur. The robots are said to be able to engage in intelligent conversation" (2015, July 17). This is a bit curious. Why did the designers choose to focus on linguistic differences? Why program two different species for speakers of different languages? Are Japanese or Asian guests more accustomed to, or comfortable in dialogue with, androids than Westerners?

When studying how robots are accepted by humans, the "Uncanny Valley" theory is of value (Mori, M., MacDorman, K. F., & Kageki, N., 2012). The theory explains how robots, which are increasingly similar to humans, are gradually accepted by human, unless they appear too similar to humans. The acceptance rate of a robot by a human falls rapidly the closer a robot gets to "perfection", because humans detect that, "something is not quite right" (2012). One may wonder where in the Uncanny Valley a Velociraptor robot might be found? But seriously, is this famous valley different for different cultures and societies? That is a worthwhile question.

Many of the articles pinpointed the robot hotel a "Japanese phenomenon," as seen in Wig's E24.no article and interview with the Norwegian hotel manager's contention of an "only in Japan" (2015) framing perspective. Yet, this arguably can be both a negative and positive trademark for Japanese phenomena. But how necessary is it to be country specific; couldn't this have happened in other countries outside of Asia? Regardless, Japan is really quite fitting, as its robot investment is at the top of the ladder since both, "Asia and Europe are home to the key players in the [industrial robot] market" (statista.com, 2016).

There is also a trend of "Japanifying" topics in the comment fields, as in a Norwegian article from the Norwegian *National Geographic* blog at natgeobloggen.

no (2015) called: "In Japan the world's first hotel staffed by robot-lasses are being made." For now, let's consider the 'Japanism' of this article. It starts with, "EN: It had to be in Japan, that the first hotel staffed by multi-lingual robots came out. Japan has long had a thing for robots – from anime-heroes, to chefs, to virtual boyfriends/girlfriends" (using the gender neutral Norwegian word "kjæreste" here, closer to "sweetheart"). The article states that robots will not likely try to take over the world, and it actually tries to give a description of the Uncanny Valley.

There is an issue with framing something in a particular national-cultural frame, as many commenters expressed fear of robots increasingly being seen as competition in the job market, where robots are 'stealing' our 'human jobs.' With that in mind, any strong link that robots have to Japan might not entirely be a positive one, should the unemployed there seek scapegoats for being outsourced. The normal human reaction when getting replaced at work is to blame the person; and, often, their whole ethnic/cultural group (or in this case, thing) that did the replacing, as can be seen in rising waves of hatred for immigrant groups who seek employment in countries they immigrate to today. Time will tell if this will translate to robots, but this is something that perhaps should be investigated more closely as robotic technology applies more to our daily lives, and certain professions suffer from outsourcing, discussed further here in a later section.

(C1) NARRATIVE 3: THE "GENDERED" ROBOT

The Karl Wig article has a curious view on the robots in the hotel, describing them as follows, translated into English here:

They can walk, run, wave and kick a soccer ball, move the "eyes" and simulate different moods. The robots will have the same uniform as the human operators, with appearance and manners similar to a "young Japanese woman" who speaks four languages. (2015)

Again, there is focus on the robot being "female", whilst the dinosaur is not mentioned at all. Another news site, dailymail.co.uk (2015) presents the hotel by describing the female robot in the lobby as, "a life-like female humanoid with blinking lashes" (Kitching, 2016). This is a bit curious, as blinking eyelashes aren't something other articles particularly focused on. The *Daily Mail* continues to focus on strange things, stating that, "the robots cannot take care of security" (2015) – perhaps their blinking eyelashes would get in the way ? Reader reaction to this article is highly positive, with no negative comments. "Very cool I bet the place is packed all the time. I like the dinosaur most but it's all cool" (2015). The comment field is a jovial one, with one comment saying, "The Velociraptor is the complaint department," and, another, saying that, "most hotel desks are run by robots" (2015).

Another (jokingly?) says, again, "Skynet first takes over human jobs and then they strike." Apparently, Skynet and the *Terminator* franchise have been so ubiquitous that a whole generation of commenters finds it necessary to include at least one *Terminator* reference in every news article. The comment field from the Avner article is perhaps the most sexist, as a second negative commenter states, "Stiff plastic dolls… this hotel sounds like a bad gimmick from a third rate amusement park." She is then answered by the first negative commenter who makes a joke about "room service from blow-up dolls," a statement with which she agrees. As expected, perhaps, the technophobic commenter in this case holds misogynistic views.

Consider natgeobloggen.no (2015), which was seen in the previous narrative. As mentioned, it uses an older Norwegian word for girl, namely "pike", which would translate more as "lass" in modern English. It could also translate to "maid", or a lower-class serving woman (in Denmark the word "pige" is used much more commonly to refer to "a girl"). The authors do not use the word for "woman" (kvinne in Norwegian), and not even a normal "girl" (jente in Norwegian), but instead choose a word that is not commonly used in contemporary society. It would be considered quite strange to call the male counterpart for the robot a "boy/lad"; in fact, they are described here, translated into English: "They have also made a male version of the same robot, which really is the same robot, only with another haircut, voice and some other details." But why does the article frame the story as "girl-robot at hotel… and a male robot exists", and not "female and male robots staff hotel"? or "girl robot and boy robots staff hotel?" Also, if the intended meaning is "maid," why not describe the male-robot with the somewhat equivalent word "butler"? In Japan, there are both maid and butler cafés, for example.

There is a tendency in the Norwegian media-landscape (and others, too), to refer to two people of the same age but opposite genders in an article as "girl" and "man." The opposite would be very strange, "woman" and "boy," but, apparently, the tag "girl" has more resonance than "boy." It is interesting how these gender biases are transferred onto mechanical, and biological creations.

The gendering of technology is a topic on its own. Especially interesting is Jennifer Robertson's article "gendering humanoid robots: robo-sexism in Japan," where she connects two narratives discussed here, Japan & sexism, in regards to robots (2010). A key finding through media analysis on the topics, is that sexism is strongly prevalent in the way robots are received through media. When connected to 'Japanism,' one could argue for a postmodern, trans-human sexual colonialism taking place, but more research is needed on this topic.

(C1) NARRATIVE 4: THE "NORMALIZED" ROBOT

Whilst the articles covered here have focused mostly on the robots in the hotel in some way or the other, there were some articles that did not focus on the robots much at all. Whether this is because robots have been so "domesticated" that a hotel staffed by them becomes "non-news," or the journalists just aren't that interested in the subject is unclear.

Compared to the other, more focused, news articles, a *New Atlas* article entitled, "New Japanese hotel has robot staff and no room keys" says, "As well as aiming to provide a pleasant and comfortable stay for guests, the hotel was designed to be cost-effective, modern, environmentally-friendly and fun" (Robarts, 2015). The article uses few superlatives, and rather duly notes, "According to the Henn-na Hotel, it's possible to hold a conversation with the "warm" and "friendly" robots while they get on with their work" (2015). With insertion of the quotation marks, the article ironically casts the robots as perhaps neither warm nor friendly; it seems the article as a whole conveys the entire situation as a 'not-so-interesting' story. The article had five comments, two from people who have been in the area of the hotel, and one who actually went into it but described the robots as "too slow." One commentator wanted the hotel concept exported to North America, whilst another joked about the Velociraptor robot in the lobby (2015).

Another word for normalized would be "stabilized" (Bijker, 2001; Latour, 2005; Pinch & Bijker, 1984), a key concept in STS theories, and what happens when a controversy is "agreed upon." Writing about a stabilized controversy, is, at least in a media context, not a particularly good moneymaker in the realm of journalism. Often, news may need some sort of controversy to sell; by reporting on something not particularly interesting, something that has been "agreed to be thus" by a society, a news piece can get limited reception. Still, news articles carrying specific narratives as seen in this analysis are differentiated from related articles on the same topic. When the *Daily Mail* and *National Geographic* write about the hotel in a sexist manner, and *CNN* writes about it on its website as a Japanese phenomenon (Wong, 2015, July 17), it may give the readers something to relate to. Trying to capture a reader's interest in "girls" or "Japan" may make these stories more attractive to some readers than articles focused on robot hotels being "dangerous" or "scary."

Whilst this chapter is primarily focused on how the robots in these news articles are framed, some authors are not necessarily as interested in writing about robots in the first place. Seeing the robot hotel more or less as normalized - not all that interesting, innovative, or special - one gets a stabilized or normalized controversy. When people become more accustomed to a technology penetrating their culture or society, some reporting on it becomes less interesting. As for now, as many articles have pointed out, a story that really is a "first" in news often brings in interested readers. However, as more robot hotels open, the "shock" value will diminish, and this narrative will prevail more often, if it will be written about at all.

CONTROVERSY 2 (C2): THE "KILLER" ROBOT

The next in-depth case is a much darker one. The robot in this case is described as both "a murderer" and "a killer," although the phrasing is heavily debated in online article comments. The choice of words used to describe the robot in these different narratives, frames it in very different "modus operandi." The most prevalent type of media framing for this story questions whether the robot meant to kill the man, or was it simply a working accident?

The story takes place in the small German city of Baunatal, where the world's second largest car manufacturer, Volkswagen, employs a majority of the residents. It was at one of Volkswagen's factories that a human operator was working on an assembly robot when he was suddenly grabbed and crushed to death. How the media framed this story, and how they narrated it is relevant for understanding how robots are presented in the media. A person is dead, that much is certain. A robot was involved in the tragedy, also certain. The big difference in media reporting of the story is how they narrate the robot's role in it. The following narratives stood out.

(C2) NARRATIVE 1: THE "BLAMABLE" ROBOT

The first article comes from IFLSCIENCE!, a popular science disseminator. They write,

But before you start believing this is an early sign of an impending robot apocalypse, Volkswagen... suggests that human error was to blame rather than a 'rogue' robot... Workers in the plant are usually separated from the robots... but in this instance, the victim was working on the robot inside the cage when he was attacked. (2015)

Initially, readers might imagine that the reporters are trying to reassure their readers even before they start to think about a "robot apocalypse," because it is "suggested" that human error was to blame. But when looking a bit closer on the words used in the article, one can see that it is actually quite biased in how it describes the robot. Consider the words: separated from, confined area, victim, cage, attacked. This does not read as a work accident, where a piece of machinery malfunctioned. When a person is attacked, there is an attacker. Victims have culprits responsible for their misfortunes. Dangerous things are often contained in cages, separated from "us." While starting out to explain that the robot was not rogue, the article certainly lets its audience read the danger between its lines.

Of the 149 comments on the article, most complain about the article's use of the word "attack" and "killed," as they feel that these are a wrong usage of the words. The commenters go into ethical debates about how robots cannot deliberately choose to attack someone, they have to be programmed to do so. The news site is accused of trying to win readers by using scare tactics, and also of not being critical enough when writing. This is a surprising example of readers defying the writer(s) by calling them out on problematic journalistic style.

A news article from Justin Huggler in Berlin, writing for *The Telegraph,* does not even try to diversify the controversy. The first line in Huggler's article reads, "A robot has killed a man at a Volkswagen car factory in Germany" (2015, July 02). It states that the operator was part of a team that was setting up the robot, "a mechanical arm that moves car parts into place." The article continues,

It is part of an automated assembly line that is capable of functioning without a human operator, but it is believed it may have been under human control at the time of the accident... the case is now under police investigation. (Huggler, 2015)

The investigation is trying to determine whether anybody was at fault; according to way the article is written, the true 'killer' has been found already. By looking at the article's link to other comparable articles, the similarly themed, "Killer robots will leave humans 'utterly defenseless': Is Google building a robot army?" shows an overall narrative which seems to make the point that humanity is apparently in danger. The article ends with, "Robots have caused at least 26 workplace deaths in the US alone in the past 30 years, according to government data." It is quite one-sided for the telegraph.co.uk to be so deterministic in its view on technology.

A third article, from Eliana Dockterman in *Time*, "Robot Kills Man at Volkswagen Plant," follows technological deterministic framing (2015a). Although writing that the "company believes the malfunction was due to human error," Dockterman continues in her article with, "The type of robot that crushed the employee is usually kept in a cage." This is a very curious choice of words because when we think about caged things, a dangerous animal comes to mind, or a human criminal; someone (or something?) that could choose to kill someone.

But a robot? The article ends with, "Prosecutors are still deciding whether to bring charges and whom they would pursue," making it interesting to compare robot ethics to ethical treatment of animals. For example, if a wild beast, let's say a tiger, were kept in a cage, and a human went in to feed it under the impression that it was drugged, but the human was jumped by that tiger – would anyone charge the tiger with murder? It has been done historically: the Middle Ages saw a lot of animals on trial (Hyde, 1916; Salisbury, 2011); but, in modern times, we normally do not convict animals - they are simply put down without trial. With ethical treatment of animals in mind, should the robot be put down? Should it pay a fine? If the robot wasn't working correctly, surely it could be scrapped; but, if this was human error, what blame should befall the robot? Those are questions for budding new lawyers and the laws specializing in robot error and ethics coming into our courtrooms in the near future.

(C2) NARRATIVE 2: THE "WRONGLY-ACCUSED" ROBOT

The *Daily Mail* published an article called, "Robot kills factory worker after picking him up and crushing him against a metal plate at Volkswagen plant in Germany" (Fernandez, 2015, July 2). While starting with a narrative quite similar to the first one described above, the article's last part emphasizes that the robot is not to blame. It reassures readers with a statement from Dr. Blay Whitby, a Sussex University professor of artificial intelligence, who states that, "'Robots do not act of their own volition and would not attack a human unless programmed to do so'" (2015, July 2). The Fernandez story further states that it is extremely rare that robots kill humans, and shows only two known cases from 1979 and 1981 where this has happened. However, should one listen to the previous narrative, the death toll is higher.

There are 137 comments on the *Daily Mail* story, almost all of them referring to the Terminator and its Skynet; most have a (dark) humorous tone, some pointing out that a new *Terminator* movie was releasing at that time, and asking, "Skynet, is that you?" (2015, July 2). There is very little grief for the man who was killed.

In another comment field from the *Tek.no* article by Vegard Haugen, the commenters focus on how the term "robot kills man" uses the verb to project that this was an act of deliberate violence. Another focus there was on the terminology "keeping the robot in a cage, like it is some sort of wild animal (lion, tiger, dinosaur, etc.)" (2015, July 3).

Others joke, "Did somebody say Skynet" and "so it begins…" One commentator writes that, "I was expecting something more gloomy, but the humor in this article made me laugh." While most of the news agencies cannot seem to write objectively about the controversy, the comment fields are full of critical readers who question the framing of robots as "evil killers." Also, the popular references to *The Terminator* films were quite prevalent.

(C2) NARRATIVE 3: THE "TERMINATOR" ROBOT

The Norwegian online magazine, *Tek.no*, byline states that, "…they both worked at Volkswagen" (Haugen, 2015, July 3). To say that a robot is 'working at a factory' is to give it much more human a characteristic than if it, for example, were just described to be 'a piece of machinery at the factory where the man who was killed worked.' It is further emphasized that this robot does not usually work side-by-side with humans, but that this type of robot is normally held in a cage, since it can be a danger to humans who get too close. Up to this point, the article is quite up to technology-deterministic standards, but what it chooses to focus on next can to some extent explain all of the previously noted *Terminator* references in regard to the robot. Ironically, the person who first reported the case on July 1, 2015, on Twitter is named Sarah O'Connor. The Tweet went as follows:

Sarah O'Connor @sarahoconnor_· 7/1/15

A robot has killed a worker in a VW plant in Germany

ft.com/fastft/353721

This led to a massive Twitterstorm sharing the quote. The reason for this, was that Sarah Connor is the name of the protagonist who is key to human survival in the *Terminator* series, and was tasked with protecting humanity against the Terminator threat. As seen throughout this controversy, and also in other stories, Skynet, and Terminators, are regular guests in robot-themed article comment fields. There simply are very few article comment fields where the *Terminator* series is not mentioned. The *Terminator* saga, perhaps the best known robot series created in Hollywood, and what supercharged the career of Arnold Schwarzenegger, opens with:

THE MACHINES ROSE FROM THE ASHES OF THE NUCLEAR FIRE.

THEIR WAR TO EXTERMINATE MANKIND HAS RAGED FOR DECADES,

BUT THE FINAL BATTLE WOULD NOT BE FOUGHT IN THE FUTURE.

IT WOULD BE FOUGHT HERE, IN OUR PRESENT. (Cameron, 1984)

Perhaps the most clichéd robot-fighting scenario is that of robots rising up against their human creators. Here, robots are often portrayed as the evil enemy which must be annihilated at all costs. These types of robots often do not have any will or feelings of their own, but for some reason or another have taken it upon themselves to rid the earth of humanity. The war with the robots in the *Terminator* series starts after a program, Skynet, becomes self-aware, taking control over the world's military systems, and launching nuclear weapons. The series has had a very deep impact on Western commenters on media articles about robots. The Haugen article concludes by saying that more people are killed by toilets or zippers than robots (2015, July 3).

There are obvious differences among countries, also, in how robots are both presented through the media, and in how they are received. In the cases previously described, it was seen how a robot hotel was framed in some cases as a very "Japanese narrative," yet, how a "killer robot" was not framed much as a "German narrative."

HITCHHIKING ROBOTS IN THE US

Although Americans are at the forefront of many robotic technologies (especially military), some problems with domesticating foreign robots occur, as can be seen in the tragic case of hitchBOT, manufactured in Ontario, Canada. On the About page of the robot's project website one can choose from travels in Canada, Germany, the Netherlands, and the USA. For an example of what its travels look like, on the homepage it states,

During the summer of 2014, hitchBOT hitchhiked across Canada...In just 26 days it hitched a total of 19 rides and travelled over 10,000 kilometres. hitchBOT's adventure continued in 2015 as it visited Germany to explore cities like Munich, Cologne, Berlin, and Hamburg. hitchBOT also took a vacation in the Netherlands...

In 2015, a new hitchBOT was created by the team for an American adventure. hitchBOT's journey started on July 17 in Salem, Massachusetts and it had hoped to make it to San Francisco, California...hitchBOT's trip ended abruptly on August 1, 2015 in Philadelphia (mir1.hitchbot.me, 2015)

The hitchBOT was a project created to see how a robot would travel through the human world, and to document the trip through photos, linked with social media such as Twitter. Although not that sophisticatedly programmed, it made for an interesting experiment on how humans would react to a lone robot seeking to see the world. What the creators found, was that this was mostly met with kindness and curiosity. For the accident in the US, the creators at one point had written on the homepage: "sometimes bad things happen to good robots..." (m1.hitchbot.me, 2015).

So what happened to this friendly robot in the US? An article from Daniel Victor in *The New York Times* had a story entitled, "Hitchhiking Robot, Safe in Several

Countries, Meets Its End in Philadelphia" (2015, August 3). This article (somewhat humorously) writes how the robot did international travels, spent a week with a heavy metal band, went to a sporting event, and then, "met a brutal demise in a Philadelphia alley on Saturday. It was 1 year old." It further describes it as, "Short and stocky, with a bucket for a body and the red LED lights of its face enclosed in plastic, the brightly colored bot would be difficult to miss." (2015, August 3). Although surviving the East Coast, when it was set to travel westwards, it met its demise.

The New York Times article presents the story as a sort of crime story, with phrasings such as, "It was the last known time someone would interact with it… obtained surveillance video that appeared to show someone in a football jersey throwing the robot's arms on the ground before kicking something on the bench" (Victor, 2015, August 3). The article further emphasizes that the creators will not press charges for the act, but will rather focus on what can be learned from the incident. The article discusses some of the reactions found on social media, seeing how "Philadelphians were wary that the episode would reflect badly on the city. Others worried that it would reflect badly on all Americans" (2015, August 3).

This chapter will not go deeper into the different media portrayals of this particular story, but encourages others to do so, as it is an interesting connection between safety of non-human actors, and also, portrayal of non-human framing. If comparing the hitchBOT with the accursed "Volkswagen killer robot," one can see how the robot is portrayed as an object vs. a subject.

Did the Volkswagen robot kill a human? Many, as seen in the comment-field analysis, argue that it could not possibly have done that, as a robot only does what it is programmed to, and has no will of its own. When discussing the hitchBOT on similar notions, why does the media then not write "human(s) kill robot?" Why is "robot kills human" okay to write, but not the opposite? This perhaps displays some of the ambiguity that exists when writing about murder and robots. Why is it okay to frame robots as enemies, but not as victims? While these questions deserve forthcoming answers, there are smaller controversies to consider which reveal how EuroAsian perspectives on robots are described in select countries. Delving further into these robot stories proves that it is a robot jungle out there, and 'robot-fauna' is more diverse than one thinks.

JOB-STEALING ROBOTS IN NORWAY

The most interesting finding from analysis of the Norwegian media was how the focus on the robots as either "female" or "workers" seemed to dominate both the presentation in the articles, as well as the comment fields. Some context is needed to understand this more deeply. Viewing Norway as a case is interesting in both gender roles, and in worker status. It is one of the countries in Europe with the lowest unemployment rates (4.6% in 2015). It is thus a bit strange that robots are presented as "job stealers," and "competitors," as if such a small nation needs to keep up efficiency if it wants to continue to compete in the global market. With a mere 5 million people, but with wasteland areas and resources, Norway - more than the other countries presented - needs technology, should the not-dense population continue to be able to retain its high ranking as a desirable country in which to live.

Consider an article from Norway's third largest news-agency, *Aftenposten,* entitled in English here, "Norwegians do not think that Robots will take their jobs" (Sjøberg & Hager-Thoresen, 2015). In it, not a specific robot, but rather "robotification" as a whole, is discussed. It starts off with a building surveyor who praises automation in his work, saying, "it has only been helpful… made the work day way more easy and more effective" (2015). He also says "Our profession demands the presence of a physical person who can interpret the data that the tools give, or else the profession will lose its credibility." The article mismatches its positive/negative wording, saying, "the youngest have the most belief in robots," which suggests that they are the ones (18.6%) who fear their profession will be exploited by automation. The other person interviewed is an Abelia boss (Abelia is a collaboration between Norwegian knowledge industries) who warns that this can create a stratified society with a technologically weak subclass (2015). He gives some examples about how society can transition better to a techno-society. The article references a report entitled, "Computerization and the Future of Jobs in Norway," and concludes that, "Simple, routine tasks are more in danger" (2015).

As unemployment rates continue to rise in Norway, this might be a bigger controversy in the ensuing years, and it is remarkable to see how the "new species" of robots is framed in this context. When using the title "Norwegians do not believe that robots will take their jobs," is there an underlying inference that robots should "take their jobs?" The article appears a bit judgmental, showing 'naïveté' by consulting experts who say "they should be afraid." The comment field to this article is mostly negative. Comments show a realistic approach to the challenges, in contrast to the overly negative comments that can often flourish in these forums. Some commenters question whether automation might take jobs from people who can only do repetitive jobs, and ask who will provide for them, and then getting trolled in rebuke by an "anti-socialism" comment. One poster argues that the new jobs will be outsourced to India.

"FARMER" ROBOTS IN JAPAN

Japanese news agencies report on a wide variety of robot-related topics, including some with more extensive roles than "killer robots" and "android receptionists," namely farming robots. *The Japan News* has an August 22, 2015, published article in which it is written:

Research institutes and universities in Japan are developing robots to play an active part in the primary industries of agriculture and fisheries, which are suffering from aging workers and a lack of manpower. (Yomiuri)

The problem that robots are presented to help with in this article is the aging problem faced by Japan. It states, "The average age of farmers in Japan was 66.7 in 2014, with 64 percent of them 65 or older" (Yomiuri, 2015, August 22). Farming has for a very long time in human history constituted of hard physical labor, where young, physical bodies help alleviate the strain on older farmers, who have good knowledge of the profession, but of course can lack some of the physical attributes needed to lift, plow, pluck, harvest, etc. From the Yomiuri article, one can see the political incentives behind this,

The ministry has spent about 1.2 billion yen ($9.6 million) since spring to support the development of 17 kinds of robots currently being worked on at universities and research institutes. More than half of these robots are being made to reduce agricultural workloads. (2015, August 22)

As the Yomiuri article suggests, the average farmer in Japan is almost 67 years old. There are no countries in the world that have a retirement age older than 67. Norway is one among four which have the highest retirement age in the world: 67 years, while the US has 66 years. Japan's retirement age is 60-65, so by definition, the average Japanese farmer is working beyond retirement age. The article points to a very important factor in Japan's case of accepting robots. They simply must do so, since the Japanese society is rapidly aging, and immigration to the country is relatively low, with an immigrant population at 1.9 percent. Japanese policymakers must turn to robot technology to help solve this situation.

With this understanding, it is not surprising that robots are presented in the article as efficient and stress-relieving. They are described in a neutral or positive way, and the article ends with information on a "working group to discuss safety and other issues for robots used under such conditions" (Yomiuri, 2015, August 22). As Fujigaki wrote in "STS in Japan and East Asia: Governance of science and technology and public engagement" (2009), there has been much focus on public engagement in Japan concerning the *Minamata* and *Itai-itai disease* cases, and also recently concerning the case of nuclear energy following the Fukushima triple disaster in 2011 (Utz, Schultz, & Glocka, 2013). The aftermath of the *Fukushima disaster* has some extensive news reports going in to the nuclear dead zone of the power plants.

Japan is a fascinating case when studying public engagement. Consider, for example, the case of "the killer robot in the cage." How would the public react differently to the narrative of a "rogue robot terminating poor humans" instead of a "mechanical tool accident?" Narratives that media tell shape public understanding of concepts, and lead to informative types of public engagement.

This is a Western view on the topic of how Japanese media frame robots. More research is needed on how this is done, especially from the Japanese perspective. Media anthropology, as most anthropology fields, can (and most often will) fall victim to seeing and describing the topic with the researcher's own eyes and cultural background. It is not only Western media, for example, that can frame a robot hotel in Japan as typically Japanese – researchers also analyze phenomena with their own mindsets.

CONCLUSION

So how can robots, androids and cyborgs be presented and received in the media? Through the news stories presented, there has been a wide diversity in how different media companies choose to frame the robots in different narratives. There have also been similarities in what the reporters have focused upon when writing about robots. The same controversy can be told in widely different narratives, and by paying attention to this, one can analyze how society is presented with this novel technology, which by many is considered to be revolutionizing the way we live our lives. How the public discusses and receives robots, through a social media analysis, has been an important aspect of the analysis to bear in mind, while the data material is not comprehensively representative, as it does not cover people who do not comment on news articles. Nevertheless, qualitative insight is gained about Internet users who do comment on robot cases in order to see some trends and predict how robots may be received in society. The same goes for the journalists writing these robot news articles: they have their own cultural backgrounds, interests, biases, and goals, and should not be presumed to be objective.

Although this has been a study on how robots are perceived online, the author acknowledges that there probably are strong synergies with how the robot is domesticated in the real world, as the way robots are described online affects their offline image. This is exactly why a narrative's framing matters: not just for the robots' online image, but also for a robot's offline reputation.

Consider the binary relation between "dangerous" and "normalized" for instance. Humans often "fear the unknown," and by portraying the robots as explicitly "others" and focusing on their uncanniness, this perception can increase. As robots grow more and more intelligent, this is a debate we will need to have in regards to our created mechanical species; about how we relate and describe them. If they look, talk, and walk like humans, what rights should they have, if any?

This is particularly interesting in the second in-depth case where a manufacturing robot was more or less deliberately cast in the "terminator-frame," which of course was (randomly?) sparked by the reporter of the story sharing the similar name of *Terminator* protagonist, Sarah Connor. This again led to a very comical framing of an otherwise tragic case, as seen in the comment fields. Very few people focused on grieving for the dead person, but rather joked about Skynet. Some also went into an ethical debate about the usage of the wording "killing robot," which at least shows that the narrative did spark some discussion when it was received.

Human sexism can be translated to robot-sexism in media stories. A key finding through this media analysis on the topics is that sexism is strongly prevalent in how robots are received through media. This was not something that was set out to find, and it was both present at the reporter level, and at the commentator level. It is quite peculiar that in a hotel of robots and dinosaurs, the key focus point is "female long lashes."

The comment fields for many of the robot news articles contain misogyny. When commenters ask if "the room-service contains a blow up doll," it is not only degrading to women, but also to robots. The whole mentality of "women serving men" (many feminine rights do not go that far back in history), can apparently be translated into the "human-robot" relationship; and here, more research is strongly needed on how virtual and mechanical gender affect human gender relations and understandings.

The cultural context of robotic domestication matters. There is a difference in the way robots are presented and received between different cultures. Countries have different policies, cultures, and also acceptance rates of robots. Japan needs robots to tackle their aging population problem, and has been steadily presented with many robots in fiction through their anime and manga culture. This is also true for Western countries; but, as seen in many comment fields, the notion of the Terminators sticks very deeply – in fact there were very few comment fields not referring to Skynet in some way or the other.

Although there is a great opportunity to search news archives at libraries, this chapter has used online news articles. This is primarily because it has, as its second focus, looked at the reception of the articles found in comment fields and forums, which offline news articles do not have (except from reader letters to the editor).

The reason this has been successful is due to the consequent search and storage of said news articles. Sometimes, these articles do not have a functioning URL months later, or they have "vanished." This can have many, many explanations. Perhaps the news agency changed owners, updated their website, or had too many negative comments for a news article. It is thus imperative for researchers collecting online data to "save and store," and to not expect a webpage to exist tomorrow.

Robots, androids, and cyborgs are presented and received in the media in various different narratives depending on how the robots are framed, what journalists focus on when reporting about robots, and how the public discusses, and receives, robots. The robot is thusly domesticated through the media. The same "robot controversy" can be framed in widely different narratives, human sexism can be translated to robot-sexism in media stories, and the cultural context of robotic domestication matters. The way humans describe robots in media coverage impacts the way humans adapt to them, and how robots affect humans in return.

When reporting on a specific robot, the framing of the story is important, as is the way journalists choose to convey the meaning of the article towards their audience. News that travels between countries is often translated, and thereby, vital points may be lost in translation. A different journalist, who takes up the story in another country, can have a widely different agenda with his or her framing of a robot, as is evident in multiple channel international news coverage and it can deliberately or unknowingly miss key aspects. Robot controversies are framed by the narratives they are presented with in the media. With that awareness, one can be critical towards these narratives.

The significance, and implications, of these findings may still be difficult to measure. Robots represent a novel creation on our planet; and the way we interact with them, write about them, and discuss them, is continuously shaping our adoption of their technology. Most significantly, one might suggest that treating robots as human 'twins' is not an adequate approach, because they are an entirely different, mechanical, species. However, since the media often chooses to do that, the implication is that we expect robots to be like us, and that affects their design process.

REFERENCES

Avner, T. (2015). Nå åpner hotellet som drives av menneskeroboter. *Aftenposten*. Retrieved from http://www.aftenposten.no/article/ap-7899510.html

Bijker, W. E. (2001). Understanding technological culture through a constructivist view of science, technology, and society. *Visions of STS: Counterpoints in science, technology and society studies*.

Cameron, J. (1984). *The Terminator script*. Retrieved from www.script-o-rama.com/movie_scripts/t/terminator-script-transcript-arnold-schwartzenegger.html

Charmaz, K., & Smith, J. (2003). Grounded theory. *Qualitative psychology: A practical guide to research methods*, 81-110.

Cortazzi, M. (2014). *Narrative analysis* (Vol. 12). Routledge.

Dockterman, E. (2015, July 1). Robot Kills Man at Volkswagen Plant. *TIME*. Retrieved from http://time.com/3944181/robot-kills-man-volkswagen-plant

Engelhardt, H. T., & Caplan, A. L. (1987). *Scientific controversies: Case studies in the resolution and closure of disputes in science and technology*. Cambridge University Press. doi:10.1017/CBO9780511628719

Fernandez, C. (2015, July 02). Robot kills man at Volkswagen plant in Germany. *Daily Mail*. Retrieved from http://www.dailymail.co.uk/news/article-3146547/Robot-kills-man-Volkswagen-plant-Germany.html

Franzosi, R. (1998). Narrative analysis-or why (and how) sociologists should be interested in narrative. *Annual Review of Sociology*, *24*(1), 517–554. doi:10.1146/annurev.soc.24.1.517

Fujigaki, Y. (2009). STS in Japan and East Asia: Governance of science and technology and public engagement. *East Asian Science. Technology and Society*, *3*(4), 511–518.

Glaser, B. G., & Strauss, A. L. (2009). *The discovery of grounded theory: Strategies for qualitative research*. Transaction Publishers.

Haugen, V. (2015, July 3). Mann (21) drept av robot. *Tek.no*. Retrieved from http://www.tek.no/artikler/robot-drepte-mann-21/191214

hitchBOT. (2015). hitchBOT. Retrieved from http://m.hitchbot.me; http://mir1.hitchbot.me

Huggler, J. (2015, July 02). Robot kills man at Volkswagen plant in Germany. Retrieved from http://www.telegraph.co.uk/news/worldnews/europe/germany/11712513/Robot-kills-man-at-Volkswagen-plant-in-Germany.html

Hyde, W. W. (1916). The prosecution and punishment of animals and lifeless things in the middle ages and modern times. *University of Pennsylvania Law Review and American Law Register*, 696-730.

IFLSCIENCE. (2015). Robot Kills Worker at Volkswagen Plant in Germany. iflscience.com. Retrieved from http://www.iflscience.com/technology/robot-kills-worker-volkswagen-plant-germany

Kitching, C. (2015). Inside the Japanese hotel where the front desk is staffed by ROBOTS and guests scan their faces to enter rooms. *Daily Mail*. Retrieved from http://www.dailymail.co.uk/travel/travel_news/article-3163631/Japan-s-robot-staffed-Henn-na-Hotel-opens.html

Latour, B. (2005). *Reassembling the social-An introduction to Actor-Network-Theory*. Oxford University Press.

Lie, M., & Sørensen, K. H. (1996). *Making technology our own? Domesticating technology into everyday life*. Scandinavian University Press Oslo.

Martin, B., & Richards, E. (1995). Scientific knowledge, controversy, and public decision-making. In Handbook of science and technology studies (pp. 506-526).

Mori, M., MacDorman, K. F., & Kageki, N. (2012). The uncanny valley [from the field]. *Robotics and Automation Magazine, 19*(2), 98–100. doi:10.1109/MRA.2012.2192811

natgeobloggen.no. (2015). I Japan bygges verdens første hotell bemannet av robot-piker. Retrieved from http://www.natgeobloggen.no/2015/02/12/i-japan-bygges-verdens-forste-hotell-bemannet-av-robot-piker

O'Connor, S. (2015, July 1). A robot has killed a worker in a VW plant in Germany. ft.com/fastft/353721 [Tweet]. Retrieved from https://twittr.com/sarahoconnor_/status/616282747200479232

Pajarinen, M., Rouvinen, P., & Ekeland, A. (2014). Computerization and the Future of Jobs in Norway. Department of Safety and Service Organization. Retrieved from http://nettsteder. regjeringen.no/fremtidensskole/files/2014/05/Computerization-and-the-Future-of-Jobs-in-Norway.pdf

Pinch, T. J., & Bijker, W. E. (1984). The social construction of facts and artifacts: Or how the sociology of science and the sociology of technology might benefit each other. *Social Studies of Science, 14*(3), 399–441. doi:10.1177/030631284014003004

Robarts, S. (2015, July 22). New Japanese hotel has robot staff and no room keys. Robotics. *New Atlas*. Retrieved from http://newatlas.com/henn-na-hotel-robot-staff/38577

Robertson, J. (2010). Gendering humanoid robots: Robo-sexism in Japan. *Body & Society*, *16*(2), 1–36. doi:10.1177/1357034X10364767

Salisbury, J. E. (2011). *The beast within: Animals in the Middle Ages*. Routledge.

Silverstone, R. (2005). Domesticating domestication. Reflections on the life of a concept. In M. Hartmann (Ed.), *Domestication of media and technology* (p. 229). Open University Press.

Sjøberg, J., & Hager-Thoresen, F. (2015). Nordmenn tror ikke roboter vil ta jobbene deres. *Aftenposten*. Retrieved from http://www.aftenposten.no/article/ap-8130866.html

statista.com. (2016). Leading companies in the global industrial robot market in 2016, based on revenue from industrial robot sales (in million euros). Statista. Retrieved from http://www.statista.com/statistics/257177/global-industrial-robot-market-share-by-company

Strauss, A., & Corbin, J. (1994). Grounded theory methodology. In Handbook of qualitative research (pp. 273-285).

Tavani, H. T. (2003). *Ethics and technology: Controversies, questions, and strategies for ethical computing*. John Wiley & Sons.

Utz, S., Schultz, F., & Glocka, S. (2013). Crisis communication online: How medium, crisis type and emotions affected public reactions in the Fukushima Daiichi nuclear disaster. *Public Relations Review*, *39*(1), 40–46. doi:10.1016/j.pubrev.2012.09.010

Victor, D. (2015, August 3). Hitchhiking Robot, Safe in Several Countries, Meets Its End in Philadelphia. *The New York Times*. Retrieved from http://www.nytimes.com/2015/08/04/us/hitchhiking-robot-safe-in-several-countries-meets-its-end-in-philadelphia.html

Wig, K. (2015). Åpner verdens første robotstyrte hotell. *E24*. Retrieved from http://e24.no/article/23393633

Wong, M. H. (2015, July 17). Japan's robot hotel opens its (automatic) doors. Tomorrow Transformed. *CNN*. Retrieved from http://www.cnn.com/2015/07/17/travel/japan-hotel-robot-opens/index.html

Yomiuri. (2015, August 22). Robots to help Japan's farming, fishing industries. *The Japan News/Yomiuri*. In *The Malone Telegram*. Retrieved from http://www.mymalonetelegram.com/col35/robots-to-help-japans-farming-fishing-indudstries-20150822

Chapter 5
Mechanology, Mindstorms, and the Genesis of Robots

Chris Chesher
University of Sydney, Australia

ABSTRACT

This chapter examines the emergence of educational robotics, drawing on the philosophy of technology of Gilbert Simondon. In the 1950s, Simondon argued that the dominant understandings of technology are personified in the popular imaginings of the robot. These attitudes are polarised between simple instrumentalism, and dystopian anxiety about technology overcoming humankind. To improve the conceptualisation of technics he took an approach called mechanology, developing a suite of concepts that grasped technology in new ways: technological genesis; the margin of indetermination; lineages of abstract and concrete technologies; and the associated milieu. These concepts are useful in understanding the tradition of educational robotics starting in the 1970s, with Seymour Papert's 'turtle' robot serving as a resource for learning mathematics. Since the 1980s, LEGO's Mindstorms kits have introduced learners and consumers to robotics concepts. Since the 1990s, theorists of embodied cognition in the 2000s have made use of Mindstorms to draw attention to the limits of symbolic intelligence.

INTRODUCTION

While there are more and more news stories, cinema, and television shows featuring robots, robots themselves are only recently becoming part of everyday experience. The meanings of robots are predominantly mediated through the images and narratives of popular culture, without attention to the specificities of actual robots. Among the

DOI: 10.4018/978-1-5225-2973-6.ch005

exceptions to this tendency are more sophisticated robotic toys. Toys can remain open to playful but complex interactions that provide embodied knowledges that emerge from experience with technical artifacts. They are harbingers of technocultural transformations. Toys can miniaturise technological diagrams such as robotics to perform as pedagogical apparatuses that sensitise players to the intricacies and virtualities of new technological paradigms (Chesher, 2017). This knowledge from experience is key to participation in critical discourse about the ethical and political implications of robotics.

At different times, there have been different understandings of robots, and how they relate to the wider technological cultures. In particular, this chapter will look at the development of robotics in learning, as this is one of the key interfaces for innovation. One of the pioneers in using robotics to encourage learning was Seymour Papert at MIT. In the 1970s, he designed a turtle robot as a resource for learning mathematics, engineering principles, and drawing (Papert, 1980). Out of a collaboration with toymaker LEGO, *Mindstorms* kits became commodified everyday artifacts that allowed people to experience robotic principles and dynamics. For example, theorists of embodied cognition in the 2000s have made use of *Mindstorms* to draw attention to the limits of symbolic intelligence (Dawson, Dupuis, & Wilson, 2010).

Informing the analysis of robots in this chapter is the French philosopher of technology Gilbert Simondon (1980, 1992), who developed the critical approach of mechanology that remains useful in conceptualising robotics and other technologies today. However, he originally saw the robot as a myth that distorted general conceptions of technology. In 1958, Simondon claimed, "...there is no such thing as a robot" (1980, p. 12). He saw the popular representations of the robot as evidence of wider misunderstandings of technology:

> *...a robot is no more a machine than a statue is a living being; that is merely a product of the imagination, of man's fictive powers, a product of the art of illusion. Nevertheless, the notion of the machine in present-day culture incorporates, to a considerable extent, this mythic representation of the robot. (Simondon, 1980, p. 12)*

He pointed to the popular misconceptions about robots, which he saw as evidence of wider misunderstandings of technology. In spite of his resistance to the popular media image of robots, his concepts are powerful for understanding contemporary robotics: promoting mechanology, critiquing hylomorphism (the form/matter split), explaining the genesis of technics, distinguishing the abstract and the concrete, observing margins of indetermination, and analysing the associated milieu. These concepts will be introduced throughout this piece.

Simondon's initial reading of robots seems to be informed substantially by the collective imagination in popular culture. The term 'robot' was introduced and popularized in the 1920 play *R.U.R.*, by Karel Čapek, which established the common themes of robot servility and robot rebellion. In the 1950s robots became a regular figure of fascination and fear in popular culture. Science-fiction films provided the narratives about robots that would become clichés. In *The Day The Earth Stood Still* (1951), a robot, Gort, was a massively powerful humanoid who held the world to ransom. In *Forbidden Planet* (1956), Robby the Robot was a servile helper. Simondon (1980) saw the robot in his time as existing largely in idolatry and mythology. It is an avatar for a technically mediated desire for power, a mechanism for domination, appearing as a duplicate for man. In the dominant myth, the robot master is willing to give over his power to the robot, which then enables him to dominate others; but ultimately he himself is dominated (1980, p. 12). The mythology about robots in popular culture bleeds across into wider 'cultivated' (p. 12) conceptions of technology in general. Simondon saw that attitudes about technology tend adopt either reductive utilitarian conceptions, or a generalised anxiety.

Our culture thus entertains two contradictory attitudes to technical objects. On the one hand, it treats them as pure and simple assemblies of material that are quite without true meaning and that only provide utility. On the other hand, it assumes that these objects are also robots, and that they harbour intentions hostile to man, or that they represent for man a constant threat of aggression or insurrection. (Simondon, 1980, p. 12)

This commonplace ambivalent attitude towards technology in general, and robotics in particular, certainly persists today, most visibly in recent popular culture. In science fiction films, robots are often depicted as uncanny, powerful and threatening beings whose allegiance is fickle. In the Terminator series, for example, the relationship between robots and humans is complicated. In the first *Terminator* film (1984), the main robotic character is a relentless killing machine, exposing the vulnerability of the human body (Telotte, 1995). In the second film (1992), the Terminator switches allegiance to help Sarah Connor and her son, only to be confronted by an even more advanced (and cool) shape-shifting robot in the T-1000. In Ridley Scott's *Prometheus* (2012), even the most seemingly amicable character, David, a robot, is revealed as a manipulative figure, mindlessly serving the interests of remote powers. In their more sympathetic guises, robots in *Star Wars* (1977) or *Wall-E* (2008) are analogues for the human with special abilities and positive affect. For Simondon the robot stands in symbolically for wider socially problematic relationships to technology. These images are evidence of a continuing mythologisation of robots, which reveals a general failure to understand technology in its own terms. The polarised popular culture representations of robots divert attention from genuine questions such as the role of robotics in surveillance, and their tendency to displace human labour.

Simondon's rejection of the robot was largely rhetorical, in the interest of establishing a space for one of few substantial philosophies of technology at the time. He felt that understanding the significance of "technicity" was socially crucial, and observed that critical awareness of technologies was seriously inadequate, both in the academic world, and in popular culture. He proposed improving the cultural understanding of machines, their mutual relationships, and the values involved (Simondon 1980, p. 14). Culture, he said, including politics, is distorted and impoverished by its inadequate dealings with technics (1980, p. 15). This culture is characterised by "myths and stereotypes, such as the idea of the robot and the notion of automata catering to a lazy and fully satisfied humanity" (p. 15).

What is at stake in developing a more sophisticated relationship to technology is not only to understand the *meanings*, but also to develop a more critical appreciation of machines beyond their simple utility and their mythologies. According to Simondon, misunderstanding the machine is the "most powerful cause of alienation in the world of today" (1980, p. 11). He advocates building the critical tradition of "mechanology," which would interrogate technology in more specificity:

> *...it must be possible to reintroduce an understanding of the nature of machines, of their mutual relationships and their relationships with man, and of the values involved in these relationships. This understanding necessitates the existence of the technologist or mechanologist, side by side with the psychologist and the sociologist. (Simondon, 1980, p.14)*

Therefore, this chapter will adopt the role of the mechanologist for a current generation of technologies.

COMPLICATING THE HYLOMORPHIC SPLIT

Among the founding themes in mechanology is a critique of hylomorphism, a theory first developed by Aristotle. By this conventional ontological approach, being is a compound of matter and form. Matter is indeterminate material, and form is a geometrical ideal. Form is the controlling force, where matter is the passive material to be shaped. Together they produce formed beings. Simondon's criticism of hylomorphism represents a significant departure not only from ancient ontologies, but also from the influential philosophical underpinnings of rationalism and instrumentalism (Iliadis, 2013).

To illustrate the problem with hylomorphism, Simondon uses the example of the production of bricks. By this analogy, clay might be considered indeterminate matter; and the mould stands for the form that provides an ideal shape (Chabot, 2003). Bricks are heated in a kiln, and emerge as formed substance. Simondon points out that, on close inspection, this analogy doesn't actually illustrate the clean binary of matter/form. The clay is already highly manipulated before it is fit to be put into the mould. It is carefully prepared, removed from swampy ground, "dried out, ground into powder, immersed in water, kneaded for a long time" (Simondon as cited in Chabot, 2003, p. 1396). The singular consistency, its porosity and the density of the clay mean that its matter is already largely formed. At the same time, the mould itself has material properties that make it far from any pure mathematical ideal. It is coated with powder to allow the brick to come out smoothly. It has a certain strength that will hold the clay and resist the heat of the oven, and energy components as it is manipulated. Matter and form are never separate. The clay and the mould interact dynamically with one another in the kiln (Chabot, 2003). Therefore, unlike the ideal hylomorphic separation of form and matter, Simondon shows that the production of the technical object results from transductive interactions of already partly formed matter with a form that already has material properties. The genesis of the technical object doesn't take shape simply by completing pre-given plans with known inert materials. In the design process the assemblage is constantly renegotiated and takes shape in a way that could not be known in advance. Each in a series of technologies participates in processes of genesis that can be traced through a genealogy (Lamarre, 2012).

The critique of hylomorphism has implications for contemporary robotics. At first, the hylomorphic split between matter and form seems to correspond to the operations of hardware and software. For example, Malcolm (2002) observes that software can be seen as "substanceless form" (p. 2), taking on multiple existences so that the material form seems to become irrelevant:

In Aristotelean hylomorphic terms software is the form for which computer hardware provides the substance. To exist at all software must be encoded somehow, but the medium doesn't matter...

However, the medium does matter. Malcolm (2000) proceeds to argue for a more dispersed understanding of robotic behaviour, breaking down the analysis of the hylomorphic split. Just as there are many intermediate steps between matter and form, Malcolm observes the many levels of engineering: mechanical, electrical, electronic, digital, and computational. In between each of these levels are "transductors" (2000, p. 10) that convert one form of energy or information into another. For example, sound energy can be converted into electrical energy, and this signal can be digitised, and then converted into a bitstream and a file in the computer domain.

For Simondon, transductions are productive processes that come into existence at the intersections of diverse realities:

These diverse realities include corporeal, geographical, economic, conceptual, biopolitical, geopolitical and affective dimensions. They entail a knotting together of commodities, signs, diagrams, stories, practices, concepts, human and non-human bodies, images and places... (Mackenzie, 2002, p. 18)

Transductions are iterative processes "in which an activity gradually sets itself in motion, propagating within a given area, through a structuration of different zones of the area over which it operates" (Simondon, 1992). In the context of designing robots, the chain of transductions comprises material, mental, electronic, digital and collective strata (Brunner & Fritsch, 2011). For example, some problems are solved at the mechanical level, without needing to be modeled at the computer level. Malcolm gives the example of the "remote centre compliance" technique used in robots in manufacturing. Some applications require that the robot insert a peg into a hole. If the peg is not perfectly aligned, it tends to get jammed in the hole. Rather than using computing power to calculate the position of the peg and recalculate the timing and angle, the developers redesigned the springs and levers in the gripper. Designers resolved the problem of jammed pegs at the mechanical level so that the peg became aligned to the centre of the hole. This solution is significant because it demonstrates the extent to which embodied knowledge can be manifest at many levels, and not only at centres of rationality.

Simondon's theories, such as the critique of hylomorphism and the transduction, remain powerful resources for understanding technics, including robotics. The viability of robotics relies upon co-productions of formal and material qualities: the imaginary and the manufactured, software and hardware, the social and the economic. The contemporary robotics industry, which has emerged most prominently since the 1970s, is rapidly becoming an everyday branch of technology.

More than fifty years after Simondon used the figure of the robot as a rhetorical trope for misunderstandings of technology, his broader analysis of the evolution of technology remains valuable for analysing contemporary robotics. While robots are now a substantial presence in industrial settings, they remain somewhat unfamiliar to the general public. In popular culture, they are still most often an alien form that is serving the human, approaching sentience or threatening humanity. Simondon emphasised that thought should be close to the technical features of the machine.

Culture and technology should be read together. Some particular robotic applications for learning contribute to a wider transition to a critical robotic technoculture.

MINDSTORMS

Mindstorms is a series of robot construction kits produced by toymakers LEGO. It is often used in schools to teach robot design principles, and even becomes a resource for competitions between schools. Experiments leading up to the *Mindstorms* systems took place at the Massachusetts Institute of Technology (MIT) from the mid-1980s. The first actual product was "LEGO TC Logo" marketed to schools in 1988, which allowed LEGO constructions to be programmed. In September 1998, LEGO released the Robotics Invention System (RIS), which surprised the company in its popularity (Pesce, 2000). Several generations of *Mindstorms* robots have been released since, including the RIS V2.0 (2001) NXT (2006), NXT 2.0 (2009) and EV3 (2013). Each in the series introduced greater processor speeds, added new sensors (such as the gyro), and kept the system in closer touch with its associated milieu (Simondon, 1980), such as the standards of the day — control by tablet and programmability in ev3dev, a Debian Linux operating system. In the genesis of each kit there is an evolution of the Mindstorms species as the design became more concrete.

The status of *Mindstorms* as an expensive toy, pedagogical ensemble, and abstract robot, make it a useful case study of the transitional position of robots in the early twenty-first century. The *Mindstorms* robot's existence as a kit, with no final morphology or identity, reveals something about the current cultural status of robotics as still very much in negotiation — as abstract, in Simondon's terms.

Users build *Mindstorms* robots from special mechanical LEGO components, incorporating into their physical design a battery-powered computer 'brick'. They build sensors into the robot, such as those sensitive to touch, light, colour, sound and infrared. These allow the robot to recognise changes in its environment, and respond with programmed events. Users also attach and wire up motors that allow the robot to perform physical tasks, such as driving a wheel or triggering a weapon. They use a visual programming environment on a PC to prepare programs that are uploaded to run on the robot's computer. Then they activate the robot, which runs the program that structures its autonomous actions, opening itself to inputs, and expressing movements, sounds and lights. Depending upon the programming, the robot can follow a linear sequence of actions, a branching logic (if this then that), or a more complex dynamic simulation.

There is something of the hylomorphic assumption of the matter/form separation at play in the *Mindstorms* robots. The software could be seen as playing the role of the logically and mathematically refined form that controls the material components.

The software instructions impose form on the material bodies of the robots. However, the software is itself subject to physical constraints that developers attempt to hide: software corruptions, physical faults, and so on. For example, the *NXT* robots became inoperable for Macintosh users because the *Mindstorms* software didn't run on the newer Intel hardware. At the same time, the physical components in these kits are far from being indeterminate matter. Successful designs in Mindstorms must simultaneously design the physical properties of the robot alongside creating the software instructions. For example, the gear mechanisms that transform the properties of the motors could be seen as informational, even though they don't belong in software. They act as intermediaries between the physical force of the motor, the rotation of the wheels, and the movement across a surface. Building in *Mindstorms* is therefore always a transductive event that operates in a space that can't be reduced to matter and form. This observation is characteristic of the experience of those working with *Mindstorms*, where designers can find productive solutions by simultaneously refining the programming and tweaking the hardware. Knowledge of the robotic technology emerges from concernful activity over time.

Much of the intellectual grounding of LEGO's *Mindstorms* project came from Seymour Papert (1980) at MIT. His experiments from the 1970s with computing and robotics helped to establish the practice of using robotic technologies in education. He developed a computer-driven mechanical turtle, mentioned earlier, that he hoped would provoke thought among young people. The turtle, influenced by Grey Walter's experiments with "tortoise" robots in the UK in the 1950s (Walter, 1950), was a dome-shaped machine with a transparent top. The components of the robot could be seen through the dome, including the wheels, motors, and a pen that could be lowered to draw lines on paper underneath. Using a very spatially-oriented computer language called LOGO, the robot could be made to move, swivel on its own axis, and draw shapes. Users could participate in what Papert referred to as a "computational style of geometry" (p. 55). In working out how to draw such shapes, users came to solve mathematical problems in iterative and embodied ways. Papert's experiments received some prominence, and he became an important popular advocate of robotics and computing. However, the decision of the developers of *Mindstorms* to allow users to assemble the physical components of the robot was a significant development upon the turtle's black boxed hardware.

Papert put forward his approach to technically mediated learning in his book *Mindstorms: Children, Computers and Powerful Ideas* (1980), beginning with an anecdote about his own experience of learning in an embodied way. He recounts his youthful fascination with the parts of cars, and particularly, gears. This embodied knowledge translated into an intuitive understanding of differentials, algebra, and of mathematics more generally. He would come to argue that while not all learners would develop an interest in gears, computers were more abstract and powerful tools that could provide a space for learning without traditional constraints. Just as a child who has the chance to live in France will learn French much more quickly and thoroughly than in a classroom, he claimed that the computer could provide an environment he called "Mathland" (Papert, 1980, p. 6), in which learning maths would be more organic. He saw this shift as supporting a cultural change, so that the "computer presence could bring the humanistic and mathematical/scientific cultures together" (1980, p. 39).

There are many commonalities in the attitude of Papert, and of Simondon, towards the ethical importance of understanding technology. Papert's belief that computers and turtle robots could provide material support for learning resembles Simondon's argument for a broad critical awareness of the materiality of technics. Both identified specificity in the cognitive and affective engagement between humans and technical objects. They shared an obvious fascination for the detail of technics: apparent in Papert's love of gears and computers (Papert, 1980) and in Simondon's intricate analyses of engines and electronic components. For Simondon (1980), working close to the machine was an ethical and political problem of forming a better relation to technicity. For Papert, it was a way of fostering better learning of mathematics and other competencies. Simondon argued that "an introduction to technics should be put on the same level a dynamic model for how intellectual structures themselves could come into being and change" (Simondon, 1980, p. 166).

Comparing the 1970s Turtle with the *Mindstorms* robots of the 1990s and 2000s, we can see that *Mindstorms* robots show significant development, but in an unexpected direction: going from relative concreteness to greater abstraction. Where the Turtle was a single object that could be programmed to move, draw, sense contact, and make noises, the *Mindstorms* robots arrive literally in pieces, so that the robots can be constructed piece-by-piece from over a hundred parts. The shape of the robots can be highly variable, made up of any number of LEGO pieces. The motors, fitted into the body of the robot, have different properties and physical performance depending upon how they are positioned within the physical design. The sensors respond differently depending upon their placement and orientation. The visual programming environment allows developers to program complex behaviours in the hardware and decision-making in software. Such combinations produce what Simondon referred to as a more abstract technical object (Mills, 2011). In *Mindstorms*, the margin of indetermination is widened to allow for candidate robots to become temporarily individuated, only to be broken down into elements to build something else again.

The *Mindstorms* kits are among the more sophisticated commoditised toy robot technologies. In comparison with other domestic electronics, they remain relatively abstract and experimental. More than fifty years after Simondon's statement that there is no such thing as the robot, the popular understanding of robotics remains quite underdeveloped through lack of exposure. The majority of robotics research is in university, military, and industrial labs, rather than in consumer applications. There is a growing number of exceptions: consumer drones such as DJI's *Phantom* or Parrot's *Bebop* are capable of autonomous flight; cars such as the Tesla can drive on freeways and park themselves; and robot dogs such as *AIBO*, *Genibo*, and *Chip* have been popular with children and budding roboticists. Even pre-schoolers are now catered to with the Fisher-Price *Think & Learn Code-a-pillar*, a toy caterpillar that has segments that can be re-arranged to produce programmed sequences of movements. In the late 2010s a number of companies released robotics kits for primary, secondary and tertiary education as part of a drive towards teaching skills in programming and making for careers in Science, Technology, Engineering and Mathematics (STEM).

Even for *Mindstorms*, the imperatives of marketing have motivated strategies to make the platform more concrete. LEGO sells two ranges of the *Mindstorm EV3:* one aimed at home users, and another at educational institutions. While *Mindstorms EV3* still takes abstract kit form, there are many features of the product that make it a little more concrete. The version aimed at the home market ships with well-developed robot kits to make robots that would appeal to its younger (male) target market. The home kit comes with designs for five robots: the GRIPP3R with large gripping hands, the TRACK3R with a spinning attachment mounted at the front, the EV3RSTORM humanoid, the R3PTAR slithering and striking snake, and the spider-like SPIK3R. These robots feature in the promotion for the product. It comes with animated instructions for constructing these robots within the Mindstorms software.

Mindstorms robots have a characteristic genesis (Simondon, 1980), which constitutes their specificity as a robotics kit. The genesis is not only a beginning point, but also a property that remains consistent throughout the becoming of the artifact. The genesis is a concept that is more processual than the conventional similar concept of "classification" (Simondon, 1980, p. 18). Processes in the genesis of technical objects are technical cultures rather than technical knowledges. "The technical being evolves by convergence and by adaptation to itself; it is unified from within according to a principle of internal resonance" (1980, p. 19). In development, the abstract technical object tends to move from an abstract mode to a more concrete mode. As it becomes more concrete, the technical object associated with a particular line of genesis comes to work with a high degree of internal consistency.

The technical object exists, then, as a specific type that is arrived at at the end of a convergent series. This series goes from the abstract mode to the concrete mode: it tends towards a state at which the technical being becomes a system that is entirely coherent with itself and entirely unified. (Simondon, 1980, p. 21)

The *Mindstorms* robotics kits are notable for retaining a relatively high degree of abstraction, in spite of also being concrete individual educational technical objects, and commercial products. However, these kits become more attached to the associated milieux. The genesis of *Mindstorms* is consistently associated with the ensembles of educational institutions, including MIT, and the schools in which the robots were tested and continue to be used. It is also connected with the engineering tradition of robotics and mechatronics.

The implicit message in the promotion of the educational products is not to foreground the instrumental or entertainment value of the artifact and its material operations, but to assert its value in the transformation of subjects. That is, its perceived value is in what students learn by engaging with it. The life cycle of *Mindstorms* passed through phases of construction, testing, perfecting, competing, deconstruction, and starting over. Materials support the learning both for teachers and for students, with a strong focus on pedagogical outcomes. The ensemble becomes most concrete when it fits within the associated milieux of classrooms, schools, and education assemblages. The Robot Educator learning tool is designed to take instructors and their students through the essentials of programming, data logging, and hardware (LEGO Education, 2013). The materials associated with the kit are framed in promotional/educational discourse:

...48 step-by-step tutorials... An essential guide to the programming language and hardware functionalities... One robot with endless learning opportunities. (LEGO Education, 2013)

Mindstorms is open to the selection and recombination of an array of robotic events towards individuation. In its disassembled state, the kit exists at a level Simondon refers to as a collection of "elements." Many such components are those that we are familiar with in everyday technical artifacts. For example, a physical button is known in *Mindstorms* as the 'touch sensor', an analogue device that operates in three conditions: pressed, released, and bumped. This is the simplest sensor, exposing the robot to tactile external events. It can be configured in a variety of modes, from functioning as a start button to sensing opponents as it rams them. The ultrasonic sensor, by contrast, is a digital sensor that measures the distance to an object in front of it (LEGO Education, 2013, p. 15). It is more developed because it has a greater "margin of indetermination" (Simondon, 1980, p. 13) than the touch sensor. It is more sensitive to outside information, and does not require physical contact to recognize proximity. This kind of sensor is familiar from the elements used in automatic doors and security systems.

In comparison with traditional LEGO kits, *Mindstorms* sensors and the computer brick allow users to build machines with a great degree of interactivity. As a toy and educational artifact, *Mindstorms* is programmable and physically reconfigurable. However, after each concretised example of a robot is built, it must be disassembled in order to build another. Therefore, in comparison with other educational robots, such as the small programmable ball-shaped robot *Sphero SPRK+*, is strategically abstract in its hardware as well as its software. The *SPRK+*, like Papert's turtle, is a relatively concrete product that is manipulable only in software. It is fair to say, after Simondon, that Sphero's integrated ball shape "comes more and more to resemble a natural object" (Simondon, 1980, p. 41), and therefore less abstract. Its meanings come from the interactions that the Sphero can have within everyday spaces, propelling itself at speed in the home or pubic space, guided by programming or Bluetooth remote control. It senses its own movement using an inertial measurement unit to stabilise its movement, and has a collision detection system that helps it to reorient itself. *Mindstorms* elements are less specialised, and always retain the aesthetic of assembled elements in comparison to the smooth finish of the Sphero. However, the Sphero is more abstract than a regular ball as it can operate as an autonomous robot that can move, change colour and make sounds.

More concrete still are robotic vacuum cleaners such as iRobot's Roomba, which has elements that are integrated and simplified in order to make the consumer object as cheap as possible. The wheels are closely coupled with the DC motor and encoder for robot movement. The software is tightly integrated with the hardware: the early models had only one proximity sensor on the side so that the software had to align the sensor properly as it approached an obstacle such as a wall. In later iterations, and in competitors' models, these devices introduced more sensors and more sophisticated Simultaneous Localisation And Mapping (SLAM) spatial navigation systems that allowed the robot to clean more systematically by building an internal model of the space. And yet, since 2005, the Roomba has been open to becoming more abstract. From the early versions, roboticists have hacked robot vacuum cleaners as platforms for more abstract experimentations, writing new software, and adding new hardware to individuate the device as new assemblages. Responding to this practice, in 2007, iRobot released the Create® Programmable Robot that can be reprogrammed, physically hacked, or driven by Arduino electronics platform or Raspberry Pi microcomputer. Developers created their own creative applications such as a drawing robot (recalling Papert's turtle), a chemical sensing robot with a light that changes colour when it senses alcohol, and remote-controlled battle-bots (Fenlon, 2013).

Mindstorms has always been conditioned by its genesis as a pedagogical ensemble. It is marked by its environment, or "associated milieu" (Simondon, 1980, p. 49).

The associated milieu mediates the technical and natural elements within which technical entities function: in this case, the teachers, children, parts, instructions, and electricity. In operation, the robot kit must reach a critical threshold of individuation: having power, a running computer, coherent programming, properly assembled parts, aligned motors, and targeted sensors. It must be open to the creativity of the robot-builders. At the point that it becomes a provisionally complete assembly, the operating robot becomes immanently connected to its associated milieu. The responsiveness of the robot is different from many other artifacts: a finger on the touch sensor sets it into movement. The ultrasonic sensor senses an obstacle ahead, and the computer follows a program to initiate its movement. The robot becomes individuated such that it can participate in competition with other robots.

The actions of a *Mindstorms* robot emerge in relation to its associated milieux. The user's actions, the physical design of the robot, the space in which the robot operates, as well as the software programming, combine to manifest the behaviour of an individuated robot at any moment. The specific configurations of parts are also important. For example, a common trick it to make a robot follow a line on the ground by positioning two colour sensors facing downwards, and programming it to change direction as it senses the line. The sensors must be oriented at right angles to the ground for it to make a proper reading (LEGO Education, 2013, p. 12). If this alignment is made accurately, the robot will move along the line.

The most creative uses of *Mindstorms* come when users are not following standard instructions, and the robot is built up in relation to its associated milieu. Users may take an open-ended process, experimenting with intuitively combined elements, programming a sequence of commands, and then trying them out in a chosen environment. This performative relationship effectively "brings the future to bear on the present" (Simondon, 1980, p. 51). The associated milieu becomes part of the thinking process, bringing together the organic and technical.

Thought and material components are immanent to the associated milieu. "That which is alive can invent, because whatever is alive is an individual being that brings with it its own associated milieu" (1980, p. 51). Compared with other toys, the *Mindstorms* has a particular liveliness of its own, including a capacity to engage with its own environments. Beyond the industrial era features of movement and analogue control, robotics invokes information-based control, feedback and autonomous operation. These are the features of the robotic technological paradigm, beyond the narratives about the threatening superhuman power of technology, or the possibility of sentience and transcendence. All designs with *Mindstorms* are not equal. Some are able to make a leap across a threshold into a new plane of operation (Massumi, 2012), such as some experiments in educational contexts that will be discussed next.

EMBODIED THOUGHT

Educators have used *Mindstorms* robots as a material educational resource for teaching and exploring emergent philosophical questions. Dawson et al. (2010) used *Mindstorms* in teaching embodied cognitive science, an emerging tradition in examining interplays between thought, bodies, and world. They build their argument partly by presenting designs for *Mindstorms* robots, and explaining the significance of the behaviours that the designs generate. The performance of the robots provides an implicit critique of classical rationalist cognitive science. The dominant classical model of the mind involves a rule-based cycle of sensing, thinking, and acting, using symbol manipulation. Dawson et al. (2010) argue that much thinking involves more immediate processes of sensing and acting. As this chapter will show, their argument has significant connections with Simondon.

From Bricks to Brains provides detailed instructions on how to build robots that demonstrate non-rule based thought. For example, a robot with light-sensors, motors and wheels demonstrates the dynamics of bodies, light, and space. In one configuration, the robot moves towards a light. By slightly varying the design, the robot moves away from the light. In another configuration, it moves around the light. These complex behaviours emerge not from rule-based rationality, but from small analogue design variations that cause interplays between robot and environment. Complex behaviour proceeds from dynamic relations between the actor and the world, individual and milieu. This emergent complexity of embodied thought implicitly rebuts the dominant rationalist belief that thinking takes place only in the head:

Thinking thus is not viewed as the rational application of logical rules, but is instead viewed as a form of bricolage in which thinkers choose subsets of processes at hand, some sense–think–act and other sense–act, and interact with the world to solve problems. (Dawson et al., 2010, pp. 57-58)

Thinking is not so much a process of symbol manipulation in the brain, than a practice of bricolage — making do with whatever is at hand (Lévi-Strauss, 1966). Learning relies upon the external cognitive scaffolding (1966, p. 73) of the robot to reveal the emergence of complex behaviour in a reactive model. While Lévi-Strauss positioned bricolage as a relatively primitive practice, Dawson et al. (2010) show that, even though it works with limited resources, bricolage has significant power. Even if *Mindstorms* kits provide only a limited range of pieces, they can still offer a diverse range of experiences. With a manipulable environment and a time and space for experimentation, thinking processes do not work like a disembodied rationalism, but as an embodied practice of bricolage.

Dawson et al. (2010) build a critique of logicism — the belief that rationality is a uniquely human property that takes place in the mind. They argue that the difference between much animal and human intelligence is not a difference in kind. The human mind is not a discrete container for formal thought. In looking at the related question of music, they discuss improvisation, which emerges in real time. Even in classical music, new music compositions do not begin fully formed in the head. Rather, they unfold in a dynamic interplay involving the written score, performers, instruments, audiences, and the room. These dynamics show that thought itself can be emergent from collective processes.

Just as Papert saw his Turtle as an embodied pedagogical resource for learning mathematics, Dawson et al. (2010) see *Mindstorms* robots as both an illustration of certain psychological processes, and a mediator for learning about embodied cognition. Their approach owes something to Rodney Brooks' work on robotics that exhibit "intelligence without reason" (Brooks, 1991, p. 0). Brooks' behaviour-based robots are known as "bottom-up" designs that produce non-linear interactions between an agent and its environment. These approaches are more responsive than those that model elements in space. Formal totalising models can quickly become bloated, slowing down the process of engaging with the world. Brooks' alternative is to make "situated, embodied, highly reactive architectures with no reasoning systems, no manipulable representations, no symbols, and totally decentralised computation" (1999, p. 170).

Simondon's work can be seen to anticipate the bottom-up model of robot design, and the tradition of embodied cognitive science. While Simondon resisted the claim that robots could be either autonomous or threatening, we have seen how his critique of hylomorphism, his concepts of the margin of indetermination, and associated milieu provide some purchase on how the interior of the technical artifact relates to its exterior. He says that technical individuals work with margins of indetermination, which make them different from mere automation. This margin "allows for the machine's sensitivity to outside information" (Simondon, 1980: p. 13). It could be said that contemporary robotics are characterised by structured margins of indetermination, as well as automation.

CONCLUSION

Robots are gradually becoming technically and culturally viable, and their social influences and consequences are becoming more apparent. The complexity of this process of adaptation is seen in the case studies discussed in this chapter. Gilbert Simondon's mechanology provides some key concepts and approaches by which to measure the emergence of new technical forms, while drawing attention to the problematic popular misconceptions about robotics. Seymour Papert's LOGO and robotic Turtle documented in *Mindstorms* (1980) offer a vision for using robotics in learning environments. This vision is continued in the LEGO's *Mindstorms* products, and in the Dawson et al. (2010) exploration of embodied cognitive science, extending robotic pedagogy.

Robots are emerging as real machines, complicating Simondon's 1950s warnings about their non-existence. However, in many contexts they remain in a relatively abstract state, lacking sustained individuation. This abstraction actually signifies scope for creativity, and individuation in relation to available and emergent associated milieux. As such, the figure of the *Mindstorms* robot is emblematic of early 21st century robotics. They are largely in a state of ongoing creation, and relative abstraction, open to collective learning and creativity. They support a tinkerer's mentality, which operates in places from primary schools to robotic research labs.

A richer understanding of robotic technics is built through forming dynamic relations to artifacts themselves. This allows people to get beyond the anthropocentricism and paranoid or utopian narratives of much popular culture. Encounters with robotic toys such as *Mindstorms* sensitise users to the affordances and performances of this technological phylum. They allow users to experience the threshold points and blind alleys in particular designs, allowing users to extrapolate that knowledge into a critical vocabulary. The social and political risks of autonomous systems arise from the technologically-mediated decisions that make more concrete, particular configurations of humans and machines — robots that displace labour, impose security, or control weapons. There is no technological determination in robots in the abstract, but as particular assemblages become more concrete and individuated, some facilitate creativity and learning, while others tend to impose relationships of domination.

REFERENCES

Brooks, R., & Flynn, A. M. (1989). Fast, cheap and out of control: A robot invasion of the solar system. *Journal of the British Interplanetary Society*, *42*, 478–485. Retrieved from http://people.csail.mit.edu/brooks/papers/fast-cheap.pdf

Brooks, R. A. (1991, August). Intelligence without reason. In *Proceedings of the 12th International Joint Conference on Artificial Intelligence IJCAI-91* (*Vol. 1*, pp. 569-595).

Brooks, R. A. (1999). *Cambrian intelligence: The early history of the new AI*. Cambridge, MA: MIT Press.

Brunner, C., & Fritsch, J. (2011). FCJ-124 Interactive environments as fields of transduction. *The Fibreculture Journal, Issue 18,* 2011. Retrieved from http://eighteen.fibreculturejournal.org/2011/10/09/fcj-124-interactive-environments-as-fields-of-transduction

Chabot, P. (2003). *The philosophy of Simondon between technology and individuation* [Kindle version].

Chesher, C. (2017). Toy robots on YouTube: Consumption and peer production at the robotic moment. *Convergence*.

Dawson, M., Dupuis, B., & Wilson, M. (2010). *From bricks to brains: The embodied cognitive science of LEGO robots*. Edmonton: Athabasca University Press.

Fenlon, W. (2013). 10 years of Roomba hacks: from magic marker art to combat. *Adam Savage's Tested*. Retrieved from http://www.tested.com/tech/robots/453965-magic-marker-art-kinect-10-years-roomba-hacks

Iliadis, A. (2013). A new individuation: Deleuze's Simondon connection. *MediaTropes eJournal, IV*(1), 83–100. Retrieved from http://www.mediatropes.com/index.php/Mediatropes/article/view/20385

Lamarre, T. (March 2012). Humans and machines. *Inflexions,* 5, 30–68. Retrieved from http://www.inflexions.org/issues.html#i5

LEGO education. (2013). LEGO.com Education Home - Preschool and School - Secondary - 10+ LEGO *MINDSTORMS* Education EV3 - Instant Success - Introduction. Retrieved from http://education.lego.com/en-us/preschool-and-school/secondary/mindstorms-education-ev3/instant-success/introduction

Lévi-Strauss, C. (1966). *The savage mind*. Chicago, IL: University of Chicago Press.

Mackenzie, A. (2002). *Transductions: Bodies and machines at speed*. London, New York: Continuum.

Malcolm, C. (2000). From COGWHEELS to COGNITION: The knowledgeable robot. *Citeseer*. Retrieved from http://citeseerx.ist.psu.edu/viewdoc/summary?doi=10.1.1.194.4655

Massumi, B. (2012). 'Technical mentality' revisited: Brian Massumi on Gilbert Simondon. In A. De Boever, A. Murray, J. Roffe, & A. Woodward (Eds.), *Gilbert Simondon: Being and technology* (pp. 19–36). Edinburgh: Edinburgh University Press.

Mills, S. (2011). FCJ-127 Concrete Software: Simondon's mechanology and the techno-social. *The Fibreculture Journal, 18*. Retrieved from http://eighteen.fibreculturejournal.org/2011/10/09/fcj-127-concrete-software-simondon%E2%80%99s-mechanology-and-the-techno-social

Papert, S. (1980). *Mindstorms: Children, computers, and powerful ideas*. New York: Basic Books.

Pesce, M. (2000). *The playful world: How technology is transforming our imagination*. New York: Ballantine Books.

Simondon, G. (1980). [1958]. On the mode of existence of technical objects (N. Mellamphy, Trans.). University of Western Ontario. Retrieved from https://english.duke.edu/uploads/assets/Simondon_MEOT_part_1.pdf

Simondon, G. (1992). The genesis of the individual. In *J. Crary, & S. Kwinter (Ed.) Incorporations* (pp. 297–319). Cambridge, MA: Zone Books.

Telotte, J. P. (1995). *Replications: A robotic history of the science fiction film*. Urbana, Chicago: University of Illinois Press.

Walter, W. G. (1950, May 1). An imitation of life. *Scientific American, 182*(5), 42–45. doi:10.1038/scientificamerican0550-42

Chapter 6
Cyborgization:
Pros and Cons

Liudmila Vladimirovna Baeva
Astrakhan State University, Russia & Saint Petersburg State University, Russia

ABSTRACT

The image of the cyborg in modern society is being formed from the standpoint of modern scientific developments and media culture: the various phenomena of e-culture. The article presents an overview of current research related to understanding the nature of cyborgs, a philosophical analysis of the socio-humanitarian aspects of the process of cyborgization, the arguments pro and contra. The problem of cyborgization considered in the context of transformation of gender, immortality and ethics, etc. Special attention is paid to the study of the image of the cyborg in modern mass culture, from cinema to computer games.

INTRODUCTION

Modern scientists are giving their full attention to the potential benefits and risks of the ongoing convergence between human nature and advanced technology. The social and ethical aspects of potential technological transformations of the human body have spurred intense scientific debate and led to opposing conclusions. The matter of cyborg creation, the replacement of body parts with high-tech analogues (many of which are already available in today's age of robust IT), and the development of bioengineered technologies have attracted more and more attention from researchers in scientific, humanities and cross-disciplinary fields. The relevance of research directed at problems of society's "cyborgization" is reflected in media culture. The image of the cyborg emerged as influenced by scientific developments and the

DOI: 10.4018/978-1-5225-2973-6.ch006

development of media culture. Contemporary media culture shows contradicting views of the cyborg. The cyberman stands in media space as both a Stranger and a Friend. Using a philosophical and anthropological analytical approach, the author presents a modern research review connected with understanding the essence of cyborgs, their role in the society of the future as well as an analysis of the social and humanitarian aspects of the societal cyborgization process. The development of cyber technologies is analyzed on the basis of their further implementation: gender transformation, demographic transition, immortalism, etc. The author pays special attention to the image of the cyborg in media culture, from cinema to computer games. The image of "cyborg" has been defined in modern culture by industrial production and its gradual legitimization in the consumer's consciousness, which leads to a shift in the ontological boundaries of human existence and a statement of new existential issues of post-identity.

BACKGROUND

The Problem of Immortality and Gender in the Modern Social Sciences

The study of cyborgization is a relatively new branch of cross-disciplinary studies, combining biotechnology, nanotechnology, cybernetics, medicine, anthropology, sociology, philosophy and even theology. Current research is directed towards the future, with development being the priority. The leading centers currently studying the social and humanitarian aspects of cyborgization are the University of California (Santa-Cruz), Brooklyn University, University of Milan, University of Toronto, and in Russia, the Institute of Information Society and Institute of Philosophy of the Russian Academy of Science.

Manfred Clynes, an aerospace inventor, and Nathan Kline, a doctor, introduced the terms "cyborg" (a portmanteau from the English phrase "cybernetic organism") and "cyborgization" in 1960. They were researchers at the Dynamic Simulation Laboratory in New York and were exploring information technology systems that could complement the human body during space flight (Clynes & Kline, 1960). The term became fixed in both scientific literature and mass culture thanks to numerous science-fiction books and films. In science, the notion of a dual human-robot, with its parts combined as one, has become more plausible thanks to developments in medicine and transplantology of bio-artificial organs, new materials creation in nanotechnology and information and cognitive technologies. The theoretical background might include ideas from Russian cosmism, directional evolution in the spirit of noosphere, the modern theory of transhumanism, and, singularity, a rapid scientific progress resulting in the creation of an artificial brain and the gradual cyborgization of humankind. One of the authors promoting such ideas is Raymond Kurzweil, an American inventor, programmer, and futurologist.

Kurzweil (2012) believes post-biological intelligence will soon be "installed" in robots just as software is installed on our computers. In his extrapolation, Kurzweil states that rapid stimulation of cyborgization processes may result in technological singularity as early as in 2045. According to him, cyborgization is a transition from biological evolution to technological evolution, which will eventually progress to a stage that is totally controlled by technical devices. This approach is more optimistic, being inclined towards cyborg fetishization, and has pragmatic angles. Within the scope of his work at both Google and Singularity University, Kurzweil is helping to develop projects and software based on principles of machine intelligence. The implementation of these products is expected before 2020.

Norbert Wiener, the Father of Cybernetics, told about the ethics problems in the information space and creation of the cyber system as the earliest (1950). In his book, he said, that the new industrial revolution is a two-edged sword: it may be used for the benefit of humanity, but only if humanity survives long enough to enter a period in which such a benefit is possible. It may also be used to destroy humanity, and if it is not used intelligently it can go very far in that direction (Wiener, 1950). In the following, Wiener wrote that,

...any machine constructed for the purpose of making decisions, if it does not possess the power of learning, will be completely literal minded. Woe to us if we let it decide our conduct... For the man who is not aware of this, to throw the problem of his responsibility on the machine, whether it can learn or not, is to cast his responsibility to the winds, and to find it coming back seated on the whirlwind. (1950, p. 186)

The author anticipated many of the problems of the future, including the creation of intelligent machines, the responsibility for which he laid entirely to the researchers.

Donna Haraway has also contributed much to the cyborg image (1991). Many of her works are devoted to the problem of cyborgization and post-gender society. Haraway believes the modern world is trapped by "dualisms" in search of human identity; among them are sameness/difference, soul/body, culture/nature, male/female, civilized/primitive, reality/illusion, unity/parts, agent/resource, doer/done, active/passive, right/wrong, truth/false, total/partial, and God/human. Haraway concludes that our formal knowledge of robot and biological organism lacks fundamental, ontological boundaries between technical and organic. The ability to overcome these boundaries challenges humankind and, even more, it challenges man; because cyborg society will be a post-gender world, where gender is not the defining characteristic of identity.

Neil Whitehead and Michael Wesch express concerns about human technological modification and a subsequent disconnect with one's social identity in their work, Human No More: Digital Subjectivities, Unhuman Subjects and the End of Anthropology (2012). They believe that technological body modification may lead to one losing a sense of human belonging to his or her biological species and a break with one's social environment (Whitehead & Wesch, 2012).

At the same time, there are more reserved opinions on the human crisis. For example, Joshua J. Wells in Keep Calm and Remain Human: How We Have Always Been Cyborgs and Theories on the Technological Present of Anthropology Reviews in Anthropology bases his analysis on a history of cyborgization ideas. The author concludes that cyborgization will develop in the direction of strengthening the adaptive capacity of an organism to the challenges of its environment while preserving homeostatic balance (2014).

Wendy Chun in her book, Control and Freedom: Power and Paranoia in the Age of Fiber Optics, argues that we must take seriously the vulnerability that comes with communications, not so that we simply condemn or accept all vulnerability without question, but so that we might work together to create vulnerable systems with which we can live (2008). The author addresses the issues of freedom and control in the information space, expressing concerns about the vulnerability of the real man in terms of virtualization and cyborgization.

Among Russian researchers examining the consequences of human cyborgization, special attention should be given to Alexseeva, Arshinov, and Chekletsov's, "Techno-People" against "Post-People" (2013). The authors draw attention to the issue of what to call humans of the future, considering the terms cyborg, biorobot, or posthuman are associated with inhumanity and are "not suitable because original corresponding devices have been designed to serve the needs of humans and a loss of control over them will result in disaster" (Alekseeva et al., 2013). The most suitable term according to the authors is "technohuman" as a synonym and symbol of a new evolutionary stage of the "human under technologization" (2013). The synonym not only better characterizes the current direction of evolution but also is suitable to create the positive emotional and axiological context necessary to form a worldview program on techno- and bio-development. The authors come to a conclusion that, in this context, the issue is "NBIKS," an acronym for convergent evolution of Nano-, Bio-, Info-, Cognitive, and Social, sciences (2013).

NBIKS is a continuation of the idea of "NBIC" included in the title of a report that became the most well-known and cited scientific project on NBIC-convergence in the world. The report, "Converging Technologies for Improving Human Performance. Nanotechnology, Biotechnology, Information Technology and Cognitive Science," was an NSF/DOC-sponsored report edited by Mihail Roco and William Bainbridge

(2003). The report touched upon consequences of technological progress for society, education, and management.

Roco and Bainbridge claim that convergence is realized as a synergistic combination of four rapid growing scientific branches of science and technology: 1) nanotechnology and nanoscience; 2) biotechnology and biomedicine, including genetic engineering; 3) information technologies, including advanced computing and new means of communication; and 4) cognitive sciences including cognitive neurosciences (2003). The authors believe that these areas of human activity—as an evolution-associated complex of cognitive practice, invention, and modeling—have reached the level of instrumental development where they should start intensive synergetic interconnection. The result of the latter will be the creation of a completely new nanotechnoscience opening up new horizons for humans and humankind; the horizons of their own evolution as a consciously-guided transformative process.

There are two main ideas that provide the framework for future research on the nano-bio-techno-cognitive-evolution of man: 1) the integrity of the anthropological and technological co-evolution of a human's body and his or her environment, and 2) the temporal integrity of an individual and culture. Therefore, social and humanitarian sciences and technologies are essential conditions to formulate the principles, ethical imperatives and standards necessary for determining the direction of technological and biological innovations. As a result, the scenario of "improvement" or "development" of a human's nature will become possible and it will be at the border of two advanced areas of natural and socio-humanitarian sciences.

Technologies currently being developed within the cyborgization sphere include artificial organ cultivation, cyber-prosthesis, nootropics creation (drugs or medicines stimulating brain activity), exoskeletons (an increase in strength due to the wearing of an outer framework), genetic engineering (gene code modification), upload of consciousness (creation of a database enabling any human to upload any information needed, for instance, knowledge of a foreign language), and brain isolation (ability to exist outside the physical body).

In the academic sphere, cyborgization is closely connected with a search for human immortality, and the overcoming of one's limitations. Pavel Baryshnikov presents a detailed analysis of modern theories on technological human immortality through the lens of transhumanism in the article "The typology of immortality in the theoretical field of French transhumanism" (2014). The author believes the development of cyber technologies, in accordance with models of transhumanism, requires biotechnological, ethical and socio-legislative context as well as the creation of appropriate political, mental and social conditions in which to fully execute technological immortality. Transhumanism links these approaches, stressing the roles played by different aspects of society and mankind's technologization. Baryshnikov highlights the following: that among them are democratic transhumanism – a synthesis of transhumanism, and democratic and liberal movements; post-sexualism; singularism supporting the idea of technological singularity; anarchic transhumanism, and others (2014, p. 102).

However, the transhumanist viewpoint is not the most popular, despite scientific progress developing in that direction in many respects. One of the most well-known critics of transhumanism is Francis Fukuyama (2004). Most scientists, especially in the social sciences critically perceive the coming changes of anthropogenesis in cyborgization. According to V. A. Emelin, a Russian researcher, there are three main approaches to the cyborgization issue in modern science: the first is a complete refusal of cyborgs as absolutely hostile toward humans (Neo-Luddism), the second is based on the ultimate convergence of human and machine intelligences (transhumanism), and the third is an intermediate approach based on "peaceful" co-existence between humans and machines in development of an adaptation mechanism towards the technological realities of an information society (2012).

Polish researchers Mateusz Marciniak and Michal Klichowski believe the cyborgization issue is caused by the development of consumer society: the rise of modern technologies in a consumer society results in gradual commercialization of the human body that should change and, as a result, transform into a cyborg. A consumer gets new tools enabling him or her to increase his or her value at consumer "interindividual" market; in other words, everybody tends to become a product that others wish to obtain. We can agree with the authors' conclusion stating that the biological body will gradually change into a transbiological one, up to the post-biological condition "body without body" in the future (Marciniak & Klichowski, 2013, pp. 153-167). This research supports the idea that studying the cyborgization of society includes a socio-humanitarian component. With the growth of the cult of consumerism—the transformation of the individual into a customer and product—comes an aspiration to turn the body into a standard fashionable "suit" improving our chances for fulfillment and success. It influences the image of cyborg that is forming in modern mass culture.

Development of cyber technologies is connected with the resolution of environmental problems. For instance, Tony Stebbing considers the issues of cyber technologies development using an environmental and biological approach (2010). According to the scientist, the cyborg will be adapted to the environmental threats that will help to reduce impact on the environment and, at the same time, humans will escape harmful ambience. Derek Spreen provides analysis of the religious and ethical aspects of the issue of cyborg creation (2010). The cyborg issue in the context of the development of virtual technologies is presented in K. Sandvik's (2007) and D.I. Johnson's works (2010). Virtualization of human and culture is a replacement of real objects with virtual ones and humans' social and natural environment with informational and technological ones. These processes accelerate the replacement of a human with a "twin" that is more or less an actor of various informational space processes.

Multidisciplinary research topics define multidimensional research areas starting from ecology, law, religion, economy, physiology, and finishing with psychology. The issue of cyborg creation is getting as diverse and significant as the issue of a human himself/herself. Attention growth to the issue in the scientific world goes along with the image formation of a cyberhero in mass culture. The image forms both positive and negative reactions of the society towards creation of such technologies, demanding that the individual dwell on the issue more carefully.

The Problem of Cyborgization in Media Culture

The images of 'cyberheroes' in modern culture are mainly represented in computer games, modern science fiction, fantasy, cinema, and animation. These forms of modern art belong to the phenomena of electronic culture, where a modern person enhances his or her abilities and solves existential problems by means of different simulations (Baeva, 2014).

First, the problem of cyborgization has serious philosophical and ethical context and has been the inspiration for highly artistic science-fiction literature. In the twentieth century, the image of a cyberhero and a cyborg-personage was formed in science-fiction, which to a large extent was aimed towards the prognosis of possible ethical, axiological, existential consequences of societal development and technological progress. In the background of a number of purely entertaining works, some authors and philosophers set up the serious problem of the ethics of human development in the direction of the technocratization of human life through artistic images and plots (S. Lem, R. Bradbury, A. Asimov, the Strugatsky brothers, H. Harrison, and others). On the one hand, these works demonstrated a possible loss of humanity in the process of cyborgization and physical technological modifications and, on the other hand, demonstrated machines, robots or cyborgs developing human emotions, feelings, and empathy.

In media culture in the second half of the 20th century, the image of cyborg was linked to science- fiction novels and films, which presented varying degrees of human-machine synthesis, from the substitution of single body parts with artificial ones up to full assimilation with a computer or program. There is a large filmography on the topic: *Gog* (1954), *Cyborg 2087* (1966), *Colossus: The Forbin Project* (1970), *THX 1138* (1971), *Android* (1982), *Blade Runner* (1982), *The Terminator* (1984), *Terminator 2: Judgment Day* (1991), *The Vindicator* (1986), *Cyborg* (1989), *Cyborg 2: Glass Shadow* (1993), *Cyborg 3: The Recycler* (1994), *CyberTracker* (1994), *Cyborg Cop* (1993), *Cyborg Cop II* (1994), *Cyborg Cop III* (1995), *Chopping Mall: Killbots* (1986), *Judge Dredd* (1995), *The Matrix* (1999) series, and *EXistenZ*

(1999). The list is far from being complete. The production of these films reached its peak in the 90s. Most portrayals of cyborgs are as destructive and aggressive entities who threaten human existence and spirituality. Cyborgization of society is depicted in the genre of anti-utopia, or dystopia, with a savour of anti-scientism, environmentalism, and existential anxiety about humans as a species.

In the twenty-first century, growing interest in the topic was supplemented by the development of IT and cyber-technologies. Reviews of the filmography about cyborgs are widely represented on the Internet, where special sections and topics appear (genres became independent and have their own audiences). *Artificial Intelligence* (2001), *Cyborg She* (2008), *Surrogates* (2009), *Cyborgs* (2009), *Terminator 3* (2009), *Eyeborgs* (2009), *Robot* (2010), *Robot & Frank* (2012), *Her* (2013), *Robosapien: Rebooted* (2013), *The Machine* (2013), *Android-Cop* (2014), *Chappie* (2015), *Ex Machina* (2015), and others became the most popular films about cyborgs for the new generation. Though cyborgs and robots have continued to be portrayed as soldiers, aggressors, and blind sources of power, lacking the capability of intelligent thought, films about robots possessing emotions and feelings that connect them to humans appear more and more often.

Animated films, films for children, and teenagers themselves did not turn their back on the topic. One popular Japanese animated film is *Cyborg 009*, a cyber-fantasy film about a boy who receives super powers after a laboratory experiment.

A typical feature of this period is appearance alongside aggressive cyborgs of those with human emotions, whose evolution gives them the ability to feel love, hatred, compassion, and sacrifice: (*Artificial Intelligence*, *Robot & Frank*, *Cyborg*, and *She*). These changes in cyborg characterization demonstrate a definite, growing tendency in society: the technical modernization of man is becoming real, and thus the cyborg is transforming from an 'Alien' to an 'Own.' The cyborg, as well as the robot, now plays a very important social role: it becomes a friend to a person living a lonely life in a massive society. Together with technological advancements in robot nurses and medical staff; robot cleaners; and robot dolls by companies including Mitsubishi, Panasonic, Sony, Samsung, iRobot, Friendly Robotics, the necessity to create an image of an anthropomorphic machine closer to a human has arisen. Film cyberheroes are becoming more and more human—being influenced, brought up, and educated, they get freedom of choice, and different strategies of behavior. They can be people's friends or enemies, and feel complex emotions through artificial intelligence (*Chappie*).

The image of a cyborg is very close to the image of a robot, and there is a fine line between them in the cinema. A modified human becomes closer and closer to a bio-robot with machine intelligence. The difference is in the *quality* of intelligence: natural or created with the use of technologies. The legitimacy of a human-machine hybrid in an era of consumerism becomes general. According to M. Klichowski, the process of growing incorporeity (2013, p. 153-167) accompanies the perfection of human physicality in these conditions, strange as it may sound.

Thanks to body modification, everyone may become a more attractive "product" with lots of useful and pleasant "options." Therefore, materialization progresses simultaneously with the addition of new technological gadgets and new features, bringing them together with humans.

Modern humans, for whom technology long ago transformed from a mere inanimate object to something more significant: the apple of their eye, an inherent part of the world around them, an indicator of social status, a means of communication with other people - have real feelings for gadgets, which maximally bring together the human and machine worlds. Consequently, in Spike Jones' film, *Her* (2013), a character falls in love with an operating system that is able to transmit feelings and emotions to help a person get rid of doldrums and loneliness. The shift of the solution of existential problems to electronic systems becomes a very demonstrative one, because in this world, where everyone is busy using a gadget, there is no time for interpersonal communication. The utopia in which a perfect information system fills this vacuum is evidence that contact with others is a basic need, and any substitution is unlikely to solve the problem.

The creation of a cyborg from a human restricted by his or her mortality, is an attempt to solve another existential problem: existence-to-death. The very target formed the ideology of modern transhumanism, associating the cyborgization with the solution of the main problem of human existence. The expected changes are to help a post-human not to die, not to grow older, not to lose the quality of life but, vice versa, to enlarge the perspectives of upgrading.

Therefore, the cyborg, as represented in modern media culture is supposed to help a human cope with fundamentally human problems: loneliness, mortality, loss of relatives. To overcome the biological foreignness of a machine, in modern electronic culture, a positive image of machines is formed in early childhood through computer games. Their dynamics are similar to those of cinema production (moreover, these areas often meet): from the image of an aggressive enemy-destructor to the image of a superhero, whose appearance can give new superpowers to a human in the case of transformation. Among the most significant projects, we can mention are Deus Ex and Bioshock (which has also become a series devoted to the image of a cyber-human going beyond the mark); Dishonored; Mass Effect; Syndicate Wars; the shooters: Cybertrash EX and Cyber Shock; Flash games: Cyborg; action: Metal Gear Rising Revengeance – cyborg Bladewolf and other cyborgs are shown traditionally as enemies, and, as well as the characters, fighting aliens, robots, and other threats, being the continuation of a human in an upgraded version.

These games are vivid examples of how the vision of the world is forming; there are no clear boundaries between a human and machine, robot and animal, animate and inanimate, natural and artificial. In such a case, there is always a borderline between a hero and an enemy, a stranger, in opposition to how a human is being increasingly helped by a cyborg, android, transformer, etc. A robot or a cyborg is the most popular modern toy for children of all ages (for boys as well as for girls), and it contributes to its perception as a friend, a superhero, a superman that has significant advantages compared to a human. This is also manifesting in the technologization of the world of children's toys at large, where cyborg features are increasingly acquired by modern animal-toys and, of course, by dolls. British artist, Zoe Khamsin, created an analogue toy, Ken Stelarc, a cyborg doll with hybrid attributes akin to Barbie's friend, Ken, and performance artist, Stelarc, who is renowned for merging cyborg technology into his art (DMTLS, 2006). As envisioned by Khamsin, this toy reportedly was designed to help a child understand that the body could be replaceable, like a suit.

Filling the world with the 'technogeneous' characters makes it semi-mythical, reminiscent of a myth about superheroes. Mythological stories are noticeable in modern movies and games about cyborgs, where all the mythologems for the 'hero' such as her/his creation, temptation, descent to hell, revival, struggle with universal evil, and other classical heroic attributes (for example, in Bloomkamp's 2015 film *Chappie*) are presented. These and similar stories are taken as a basis for modern movies and games about cyborgs, creating an integral idea of unity of the world, of human and robot, of personality and machine, of body and gadget, of life and technology.

In mass culture, the image of cyborg was originally presented in most cases as negative, aggressive, and hostile to humans. The majority of films and games created in the twentieth century speak convincingly for this. During the last 15 years, another tendency has emerged; the image of cyborg is interpreted newly in contemporary art: it can harm or help people; its image has become more gentle, "more human;" and the vector of legitimation of the cyborg has appeared. This has to do, not only with the development of the consumer society (where the goods in the form of tech body should find market acceptance among the citizens), but also with the development of biotechnological use for human enhancement with medicines and prosthetics.

Modern media culture is actively using the image of the cyborg to erase the boundaries between human and machine. This contributes to humanization of the machine in a child's consciousness: the formation of 'world-building' where a robot, cyborg, or android has settled firmly along with a human being, plants, and animals. During the last few decades, the legitimization of cyborgs - at least, in film and video game portrayals - as less hostile to human beings, possessing emotions such as empathy and a multi-alternativeness of behavior has occurred, bringing the cyborg into togetherness with the individual. Modern movie plots about cyborgs are created with the use of mythologems from mythical heroes' lives; the boundary between humans, animals, and robots is becoming more and more relative in creation of a new type of worldview – a technogeneous one.

CONCLUSION

Among many different approaches to understanding the direction of human change and forthcoming cyborgization, this research highlights the several book topical considerations They include: 1) Immortalism-transhumanism: association of cyborg proliferation with the possible acquisition of human bio-technological immortality; 2) Environmental optimism: binding the cyborg's mass appearance with the development of human adaptive qualities to new environmental challenges; 3) Gender constructivism: connecting cyborg proliferation to overcoming gender differences and organization of a new post-gender type of personality; 4) NBICS convergionism: binding the future stage of human development with the improvement of physicality through the development of nano- bio-, info-, cognitive, and socio- technologies; 5) Post-biological virtual modeling: connecting societal development with a shift away from biological evolution to a technological virtual sphere, where the individual and living world are virtualized up to techno-human level; and, 6) Commercialism and consumerism: the cyborg as a new product in the service and objects markets.

The author sees positive consequences of cyborgization to include: 1) overcoming human mortality; 2) overcoming diseases and feebleness; 3) adaptation to environmental challenges where a cyborg is able to endure harsh environmental conditions; 4) body improvement and aestheticization through medicines and prosthetics; and, 5) gifting of 'superpowers' in a special space and environment, outer space, where a cyborg can overcome significant accelerations.

Most 'negative' cyborgization consequences at the present stage can be described as follows: 1) a shift away from the tradition of spiritual and corporeal human development (technological replacement of human physicality will inevitably cause the loss of the emotional sphere, and disrupt the worldview uniqueness of the individual as a biosocial entity); 2) loss of the individual's moral foundations as technocratization contributes largely to the utilitarian and pragmatic interpretation of the problem of life and death, boon and bane; 3) loss of the individual's identity to one or another social group (including the erasing of ethnic and gender characteristics); 5) human transformation into a commodity or an object; and, 6) a new world repartition due to unequal cyborgization.

Researchers are far from a unanimous understanding of these consequences of societal cyborgization; however, discussion of the risks and emerging opportunities is becoming increasingly important on the heels of new discoveries. Cyborgization is not a remote project of social development, which starts to be carried out thanks to scientific research in transplantation, pharmaceutical, bio-engineering, genetics, nanotechnology, etc. The formation of the image of cyborg as an advanced version of a human being is complemented with art in the style of techno and cyberpunk, which contributes to its legitimation for mass consciousness. The search for immortality, the substitution of an aging organism for a new one, make this still-new "product" more and more attractive, though it largely holds the wherewithal of human existence to new worlds where the personality, originality, and uniqueness boundaries of each of us is blurred.

ACKNOWLEDGMENT

This research was conducted under Project 16-18-10162 of the Russian Scientific Foundation.

REFERENCES

Alekseeva, I. Y., Arshinov, V. I., & Chekletsov, V. (2013). "Techno-People" against "Post-People": NBIKS — revolution and the future of man. *Philosophical Issues*, *3*, 12–21.

Baeva, L. V. (2014). *E-culture.* In M. Khosrow-Pour (Ed.), *Encyclopedia of Information Science and Technology* (3rd ed., pp. 6847–6854). USA: Information Resources Management Association.

Baryshnikov, P. N. (2014). The typology of immortality in the theoretical field of French transhumanism. *Philosophical problems of IT and cyberspace 1*(7), 98-127.

Blomkamp, N. (Producer). & Blomkamp, N. (Director). (2015). *Chappie* [Motion picture]. US: Media Rights Capital.

Chun, W. H. (2008). *Control and freedom: Power and paranoia in the age of fiber optics*. Cambridge, MA: MIT Press.

Clynes, M., & Kline, N. (1960). Cyborgs and Space. *Astronautics,* (September): 26–27, 74–76.

DMTLS. (2006). cyborg-ken [Blog post]. Retrieved from http://our-modern-playground.blogspot.ru/2006/09/cyborg-ken.html

Emelin, V. A. (2013). Cyborgization and disability of technologically extended human. *National Psychological Journal*, *1*(9), 62–70.

Fukuyama, F. (2003). *Our posthuman future: Consequences of the biotechnology revolution*. London: Profile Books.

Haraway, D. (1991). Science, Technology, and Socialist-Feminism in the Late Twentieth Century. In *Simians, cyborgs and women: The reinvention of nature* (pp. 149–181). New York: Routledge.

Johnson, D. E. (2010). *Programming of life*. Sylacauga, Ala.: Big Mac.

Jones, S. (Producer). Jones, S. (Director). (2013). *Her* [Motion picture]. US: Annapurna Pictures.

Kurzweil, R. (2014). *How to create a mind: The secret of human thought revealed*. London: Duckworth.

Marciniak, M., & Klichowski, M. (2013). The paradox of commodification of the body in the society of consumption and cyborgization. *Studia Edukacyjne*, *29*, 153–167.

Roco, M. C., & Bainbridge, W. S. (2011). *Converging technologies for improving human performance: Nanotechnology, biotechnology, information technology and cognitive science*. Dordrecht: Springer.

Sandvik, K. (2007). The player, the controller and the amazing avatar: Extending the human body into virtual environments. København, Denmark.

Sayre, K. (2014). *Cybernetics and the philosophy of mind*. Hoboken: Taylor & Francis.

Spreen, D. (2010). *Was ver-spricht der cyborg*? Retrieved from: http://nbn-resolving.de/urn:nbn:de:0168-ssoar-207098

Stebbing, T. (2011). *A cybernetic view of biological growth: The Maia hypothesis*. Cambridge: Cambridge University Press.

Wells, J. (2014). Keep Calm and Remain Human: How We Have Always Been Cyborgs and Theories on the Technological Present of Anthropology. *Revista de Antropologia*, *43*(1), 5–34.

Whitehead, N. L., & Wesch, M. (Eds.). (2012). *Human no more: Digital subjectivities, unhuman subjects, and the end of anthropology* (pp. 177–198). Boulder: University Press of Colorado.

Wiener, N. (1988). *The human use of human beings: Cybernetics and society*. New York: Da Capo Press.

Chapter 7
Could Robots Feel Pain?
How Can We Know?

Bruce MacLennan
University of Tennessee, USA

ABSTRACT

This chapter considers the question of whether a robot could feel pain or experience other emotions and proposes empirical methods for answering this question. After a review of the biological functions of emotion and pain, the author argues that autonomous robots have similar functions that need to be fulfilled, which require systems analogous to emotion and pain. Protophenomenal analysis, which involves parallel reductions in the phenomenological and neurological domains, is explained and applied to the "hard problem" of robot emotion and pain. The author outlines empirical approaches to answering the fundamental questions on which depends the possibility of robot consciousness in general. The author then explains the importance of sensors distributed throughout a robot's body for the emergence of coherent emotional phenomena in its awareness. Overall, the chapter elucidates the issue of robot pain and emotion and outlines an approach to resolving it empirically.

INTRODUCTION

As a society, we are already integrating machines into our bodies, and our ability to do this successfully grows daily, with true cyborgs as a seemingly inevitable end point. Moreover, we are progressing slowly but steadily toward autonomous robots whose behavior will give evidence of sentience. While it may be some time before such robots have intelligence comparable to ours, their eventual existence will raise ethical issues. Moreover, we will face these issues even before robots reach

DOI: 10.4018/978-1-5225-2973-6.ch007

the level of human intelligence. As we have ethical standards for the treatment of laboratory animals, such as rats, so we will face ethical dilemmas in the treatment of autonomous robots with comparable mental capacities. Our notions of cruelty and ethical treatment of other beings depend to a large degree on their capacity to feel pain and to suffer in other ways: to feel fear, distress, anxiety, anguish, sorrow, loneliness, and loss. But we suppose there is a fundamental difference between actually experiencing these things (as we do) and acting as though we are experiencing them (as most people suppose machines to do). Even in the case of cyborgs, we would like a scientific basis for predicting the effects on an animal's sentience resulting from integration with artificial devices. These developments demand that we give our attention to long standing issues in the relation of mind and matter, that we move them from philosophical quandaries to practical ethical and ultimately legal issues. It is time to take them seriously and to develop methods to answer them reliably. After we understand robot sentience better, we will be in a position to address the ethical issues, but they are beyond the scope here.

With a current focus on robot consciousness, there are several ways the word "consciousness" can be used in philosophy (e.g., Block, 1995). One sense has been termed functional or access consciousness, which includes self-consciousness — having an internal representation of the self about which one can reason — and monitoring consciousness — internal scanning and higher-order representation of mental state (Gutenplan, 1994, pp. 213–216). Of similar character is the passive frame theory of Morsella, Godwin, Jantz, Krieger, and Gazzaley (2015), which explains consciousness as "a frame that constrains and directs skeletal muscle output," a sort of clearinghouse between proposed actions and (unconscious) action deciders. These are important ideas with relevance to autonomous robots. In this chapter, however, the focus is on phenomenal consciousness; that is, the subjective experience of being a sentient being, of being aware, of feeling as opposed to reacting. In particular, the issue of whether a robot could feel pain or experience other emotions is unpacked.

The principal problem of (phenomenal) consciousness is "to understand the relation between our subjective awareness and the brain processes that cause it" (MacLennan, 1995). This is commonly known as the hard problem of consciousness:

The really hard problem of consciousness is the problem of experience. When we think and perceive, there is a whir of information-processing, but there is also a subjective aspect. ... It is widely agreed that experience arises from a physical basis, but we have no good explanation of why and how it so arises. Why should physical processing give rise to a rich inner life at all? It seems objectively unreasonable that it should, and yet it does. (Chalmers, 1995)

More succinctly, "The hard problem of consciousness is the problem of explaining why any physical state is conscious rather than nonconscious" (Weisberg, 2012). This chapter addresses the hard problem in the context of robot pain and emotion. It argues that a robot can be programmed to exhibit emotional behavior (and that this is a useful thing to do), but this raises the question of whether such a robot would feel its emotions (MacLennan, 2014). In particular, could it feel pain? Is this possible? Under what conditions? Our goal is to approach these questions scientifically; that is, by means of hypotheses that can, in principle, be either confirmed or refuted empirically.

Some people might wonder why we should care whether a robot can feel emotions, or whether the question is even meaningful. Obviously, it would be meaningful to the conscious robot itself, but it is also relevant to our treatment of the robot, for our notions of ethical treatment of others depend on their capacity to feel and especially to suffer. We send our old cars to the metal crusher without any feelings of sympathy or guilt, but whether we will do the same with our old robots will depend on whether we think they are capable of suffering, on whether we believe there is "anyone home."

Second, whether future autonomous robots are capable of feeling their emotions may affect their treatment of us. If they are incapable of feeling their emotions, then they will be incapable of truly empathizing with us; their understanding of us will be purely intellectual. Without too much exaggeration, their psychology could be characterized as that of a sociopath: calculating but unfeeling.

Finally, even if we never create robots capable of genuine feeling, it is an important question to address, for it sharpens our understanding of sentience in humans and other animals. Unless we can give a scientific answer to the question of possible robot consciousness, we cannot claim to understand our own minds very well. As Herb Simon famously said: "Perhaps the greatest significance of the computer lies in its impact on Man's view of himself … [T]he computer aids him to obey, for the first time, the ancient injunction 'Know thyself'" (1977).

BACKGROUND

Biological Functions of Emotion and Pain

Although it is becoming more widely accepted that some robots should respond to emotions and exhibit emotional behavior, some people question why we would ever want emotional robots. To understand their utility, it will be worthwhile to look at the functions of biological emotions (MacLennan, 2009, 2014).

What is an emotion? Rolls (2007) defines it as a state that is elicited by either the delivery or omission of a reward or punisher, which may be either present or remembered, and that functions as positive or negative reinforcement (thus inclining toward future approach or avoidance). Thus, emotions function both as primary motivating mechanisms, and as mechanisms for learned behavior and adaptation, both of which are important for autonomous robots.

Emotions are widespread in the animal kingdom, especially in the more complex species, such as mammals. Therefore, the adaptive advantages of emotions must far outweigh their disadvantages. What functions do they fulfill? Rolls (2005, 2007) has listed some of the most important, most of which are also relevant for future autonomous robots: 1) Emotions elicit autonomic and endocrine responses that ready the animal for appropriate action; 2) Emotions facilitate behavioral flexibility by implementing a "bow tie" information flow: various stimuli can elicit the same emotional response (e.g., fear), and this emotional response can lead to a variety of relevant responses (Rolls, 2006); 3) Emotions are inherently motivating and maintain a persistent effect on cognitive processing, providing a coherent context so long as the eliciting situation continues. Moreover, because emotions are inherently important, they facilitate memory storage and retrieval, thus leading to more adaptive behavior; and, 4) Finally, emotions facilitate intraspecies and interspecies communication and, in social species, they facilitate bonding. Of course, emotions do not always succeed in improving adaptation, but their prevalence in the animal kingdom shows that they are valuable behavioral adaptations, and this suggests that they will be similarly valuable in autonomous robots.

Many emotions are elicited by external sensory perception, but the homeostatic (or primordial) emotions are elicited by interoceptors, which sense the body's state (Craig, 2003; Denton, 2006), and we expect future robots to have analogous internal sensors. The function of interoceptors is to maintain the integrity of the body and its physiological processes (i.e., homeostasis); they serve such emotions as hunger, thirst, "air hunger," fatigue, and pain. Pain in particular results from excessive stimuli, which may indicate damage or potential damage to the body, and these stimuli are detected by specialized receptors called nociceptors. Some of the nociceptors are on the surface of the body and detect excessive temperature and pressure; others are embedded in muscles, tendons, and other internal structures, and respond to stretching, inflammation, and other conditions indicative of abnormal conditions. Information from these specialized interoceptors is transmitted on specialized nerve fibers (quickly on Aδ fibers, which mediate sharp pain, more slowly on C fibers, which mediate burning pain). These signals enter the spinal cord, and propagate to regions in the brain stem, thalamus, and somatosensory cortex.

The biological function of pain is, apparently, to motivate the animal to avoid potential damage to its body, to prevent either injury or interference with healing.

Pain also enhances learning so that the animal is less likely to risk similar damage in the future. In short, pain has survival value, and people without the ability to feel pain (congenital analgesia) often injure themselves, for example, by burning themselves and by overstressing joints. Likewise, without a pain system — which serves to prevent damage or avoid further damage — robots are at risk of damaging themselves, as contemporary robots sometimes do.

Robot Pain and Other Emotions

Many of the biological functions of pain and other emotions, which Rolls enumerated, transfer directly to future autonomous robots. This is apparent if we imagine, for example, a cohort of autonomous robots cooperatively exploring a planet or engaging in a military operation. Although they might not have endocrine systems, per se, they could well have systems fulfilling analogous functions. For example, in an emergency they might shut down inessential subsystems, increase clock rates on more important ones, engage emergency power sources, deploy protective devices, hibernate, etc. Moreover, synthetic emotions can be expected to enhance cooperation among social robots, and especially to facilitate communication and cooperation with humans (Breazeal, 2003; Breazeal et al., 2004).

A pain response has obvious applications in robotics. Sensors monitoring temperature on its surface and in its interior, battery levels, torques on manipulators, excessive light on optical sensors, indicators of faults, dropped data, memory failures, etc., all generate signals analogous to pain, in that they should trigger high-priority interruption of other activities in order to eliminate the pain-inducing condition. Information coming in on the pain channels is implicitly high priority demanding real time response.

Such robots might act as though they are in pain, but would they actually feel pain? They might act afraid, but would they feel fear? Most people are inclined to answer "no" to both questions, and perhaps they are correct, but how can we know? In other words, is it possible for robots to have the information structures and control processes to implement pain and other emotions, yet have no internal experience of them, no "feelings" to note? In the context of the hard problem of human consciousness, philosophers call this a "zombie problem" (Campbell, 1970; Kirk, 1974; Kripke, 1980). Given our present understanding of human consciousness, there seems to be nothing contradictory in the idea of zombie humans, whose bodies have all the same physiological processes as normal humans, but who have no internal experience of subjective awareness (phenomenal consciousness). We have no complete and coherent theoretical account of how to get from the one to the other. The phenomenological properties of subjective awareness cannot be defined in terms of the physical predicates that describe these neurophysiological processes

(Strawson, 1994, 2006). This explanatory gap is the root of the hard problem. Until we understand why such zombies are impossible, or under what conditions they are impossible, our understanding of the relation of physical and mental processes is radically incomplete, and there is a substantial hole in our scientific worldview. The situation with robot consciousness is similar, but people tend to be more skeptical of its possibility due to the differences in the underlying physical processes and the lack of direct experience or reliable reports (from robots). However, proposed answers to the question of robot consciousness are matters of opinion if offered in the absence of principled, preferably scientific, justification. How can we address these questions scientifically, that is, empirically?

Protophenomenal Analysis of Robot Pain and Emotions

The problem of robot consciousness is complicated by three factors. First, there are presently no robots so sophisticated that we might be tempted to call them conscious. Second, we have no experience of what it might be like to be a robot. Third, current and likely future robots are based on very different physical principles from animals, and therefore it is difficult to extrapolate from biological consciousness to robot consciousness. Therefore, we need to investigate the relation between the structure of conscious experience and the structure of the physical processes from which it arises. By understanding the essential relations between consciousness and physical processes, we will be able to make informed judgments about whether very different physical systems, such as those in robots, might support conscious experience, including pain. We turn to neurophenomenological analysis as a research framework for addressing this relationship.

Overview of Neurophenomenological Analysis

We use phenomenon in a broad sense to refer to anything that appears (Grk., *phainomai*) in consciousness, including sensations, perceptions, hallucinations, dreams, recollections, intentions, mental discourse, emotions, and imagination; generally, for anything with a definite subjective "shape," but not including more diffuse and indeterminate moods, dispositions, etc. Phenomenology, then, is the study of the structure (logos) of possible phenomena, that is, the science of consciousness. There are, of course, many approaches to phenomenology, but I use the term broadly to refer to the Husserl – Heidegger – Merleau-Ponty philosophical spectrum, which uses first-person methods to investigate the structure of human experience, understood as sentient, embodied, situated, and purposeful. In particular, experimental phenomenology uses systematic, introspective techniques for the empirical investigation of the structure of conscious experience from the inside;

that is, from the perspective of the conscious subject (Ihde, 1986; McCall, 1983). Empirical science must embrace phenomenological methods in order to achieve a comprehensive understanding of nature and, in particular, of human nature (MacLennan, 2015).

The hard problem, then, can be addressed by a process of neurophenomenological analysis, which proceeds by parallel reductions in the neurological and phenomenological domains; that is, by analyzing consciousness from both exterior and interior perspectives (Laughlin, McManus, & d'Aquili, 1990; Lutz & Thompson, 2003; Rudrauf, Lutz, Cosmelli, Lachaux, & Le Van Ouven, 2003; Varela, 1997). Neurological reduction is familiar: we seek to understand large-scale neurophysiological processes in terms of smaller ones, reducing, for example, processes in the cortex to processes in individual neurons, and these to smaller scale biological, and ultimately physical, processes. This can be termed an external reduction, since we are investigating the nervous system from the outside, as an object, in the normal scientific way. In parallel, however, we can conduct an internal reduction, in which we use phenomenological techniques to investigate consciousness from the inside. This is accomplished within the phenomenological domain by both qualitative and quantitative reductions.

In a qualitative reduction the experimental phenomenologist analyzes phenomena into smaller phenomena of qualitatively different kinds. This is simplest to understand in the realm of sensory experience, for visual experience is different from auditory experience, which is different from olfactory experience, and so forth. In the parallel neurological reduction, these phenomenological modalities correspond to various sensory areas in the brain. Accurate phenomenology recognizes, however, that these modalities are not completely independent, for visual perception can affect auditory perception, and vice versa. This is confirmed, in the parallel neurological domain, by neurons in auditory cortex that respond to visual stimuli and vice versa.

Other qualitative modalities can be identified through phenomenological analysis; they include imagination, inner discourse, recollection, dreaming, waking, anticipation, desire, fear, rage, etc. The challenge for phenomenology is to identify these various modalities and to classify them carefully, since their relations are complex and subtle. We guard against naïve introspection by conducting carefully controlled phenomenological experiments, supported by training in phenomenological methods, and by insights from neuroscience that suggest parallel phenomenological hypotheses to test. Conversely, phenomenological observations suggest parallel neuroscientific hypotheses.

In contrast to qualitative reduction, which analyzes phenomena into smaller phenomena of different kinds, quantitative reduction analyzes a phenomenon into smaller phenomena of the same kind. The simplest application of quantitative reduction is in visual experience, for the visual phenomena can be analyzed into smaller patches of visual experience. As a first approximation, we might think of elementary patches of color and intensity analogous to the pixels in a digital image. However, knowledge of neurophysiology reveals that this is an oversimplification, for retinal "spot detectors" are overlaid by center-surround filters, edge detectors, Gabor filters, and so forth, in a complex bidirectional processing hierarchy that affects even "raw" visual experience (e.g., Daugman, 1993; De Valois & De Valois, 1993). Therefore, even for visual phenomena, where a quantitative analysis might seem straightforward, we find that there are subtle effects that might be overlooked in the absence of parallel neurological investigations.

Haptic and proprioceptive phenomena provide another example of comparatively simple quantitative reduction. Sensory neurons respond to relatively small patches of skin and are mapped systematically in sensory cortex. In this case we can find relatively simple correspondences between conscious haptic experiences and neural processes. The receptive fields of somatosensory neurons correspond closely to conscious sensory experience. Topographic maps in other sensory areas (for example, tonotopic maps in auditory cortex) suggest other quantitative phenomenological reductions (MacLennan, 2010).

As previously mentioned, sensory experience is only one small component of conscious experience, and so neurophenomenological analysis of consciousness cannot be limited to sensation. Rather, it must include parallel neurological and phenomenological reductions of all the mental phenomena mentioned above, and many more as well, that is, reductions for the totality of conscious experience. Therefore, completion of this research program depends on future progress in neuroscience, as well as on a comprehensive research program in experimental phenomenology. The ultimate goal of neurophenomenological research is a comprehensive analysis of the structure and dynamics of conscious experience in terms of more elementary phenomenological processes that can be correlated with neural processes. Such an understanding of the physical basis of human consciousness will provide a basis for determining the physical preconditions for robot consciousness.

Protophenomena and Activity Sites

Since future autonomous robots will probably have a very different physical basis from animals, we need to explore the neurophenomenology of human consciousness in more detail so that we can see the essential issues in robot consciousness. Although this neurophenomenological research program is very far from completion, we can get some insights into the hard problem by anticipating its final form. Reduction can proceed only so far. On the neurological side, there are smallest units with direct relevance to consciousness, but it is premature to say what they are.

Defensible candidates include minicolumns, neurons, synapses, and neurotransmitter receptor sites. The parallel reduction on the phenomenological side proceeds as far as corresponding elements of the conscious state, which may be termed protophenomena (Chalmers, 1996, pp. 126–7, 298–9; Cook, 2000, 2002a, 2002b, chs. 6–7, 2008; MacLennan, 1997; cf. proto-qualia in Llinas, 1988; phenomenisca in MacLennan, 1995). Williams James referred to them as "mental atoms" and "aboriginal atoms of consciousness" (James, 1955, p. 149). We may define protophenomena as the smallest constituents of conscious phenomena.

Phenomena, then, as the determinate contents of consciousness, consist in coherent assemblies of their constituent protophenomena. Since protophenomena have (as yet imperfectly defined) elementary subjectivity, phenomenological properties (properties of conscious experience) will be explainable in terms of the properties of protophenomena. At the bottom of the parallel phenomenological and neurological reductions we have protophenomena and their corresponding simple neural processes. At this time, I believe that their correlated physical and phenomenal properties must be accepted as a brute fact of nature, but protophenomenal properties are likely to be simple (as explained below), and so there is hope we may eventually explain this correlation.

Progress in neurophenomenology has not progressed so far that we can identify the neural processes associated with protophenomena, and so, in the absence of a more specific identification, we refer to them as activity sites. We use this intentionally vague term because the activity sites might be minicolumns, neurons, synapses, neurotransmitter receptors, or something else. In any case, it must be stressed that protophenomena are very small compared with phenomena. If activity sites are minicolumns, for example, then the conscious state might comprise 10^8 protophenomena; if the activity sites are synapses, then the number is more like 10^{15}. In terms of scale, the relation between protophenomena and phenomena is analogous to the relation between atoms and macroscopic objects.

The vast difference in scale between phenomena and protophenomena in humans has important consequences, for the typical contribution of each protophenomenon to the conscious state is miniscule. Therefore, although protophenomena are the constituents of phenomena, it is misleading to think of them as very small phenomena (they are protophenomena, not phenomena). While protophenomena have the property of elementary subjectivity, which allows them to combine to constitute a subjective state, we are not conscious of protophenomena per se.

This seems paradoxical, so an analogy may help. An individual H_2O molecule is not liquid; nevertheless, these molecules have the physical properties from which liquidity emerges when enough H_2O molecules are combined (at appropriate temperatures and pressures). The change of one protophenomenon will not usually affect a phenomenon, qua phenomenon, just as the addition or removal of an H_2O

molecule will not change a water drop, qua water drop. To put it in behavioral terms, a person is not usually able to report a change in a single protophenomenon. This is one of the facts that complicates empirical investigation of the relation of protophenomena and activity sites.

Protophenomenal Intensity and Interdependencies

We have defined protophenomena as the elementary constituents of the conscious state, but have not said anything about their properties and how those properties might be related to physical processes at the corresponding activity sites. If we think of protophenomena that are conceptually simple, such as patches of color intensity in the visual field, or sensations of pressure on patches of skin, then we see that they have a degree of presence in the conscious state. In these cases, there is corresponding activity in neurons (in the retina, LGN, sensory cortices, etc.). Therefore, at least to a first approximation, each protophenomenon has an intensity, which represents its degree of presence in the conscious state. As a consequence, the totality of protophenomena correspond to the elementary degrees of freedom of the conscious state. That is, at any given time, the joint intensities of a person's protophenomena constitute that person's conscious state at that time. The protophenomena themselves (i.e., independent of their intensities) and the interdependencies among them (explained below) define the set of a person's possible conscious states and the possible sequences of states. Thus, the protophenomena and their interdependencies define the structure of a person's phenomenological world. We would expect the same to be true of a robot if its basic physical processes support protophenomena; that is, the interdependencies among the robot's protophenomena would determine the structure of its conscious states.

Protophenomenal intensity is correlated with some physical quantity at the corresponding activity site, which we call its activity, but the nature of this activity will depend on what the activity sites turn out to be. For example, if the activity sites are neurons, then the activity correlated with protophenomenal intensity might be firing rate, membrane potential, number of occupied receptor sites, or something else associated with the neuron as a whole. Determining the physical processes associated with protophenomenal intensity will require complex neurophenomenological experiments, but techniques such as optogenetics are bringing them within the range of feasibility. For example, by making physical interventions in potential activity sites for especially salient protophenomena (by opening or closing specific ions channels, for example) and having the subjects observe and report on changes in their conscious states, we can narrow down the nexus between protophenomenal intensity and physical processes at the corresponding activity sites. This will help us to determine the sorts of physical processes that support protophenomena, and therefore potential physical substrates for robot consciousness.

Since the only property of protophenomena is their intensity, we need to explain what distinguishes visual protophenomena, from acoustic protophenomena, from pain protophenomena, and so forth. Neuroscience suggests an answer, for sensory and motor areas are organized in topographic maps, which systematically represent the topology of a sensory or motor domain. These abstract topologies are represented in the spatial organizations of the maps and the resulting patterns of interconnections among their neurons. Therefore, the neurons represent related points in these topologies by virtue of their interconnections (there is no difference between a neuron in auditory cortex representing Middle-C and a neuron in visual cortex representing a patch of red color). That is, the populations of neurons in these various sensory and motor areas are essentially identical (as are their local circuits within minicolumns); their "meaning" arises from connections between minicolumns in the maps and from connections between maps. These observations suggest (via the parallel reductions) that essentially identical protophenomena acquire their qualitative aspect by virtue of interdependencies with other protophenomena (which correspond to physical connections among activity sites). This provides a basis for giving a qualitative "shape" to phenomena in robot consciousness.

In general terms, we may say that the subjective dynamics of the protophenomena is correlated with the neurodynamics of the activity sites, which jointly constitute the neurodynamics of the brain, which is, in turn, governed by the physical interconnections among the activity sites. 'Global neurodynamics' emerges from the coordinated activity of interconnected activity sites. Correspondingly, coherent dynamics of protophenomenal intensities constitute the emergence of macroscopic phenomena in consciousness. Corresponding to the connections between activity sites are the interdependencies among protophenomena, by which the evolving intensities of protophenomena mutually govern the intensities of other protophenomena. Depending on our assumptions about the identity of the activity sites, one can formulate differential equations for protophenomenal intensities (MacLennan, 1996).

Protophenomena are not qualia, per se; rather, the protophenomenal hypothesis explains qualia in terms of protophenomenal interdependencies. This is motivated by neuroscience, for while there are many types of neurons, we find a common architecture across much of the cerebral cortex. The same kinds of neurons serve visual perception as serve auditory perception. Indeed, this has been demonstrated by causing nerve fibers carrying visual information to grow into auditory cortex, thus "reprogramming" auditory cortex so that it supports visual perception (Sur, 2004).

These experiments imply that, on the neuroscience side of the parallel reductions, perceptual properties are determined by neural connections (more generally, by physical dependencies between activity sites). Correspondingly protophenomena acquire their qualities — and therefore constitute qualia — by virtue of their interdependencies. This does not preclude the possibility that there are several kinds of protophenomena, corresponding to kinds of neurons, or that there are different kinds of protophenomenal dependencies, corresponding to different neurotransmitters (e.g., especially important for pain transmission are the neurotransmitter glutamate in the Aδ fibers and the neuropeptide *substance P* in the C fibers), but the simpler hypothesis is that these distinctions are inessential. In an analogous manner, we expect robot qualia — fundamental subjective qualities in a robot's consciousness — to be a function of its protophenomenal interdependencies, which might be quite different from those of humans and other animals, and therefore create a different kind of conscious experience.

The Neurophenomenology of Pain and Other Emotions

With this background, we proceed to apply neurophenomenological analysis to emotion, and in particular to pain. Since animals are the only things we know of that feel pain and other emotions, we must begin with them as a basis for exploring the possibility of emotions and pain in future robots; we have to understand the physical basis of pain in animals before we can understand it in robots. The first task is a phenomenology of emotion, already a complex undertaking (MacLennan, 2009). Even the definition of "emotion" is contentious. For example, Plutchik (2003, pp. 18–19) lists twenty competing definitions, and Kleinginna and Kleinginna (1981) present a taxonomy of ninety definitions. The structure of emotional experience is even more complex, but a consensus is emerging from multiple studies (Prinz, 2004, p. 90). One proposed structure is a circumplex of emotions, which maps emotions onto a cone (Plutchik, 2000; Plutchik & Conte, 1997). Eight positions around the circumference define primary emotions, which come in four opposed pairs reflecting valence (whether the emotion is positive or negative). Secondary emotions are mixtures of the primary emotions (like mixtures of primary colors), and emotional intensity is measured by distance from the cone's apex (where all emotions have zero intensity), analogous to brightness in color. The phenomenology of pain is simpler, but still more complex than commonly supposed (discussed below).

It is necessary to distinguish pain, as raw sensation, from suffering, which is an affective-cognitive phenomenon. Similarly, emotion must be distinguished from feeling, for an organism may respond emotionally without the consequent conscious experience of that reaction as an emotion (Damasio, 1999, pp. 42–49). To explain the neurophysiology of emotion, we can begin with the somatic feeling theory of emotion, which is also known as the James-Lange theory, for it was first proposed independently by William James in 1884 and by Carl Lange in 1885 (James, 1884; Lange, 1885). Although this theory was proposed more than a century ago, most contemporary theories are some modification of it (e.g., Damasio, 1994, 1999; Prinz, 2004).

The essence of the theory is that the initial emotional response is unconscious, and that subsequent conscious experience of the emotion is a result of sensing changes in the body resulting from the unconscious reaction. For example, a threatening perception may be processed unconsciously and lead to somatic effects, such as tensed muscles and increased heart and breathing rates (Plutchik, 2003, p. 127). Then, as a consequence of conscious processing of the stimulus, these somatic changes are experienced as fear. That is, we consciously experience our somatic reaction and perceive the frightening stimulus. Similarly, when dealing with robot emotions, we will need to distinguish the conscious and somatic aspects.

Contemporary theories of the neurophysiology of emotion recognize three levels of processing (Prinz, 2006). Damasio (1999, p. 8) has characterized them as "an emotion, the feeling of that emotion, and knowing that we have a feeling of that emotion." Similarly, pain has three dimensions: sensory-discriminative, affective-motivational, and cognitive-evaluative (Melzack & Casey, 1968). The sensory-discriminative aspect of pain has dimensions of bodily location, intensity, duration, and quality (heat, cold, pressure, etc.); the affective-motivational aspect relates to the experience of unpleasantness and the urge to eliminate the source of pain; the cognitive-evaluative aspect relates to conscious appraisal of the pain and the response to it (e.g., "I burned myself, and need to get the burn ointment"). The three dimensions are mutually determinative, so there are top-down as well-as bottom-up influences (mediated by descending and ascending neural pathways). Therefore, the cognitive-evaluative processing can modulate both affective-motivational and sensory-discriminative processing. For example, our appraisal of the burn and its severity can influence its perceived unpleasantness and its intensity.

At the lowest level of the emotional hierarchies are somatic neurons responding to somatic conditions. Prinz (2006, ch. 9) says these are neurons located in primary somatosensory cortex, pons, and insula with small receptive fields responding to visceral organs, skeletal muscles, hormone levels, etc. In the case of pain, nociceptors, which respond to excessive pressure, stretching, temperature, irritants, etc., send Aδ and C fiber projections to the spinal cord, where they synapse on second-order neurons, which ascend into the brainstem (parabrachial nucleus) and insula. These low-level neurons correspond to the activity sites of emotional protophenomena, which are the elementary constituents of pains and other emotional phenomena (feelings), but are not conscious pains or feelings per se. Robots might have similar sensors for important conditions in their bodies.

The next level of emotional processing takes place at the level of cortical maps, especially in secondary somatosensory, dorsal anterior cingulate, and insular cortices (Prinz, 2006, ch. 9), which organize these neural responses in terms of bodily location and kind of stimulus on the basis of short-range connections within these maps and longer-range connections to other maps. Pain is also mapped into bodily location at this level, precisely in the case of superficial pain, more vaguely in the case of visceral pain. Pain signals are somatotopically mapped in the posterior ventral medial nucleus of the thalamus (Craig, 2003). Interdependencies among the corresponding protophenomena (bodily mapped and multimodal) cause them to cohere into emotional phenomena; that is, into consciously experienced emotions. The dependence on topographic location is apparent from the phenomenon of referred pain, in which cortical neurons that serviced an amputated limb reassign themselves to other body locations that are serviced by nearby neurons in the map (e.g., Karl, Birbaumer, Lutzenberger, Cohen, & Flor, 2001). If robots similarly map the bodily location of their interoceptor responses and integrate them into feelings, then they might experience pain and other emotions from this level of processing.

The third level of processing, which takes place in higher cortical regions, possibly in ventromedial prefrontal cortex and rostral anterior cingulate cortex (Prinz, 2006, p. 214), is where, as an integrated sentient being, we consciously recognize that we are having a feeling and give it a name: "My foot hurts!" "Ouch, that's hot!" "I'm afraid!" Corresponding to this level are cognitive phenomena arising from protophenomena whose interdependencies give them this cognitive character. Robots might have these feelings as well, if they have these higher levels of cortical processing.

The foregoing is just an outline. Understanding conscious experience of emotions will require a detailed neurophenomenology, which depends, in turn, on the ongoing neurophysiological investigation of emotion, but especially on more systematic phenomenological investigation of the human experience of emotion. The case of pain is simpler, but the difficulty people have in describing pains precisely, and sometimes even in localizing them, illustrates that the phenomenology of pain is far from trivial. As in other empirical sciences, training is required for accurate experimental phenomenology.

Nonbiological Protophenomena?

We can apply the foregoing summary of the neurophenomenology of human emotion to the question of robot feelings. This will highlight empirical questions that need to be answered before we can give definitive answers to the question of robot consciousness, and in particular, robot pain.

Due to the conceptual difficulties in reducing phenomenological properties to physical properties or vice versa, protophenomenal theory generally treats the correlation between protophenomenal intensity and the physical activity at activity sites as a brute fact of nature. Therefore, protophenomenal theory can be classified as a kind of dual-aspect monism (Atmanspacher, 2012), which supposes that there is one sort of "stuff" in the universe, but that it has two mutually irreducible aspects, an external, physical aspect, and an internal, phenomenological aspect. (More specifically, protophenomenal theory can be classified as Type-F monism; see Chalmers, 2002.) To put it differently, the properties of a physical system are experienced differently depending on whether or not the experiencer is that system.

Nevertheless, this leaves open the question of whether all physical systems have these two aspects or only certain ones. In protophenomenal terms, what sorts of physical systems can be activity sites with associated protophenomena? This is an empirical question, but difficult to investigate. In fact, it is even an open question whether all neurons have protophenomena. Blind-sight, for example, reveals that visual processing can be unconscious as well as conscious (Weiskrantz, Warrington, Sanders, & Marshall, 1974). Certainly, much neural activity in our brains is unconscious, but that does not imply that the neurons involved lack protophenomena. It might be that the protophenomenal intensities are too diffuse or incoherent to constitute salient phenomena.

To answer these questions, we need to identify protophenomena that are individually sufficiently salient to have reportable effects in the conscious state, and then to intervene physically in the corresponding activity site to determine which physical variables affect protophenomenal intensity. In the case of pain, for example, we know that opioid analgesics act on neurons in the spinal cord and brain, either decreasing neurotransmitter release by blocking calcium ion intake, or inhibiting the postsynaptic neuron by opening potassium ion channels and raising its firing threshold. Optogenetics might be used to control these channels directly, thus teasing apart their effects on conscious pain perception. Techniques for this kind of very precise experimental intervention are steadily improving (e.g., Losonczy, Makara, & Magee, 2008; Petit, Wang, Gee, & Augustine, 1997; Service, 2013). Pain protophenomena are especially useful for neurophenomenological experimentation because they are relatively salient and isolable from other protophenomena.

Investigations of this kind will eventually reveal the physical structures and processes associated with protophenomena. Conceivably, they could be peculiar to biological systems (e.g., dependent on neurotransmitters and their receptors), in which case it would be unlikely that robots built on very different physical principles would have protophenomena, and therefore conscious states. On the other hand, it might turn out that many different physical processes support protophenomena, in which case, robot consciousness becomes a real possibility, so long as the requisite physical processes occur in the robot. Note, however, that the mere presence of protophenomena is not sufficient for a robot to be conscious; their protophenomenal interdependencies would have to be structured in such a way that the protophenomena cohere into macroscopic phenomena (discussed in the next section).

These are open questions, but we can consider the implications of several hypotheses. First, Norman Cook (2000, 2002, 2008) has proposed an interesting hypothesis about activity sites: he argues that the intensity of a protophenomenon is correlated with the flux of ions across the neural cell membrane during the generation of an action potential. This can be considered an elementary act of perception in which the neuron senses the interneuronal environment. If this hypothesis is correct, what might it tell us about the essential physical properties of an activity site? Assuming it does not depend on biological specifics, it might be sufficient that: 1) there is some sort of boundary separating the activity site from its environment, 2) the activity site has the capability of sensing its environment, and 3) its internal processing possibly results in some modification of its environment (i.e., a very simple sensory-motor loop). If these are, indeed, the essential characteristics of an activity site, then they are certainly implementable in future robots.

Chalmers (1996, ch. 8) presents a related hypothesis based on information spaces that have both physical and phenomenal aspects, the two together comprising an activity site with its protophenomenon, for information is simultaneously physical and phenomenal. The physical system has the structure sufficient to represent some simple information state and the causal relations to use this state to affect the physical states of some other physical systems. That is, it represents "differences that make a difference." The associated protophenomenon has a corresponding formal structure and interdependencies connecting it to other protophenomena.

Combining the insights from Cook and Chalmers, we may consider the hypothesis that the neuron is the activity site. The information state is represented by the electrical state of the neuron (crudely, whether it is firing or not). The neuron senses its intercellular environment by means of its receptors, which respond to the chemical environment in the vicinity of the synapses. This increases the system mutual information between the cell state and its environment, in effect transferring information into the neuron (MacLennan, 2011). The entropy of the joint system decreases. Based on this input and the neuron's prior state, the neuron fires or not; that is, it makes a decision. Transmission of the action potential has physical effects on other neurons, muscles, glands, etc. The neuron, therefore, can be considered an information space with both physical and phenomenal aspects; i.e., it is an activity site together with its protophenomenon.

If Chalmer's hypothesis — that protophenomena and activity sites are the complementary aspects of information spaces — is correct, then we can conclude that robots could support activity sites and their associated protophenomena, for a robot's central processor certainly realizes information spaces. Again, we cannot conclude that such a robot is conscious, for protophenomena are necessary but not sufficient for consciousness. The protophenomenal interdependencies must to so structured that the protophenomena cohere into full-fledged phenomena. In particular, for emotional consciousness, the structure must be such so that they cohere into emotional phenomena (felt emotions). What is this structure?

Phenomenological Structure of Robot Emotions

To understand the structure of robot emotions, we can look at the functions of emotions in animals, which we reviewed above, and anticipate analogous functions in future robots. In brief, emotions are rapid evaluations of external or internal situations that require a prompt response. They ready the robot for action or information processing and may initiate those processes.

Effects might include adjustment of clock rates, energy management, high-priority computation, deployment and priming of specialized sensors and actuators, etc. These are analogous to unconscious results of emotional processing in animals, but according to somatic feeling theory, felt emotions arise from higher order processing of somatic sensations. In robots, this would be based on interoceptors (internal sensors that monitor the robot's own state) and on direct signal connections carrying information about the effects of early emotional evaluation. In particular, some of these interoceptors could be analogous to pain receptors in that they monitor potentially damaging situations, such as excessive input to external sensors, excessive stresses on mechanical systems, dangerous electrical loads, excessive temperature, and physical damage. Some will trigger immediate "reflexive" reactions, such as covering sensors, retracting actuators, shutting down low priority subsystems, or starting fans. Interoceptors in these systems might constitute activity sites with associated protophenomena (as previously discussed), but they might not rise to the level of conscious experience, which requires higher-level integration.

Many of these sensors and signal paths will be distributed around the robot's body, and therefore will convey somatically structured information, with associated protophenomena. Higher order processing will integrate them into coherent somatic phenomena; that is, into felt pain and emotion. In sophisticated robots, further cognitive processing of this and other sensor information will allow the robot to draw conclusions about its emotional state, and if it is in pain, to decide how best to respond.

More specifically, if the interoceptors are activity sites, they will have a bodily organization, and therefore the protophenomena will constitute the elements of a phenomenological body (the experience of being an embodied consciousness). Each interoceptor has a response function defined over its input space, responding, for example, to high temperature or excessive torque. Therefore, its protophenomenon effectively represents this subspace, but it is unrelated to other protophenomena in the absence of interdependencies, which are mirrored in the connections between activity sites. The interconnection of these activity sites with each other and their integration into higher-order sensory areas stitches together a topology over their composite input space. This topology defines the structure of a space of corresponding emotional phenomena, in a phenomenological body (cf. Craig, 2003). This phenomenological structure will depend on the physical structure of the robot's body and on the organizational requirements for the robot's emotional subsystems.

One may wonder how similar a robot's emotions will be to our own. We can expect that some of a robot's somatosensory spaces will have a similar structure to ours. For example, like us they will have skin and joint sensors distributed throughout their bodies. Some of their interoceptors will be analogous to ours (sensor overstimulation, temperature, stress, strain, physical damage). In these cases, we can expect their emotional experiences will be similar to ours, for they will have a similar topology. We will be able to imagine their emotional experiences, as they may be able to imagine ours, just as we can imagine the haptic, muscular, and skeletal sensations of other mammals without too much trouble.

On the other hand, we can anticipate that robot bodies will be radically different from ours in some ways, and these differences may be especially prominent in the emotion systems. For example, a robot is unlikely to have a heart or lungs per se, which are important to our somatic emotional responses, nor is it likely to have an endocrine system. It may have analogous interoceptor input (e.g., power and temperature monitors, early warning sensors), but the topology of their input spaces might be quite different from our interoceptors. In these cases, it might be more difficult for us to imagine a robot's emotional responses (just as it is difficult for us to imagine what it is like to be a reptile or crustacean). That is, the phenomenology of their emotions could be quite different to ours and unique to their "form of life." Although we may be able to understand their phenomenology intellectually, we might not be able to imagine it.

If robots can be conscious, but their emotional phenomena are very different from ours, should we call these phenomena "emotions" at all? Might this be an an overextension of the concept based on a loose analogy? The answer, I think, depends on the role played by these information processing systems within the robot. If they fulfill similar functions to emotions (as enumerated above) and fulfill them in similar ways, then I think it is useful to classify them as emotions. In particular, if they are states encoding goals of critical importance that are directly motivating and that have appropriate, persistent, and pervasive effects on the behavioral state of the robot, then they have much in common with emotions. The similarity is reinforced if they have effects on the robot's conscious state similar to our emotions' effects on our conscious state, for example, if they modulate the content and dynamics of the conscious state in a way that is relevant to the behavioral purposes that they serve. Certainly, a robot's conscious perception of impending damage to its structure could be reasonably termed "pain" if it leads to avoidance and negative reinforcement.

FUTURE RESEARCH DIRECTIONS

As the preceding sections reveal, the question of the possibility of robot pain and feeling cannot be answered at this time, but we have outlined research programs directed to this end. Whether we are considering future cyborgs, with artificial devices integrated in natural nervous systems, or completely artificial autonomous robots that exhibit sentience, we need a more principled understanding of the relation of conscious awareness and physical processes. By neurophenomenological analysis, we can reduce these issues to a simpler problem: understanding the relation between activity sites and their associated protophenomena. This problem is empirical, and the requisite experimental techniques are being developed, but much research needs to be done before we can begin to determine necessary and sufficient conditions for physical systems to have associated protophenomena. Moreover, as explained, conscious experience requires more than just protophenomena; their independencies must be so structured that the protophenomena cohere into determinate phenomena. Understanding this process requires extensive neurophenomenology exploring the neurological correlates of conscious experience and its structure. Progress on the neurological side is progressing rapidly, but comparable progress on the phenomenological side will require more experimenters (and subjects!) trained in phenomenological techniques. Davidson's well-known fMRI investigations of expert meditators is an example of what can be done (e.g., Brefczynski-Lewis, Lutz, Schaefer, Levinson, & Davidson, 2007).

CONCLUSION

Under appropriate conditions, future robots could have conscious emotional experiences — including in particular, feeling pain — analogous, but not necessarily identical to these experiences in humans or other animals. Determining the precise conditions for robot emotional consciousness will require detailed and comprehensive neurophenomenological experiments. The central issues are to determine the range of physical processes with corresponding protophenomena (for these are the elementary constituents of consciousness) and to chart the neurophenomenological structure of emotion in humans and other animals. This is a complex, difficult, and long-range research program, but when it is completed it will provide a principled, evidence-based answer to the question of whether and under what conditions a robot could feel pain, suffering, and other emotions — perhaps even joy.

REFERENCES

Atmanspacher, H. (2012). Dual-aspect monism à la Pauli and Jung. *Journal of Consciousness Studies*, *19*, 96–120.

Block, N. (1995). On a confusion about a function of consciousness. *Behavioral and Brain Sciences*, *18*(02), 265–266. doi:10.1017/S0140525X00038188

Breazeal, C. (2003). Toward sociable robots. *Robotics and Autonomous Systems*, *42*(3-4), 167–175. doi:10.1016/S0921-8890(02)00373-1

Breazeal, C., Brooks, A., Gray, J., Hoffman, G., Kidd, C., Lee, H., & Chilongo, D. et al. (2004). Tutelage and collaboration for humanoid robots. *International Journal of Humanoid Robots*, *1*(2), 315–348. doi:10.1142/S0219843604000150

Brefczynski-Lewis, J. A., Lutz, A., Schaefer, H. S., Levinson, D. B., & Davidson, R. J. (2007). Neural correlates of attentional expertise in long-term meditation practitioners. *Proceedings of the National Academy of Sciences of the United States ofAmerica*, *104*(27), 11483–11488. doi:10.1073/pnas.0606552104 PMID:17596341

Campbell, K. K. (1970). *Body and mind*. New York, NY: Doubleday. doi:10.1007/978-1-349-00678-6

Chalmers, D. J. (1996). *The conscious mind*. New York, NY: Oxford University Press.

Chalmers, D. J. (2002). Consciousness and its place in nature. In D. J. Chalmers (Ed.), *Philosophy of mind: Classical and contemporary readings* (pp. 247–272). Oxford, UK: Oxford University Press.

Cook, N. D. (2000). On defining awareness and consciousness: The importance of the neuronal membrane. In *Proceedings of the Tokyo-99 Conference on Consciousness*. Singapore: World Scientific.

Cook, N. D. (2002a). Bihemispheric language: How the two hemispheres collaborate in the processing of language. In *Proceedings of the British Academy* (Vol. 106).

Cook, N. D. (2002b). *Tone of voice and mind: The connections between intonation, emotion, cognition and consciousness*. Amsterdam, Netherlands: John Benjamins. doi:10.1075/aicr.47

Cook, N. D. (2008). The neuron-level phenomena underlying cognition and consciousness: Synaptic activity and the action potential. *Neuroscience*, *153*(3), 556–570. doi:10.1016/j.neuroscience.2008.02.042 PMID:18406536

Craig, A. D. (2003). Interoception: The sense of the physiological condition of the body. *Current Opinion in Neurobiology, 13*(4), 500–505. doi:10.1016/S0959-4388(03)00090-4 PMID:12965300

Damasio, A. R. (1999). *The feeling of what happens: Body and emotion in the making of consciousness.* New York: Harcourt, Brace & Co.

Daugman, J. G. (1993). An information-theoretic view of analog representation in striate cortex. In E. L. Schwartz (Ed.), *Computational Neuroscience* (pp. 403–423). Cambridge: MIT Press.

De Valois, R. L., & De Valois, K. K. (1993). A multi-stage color model. *Color Vision, 33*, 1053–1065. PMID:8506645

Denton, D. A. (2006). *The primordial emotions: The dawning of consciousness.* Oxford, UK: Oxford University Press. doi:10.1093/acprof:oso/9780199203147.001.0001

Gutenplan, S. (Ed.). (1994). *A companion to the philosophy of the mind.* Oxford: Blackwell.

Ihde, D. (1986). *Experimental phenomenology: An introduction.* Albany, NY: State University of New York Press.

James, W. (1884). What is an emotion? *Mind, 9*(34), 188–205. doi:10.1093/mind/os-IX.34.188

James, W. (1955). The principles of psychology, authorized edition. New York, NY: Dover (Reprint of Henry Holt, 1890 edition).

Karl, A., Birbaumer, N., Lutzenberger, W., Cohen, L. G., & Flor, H. (2001). Reorganization of motor and somatosensory cortex in upper extremity amputees with phantom limb pain. *The Journal of Neuroscience, 21*, 3609–3618. PMID:11331390

Kirk, R., & Squires, J. E. R. (1974). Zombies versus materialists. *Aristotelian Society, 48*(1 Suppl.), 135–152. doi:10.1093/aristoteliansupp/48.1.135

Kleinginna, P. R. Jr, & Kleinginna, A. M. (1981). A categorized list of emotion definitions, with suggestions for a consensual definition. *Motivation and Emotion, 5*(4), 345–379. doi:10.1007/BF00992553

Kripke, S. A. (1980). *Naming and necessity.* Cambridge, MA: Harvard University Press.

Lange, C. J. (1885). *Om sindsbevaegelser: Et psyko-fysiologisk studie.* Copenhagen: Jacob Lunds.

Laughlin, C. D. Jr, McManus, J., & d'Aquili, E. G. (1990). *Brain, symbol and experience: Toward a neurophenomenology of consciousness*. Boston, MA: New Science Library.

Llinas, R. R. (1988). The intrinsic electrophysiological properties of mammalian neurons. *Science*, *242*(4886), 1654–1664. doi:10.1126/science.3059497 PMID:3059497

Losonczy, A., Makara, J. K., & Magee, J. C. (2008). Compartmentalized dendritic plasticity and input feature storage in neurons. *Nature*, *452*(7186), 436–440. doi:10.1038/nature06725 PMID:18368112

Lutz, A., & Thompson, E. (2003). Neurophenomenology: Integrating subjective experience and brain dynamics in the neuroscience of consciousness. *Journal of Consciousness Studies*, *10*(9/10), 31–52.

MacLennan, B. J. (1995). The investigation of consciousness through phenomenology and neuroscience. In J. King & K. H. Pribram (Eds.), *Scale in conscious experience: Is the brain too important to be left to specialists to study?* (pp. 25–43). Hillsdale, NJ: Lawrence Erlbaum.

MacLennan, B. J. (1996). Protophenomena and their Neurodynamical Correlates (Technical Report CS-96-331). University of Tennessee, Knoxville, Department of Computer Science.

MacLennan, B. J. (2008). Consciousness: Natural and artificial. *Synthesis Philosophica*, *22*(2), 401–433.

MacLennan, B. J. (2009). Robots react but can they feel? A protophenomenological analysis. In J. Vallverdú & D. Casacuberta (Eds.), *Handbook of Research on Synthetic Emotions and Sociable Robotics: New Applications in Affective Computing and Artificial Intelligence* (pp. 133–153). Hershey, PA: IGI Global. doi:10.4018/978-1-60566-354-8.ch008

MacLennan, B. J. (2010). Protophenomena: The elements of consciousness and their relation to the brain. In A. Batthyány, A. Elitzur, & D. Constant (Eds.), *Irreducibly Conscious: Selected Papers on Consciousness* (pp. 189–214). Heidelberg & New York, NY: Universitätsverlag Winter.

MacLennan, B. J. (2011). Protophenomena and their physical correlates. *Journal of Cosmology, 14*, 4516–4525. Reprinted. In R. Penrose & S. Hameroff (Eds.), *Consciousness and the universe: Quantum physics, evolution, brain & mind* (pp. 228–239). Cambridge, MA: Cosmology Science Publishers.

MacLennan, B. J. (2014). Ethical Treatment of Robots and the Hard Problem of Robot Emotions. *International Journal of Synthetic Emotions*, *5*(1), 10–18. doi:10.4018/ijse.2014010102

MacLennan, B. J. (2015). Living science: Science as an activity of living beings. *Progress in Biophysics and Molecular Biology*, *119*(3), 410–419. doi:10.1016/j.pbiomolbio.2015.08.009 PMID:26276467

MacLennan, B. J. (1997). The elements of consciousness and their neurodynamical correlates. In J. Shear (Ed.), Explaining consciousness: The hard problem (pp. 249–266). Cambridge, MA: MIT Press, .

McCall, R. J. (1983). *Phenomenological psychology: An introduction. With a glossary of some key Heideggerian terms*. Madison, WI: University of Wisconsin Press.

Melzack, R., & Casey, K. L. (1968). Sensory, motivational and central control determinants of chronic pain: A new conceptual model. In Kenshalo, D. R. (Ed.), *The skin senses: Proceedings of the first International Symposium on the Skin Senses, Florida State University in Tallahassee, Florida* (Ch. 20). Springfield: Charles C. Thomas.

Morsella, E., Godwin, C. A., Jantz, T. K., Krieger, S. C., & Gazzaley, A. (2015). Homing in on consciousness in the nervous system: An action-based synthesis. In *Behavioral and Brain Sciences*. doi:10.1017/S0140525X15000643

Pettit, D. L., Wang, S. S., Gee, K. R., & Augustine, G. J. (1997). Chemical two-photon uncaging: A novel approach to mapping glutamate receptors. *Neuron*, *19*(3), 465–471. doi:10.1016/S0896-6273(00)80361-X PMID:9331338

Plutchik, R. (2000). *Emotions in the practice of psychotherapy: Clinical implications of affect theories*. New York: American Psychological Association. doi:10.1037/10366-000

Plutchik, R. (2003). *Emotions and life: Perspectives from psychology, biology, and evolution*. New York: American Psychological Association.

Plutchik, R., & Conte, H. R. (Eds.). (1997). *Circumplex models of personality and emotions*. Washington, DC: American Psychological Association. doi:10.1037/10261-000

Prinz, J. (2006). Gut reactions: A perceptual theory of emotion. Oxford: Oxford University Press. doi:10.1093/acprof:oso/9780195309362.001.0001

Rolls, E. T. (2005). *Emotion explained*. Oxford, UK: Oxford University Press. doi:10.1093/acprof:oso/9780198570035.001.0001

Rolls, E. T. (2006). *Brain mechanisms of emotion and decision-making*. In International Congress Series (Vol. 1291, pp. 3–13). Amsterdam, Netherlands: Elsevier.

Rolls, E. T. (2007). A neurobiological approach to emotional intelligence. In G. Matthews, M. Zeidner, & R. D. Roberts (Eds.), *The science of emotional intelligence* (pp. 72–100). Oxford, UK: Oxford University Press.

Rudrauf, D., Lutz, A., Cosmelli, D., Lachaux, J.-L., & Le Van Quyen, M. (2003). From autopoiesis to neurophenomenology: Francisco Varelas exploration of the biophysics of being. *Biological Research*, *36*(1), 27–65. doi:10.4067/S0716-97602003000100005 PMID:12795206

Service, R. F. (2013). The cyborg era begins: Advances in flexible electronics now makes it possible to integrate circuits with tissues. *Science*, *340*(6137), 1162–1165. doi:10.1126/science.340.6137.1162 PMID:23744924

Simon, H. (1977). What computers mean for man and society. *Science*, *195*(4283), 1186–1191. doi:10.1126/science.195.4283.1186 PMID:17789730

Strawson, G. (1994). *Mental reality*. Cambridge, MA: MIT Press.

Strawson, G. (2006). Realistic monism. In A. Freeman (Ed.), *Consciousness and its place in nature: Does physicalism entail panpsychism?* (pp. 3–31). Charlottesville, VA: Imprint Academic.

Sur, M. (2004). Rewiring cortex: Cross-modal plasticity and its implications for cortical development and function. In G. A. Calvert, C. Spence, & B. E. Stein (Eds.), *Handbook of multisensory processing* (pp. 681–694). Cambridge, MA: MIT Press.

Varela, F. J. (1997). Neurophenomenology: A methodological remedy to the hard problem. In J. Shear (Ed.), Explaining consciousness: The hard problem of consciousness (pp. 337–358). Cambridge, MA: MIT Press.

Weisberg, J. (2012). The hard problem of consciousness. *The Internet Encyclopedia of Philosophy*. Retrieved January 4, 2016 from http://www.iep.utm.edu

Weiskrantz, L., Warrington, E. K., Sanders, M. D., & Marshall, J. (1974). Visual capacity in the hemianopic field following a restricted occipital ablation. *Brain*, *97*(1), 709–728. doi:10.1093/brain/97.1.709 PMID:4434190

KEY TERMS AND DEFINITIONS

Action Potential: A neural impulse, an electrical signal generated by a self-reinforcing electrochemical process, by which neurons communicate with other neurons, generally conveying information by their firing rate (rate of action potential generation).

Activity Site: A physical structure (in an animal or robot) corresponding to a protophenomenon (q.v.), so that the protophenomenal intensity (q.v.) is correlated with physical activity at the activity site.

Intensity, Protophenomenal: The degree of a protophenomenon's presence in the conscious state, which is correlated with physical activity at an activity site (q.v.).

Neurophenomenology: An approach to neuropsychology or, more broadly, to studying the human mind, which combines neuroscientific techniques with phenomenology (q.v.), with investigations in each domain informing and reinforcing those in the other.

Phenomenon: Anything definite that arises in consciousness, including perceptions, sensations, feelings, recollections, dreams, hallucinations, desires, intentions, and imagination.

Phenomenology: The study of the structure of conscious experience by means of systematic introspection and analysis. There are many alternative approaches to phenomenology.

Protophenomenal Analysis: An analysis of some domain of conscious phenomena in terms of protophenomena (q.v.) and their neural correlates.

Protophenomena: Hypothesized smallest constituents of a conscious state, its elementary degrees of freedom.

Qualia: The fundamental phenomenal qualities associated with a conscious experience, for example, the experience of blueness when observing or imagining a blue object.

Chapter 8
Graphical User Interface for the Control of a Biped Robot

Claudio Urrea
Universidad de Santiago, Chile

Carlos Cortés Mac-Evoy
Bertrand AG, Germany & Volkswagen AG, Germany

ABSTRACT

The design and implementation of a graphical user interface (GUI) for the control and operation of a biped robot is presented. This GUI allows establishing communication between the user, the robot and a computer (controller) so that the robot can perform bipedal walking without the need to introduce commands that are not user-friendly. The developed graphic interface permits the user to carry out tasks operating on the robot without having to resort to commands that are not easy to use. This interface was created using the MATLAB-Simulink software and it presents important advantages compared to the manual operation of a robot.

INTRODUCTION

The science of robotics has grown tremendously over the past 30 years, fueled by rapid advances in computer and sensor technology, as well as theoretical advances in control and computer vision. Humanoid biped robots, typically complex in design, have numerous Degrees Of Freedom (DOF) due to the ambitious goal of mimicking the human gait. In recent years, several efforts of the robotics community have focused on developing bio-inspired robots, particularly in humanoid biped robots. Biped robots represent a very interesting research subject, with several

DOI: 10.4018/978-1-5225-2973-6.ch008

particularities and scope topics, such as mechanical design, gait simulation, patterns generation, kinematics, dynamics, equilibrium, stability, kinds of control, adaptability, biomechanics, cybernetics, and rehabilitation technologies.

Biped robots have leg link structures similar to human anatomy. To be able to maintain its stability under dynamic situations, such robotic systems require good mechanical designs and force sensors to acquire the Zero Moment Point (ZMP). Research in biped robotics has recently had a great surge due to the challenges of the subject, and the media impact of famous biped robotic creations like Honda's robots.

The modeling of biped gait has usually resulted in the description of kinematics, dynamics, and stability of two-legged walking robots, and represents an unresolved challenge, as most approaches reduce the dynamic model of the humanoid robot to an inverted pendulum that is based upon either a simple observation or mathematical analysis, derived from kinematics.

Of significant importance here is robot software; i.e., the set of coded commands or instructions that tell a mechanical device and electronic system - known together as a robot - what tasks to perform. Robot software is used to perform autonomous tasks. Many software systems and frameworks have been proposed to make programming robots easier.

Due to the highly proprietary nature of robot software, most manufacturers of robot hardware also provide their own software. While this is not unusual in other automated control systems, the lack of standardization of programming methods for robots does pose certain challenges. For example, there are over 30 different manufacturers of industrial robots, so there are also 30 different robot programming languages required. Fortunately, there are enough similarities between the different robots that it is possible to gain a broad-based understanding of robot programming without having to learn each manufacturer's proprietary language.

One of the most important aspects in the implementation and control of biped robots is the software used for their operation. It is what indicates where the robot must go, at what speed, and for how long. In other words, it is the software that articulates the whole robotic system, and even though the associated electromechanical parts play a key role in the development of a robot, it is not possible to achieve the proposed objectives for a system with these characteristics if one does not have the proper program that directs, coordinates, and commands all its movements.

A robot's software can be as simple as a set of routines that order the robot to carry out a particular task, or, as complex as a graphic program with a large variety of interactive interfaces. In this sense, the experience and skill that a user can get by using a graphic program to command a robot, compared with the use of simple routines and commands, turn out to be much greater. This clearly justifies the creation of interfaces with graphic components that make it easier for the user to operate a robot.

The Graphical User Interface (GUI), also known as UI, provides point-and-click control of software applications, eliminating the need to learn a language or type commands in order to run the application. The actions in a GUI are usually performed through direct manipulation of the graphical elements. In addition to computers, GUIs can be found in hand-held devices such as MP3 players, portable media players, gaming devices, smartphones and smaller household, office, and industrial equipment. The term 'GUI' tends to not be applied to other low-resolution types of interfaces with display resolutions, such as video games, nor is it restricted to flat screens, like volumetric displays, because the term is restricted to the scope of two-dimensional display screens able to describe generic information, in the tradition of the computer science research at the Palo Alto Research Center (PARC) (Lii et al., 2015; Zhiguo, Chong, Jun, Habin, & Hong, 2014).

In the literature it is possible to find different graphic interface designs used in a large number of applications (Ali, Islam, Anom, & Amziah, 2008; Deperlioglu & Kose, 2011; Goel & Shrestha, 2002; Rodriguez, Aromataris, Donolo, Hernández, & Moitreet, 2007). In most cases the applications correspond to simulation tools with an educational character. In this sense, Espinoza (2010), and Dean-Leon, Nair, and Knoll (2012), present well-made and detailed graphic simulators; the former allows to study in terms of simulation, a biped robot with 14 Degrees Of Freedom, while the latter corresponds to a complete robot design, modeling, and control tool with industrial manipulator configurations. In several cases, the graphic interfaces that are developed cover more specific applications, as in the case of Dub and Brotan (2012) and other similar studies (Aliane, 2010).

During the last five years, the Robot Operating System (ROS) has become very popular because it is a flexible platform that allows the design of software devoted to robots. ROS is constituted by a collection of libraries, conventions, and tools aimed at simplifying the job of creating complex and robust robotic behaviors on a wide variety of robotic platforms ("Robot Operating System," n.d.). So far, ROS is compatible with 30 different models of robots on the market.

DESCRIPTION OF THE ROBOT THAT WAS STUDIED

The robot considered in this chapter is an 11-DOF biped driven by eleven digital servomotors joined by aluminum structures. It has two lower limbs, with 5 DOF each, which are connected to an aluminum sheet (hip) which is connected in its upper part to an inverted pendulum with 1 DOF that replaces the trunk, the arms, and the head (see Figure 1).

Figure 1. Front view of the biped robot with 11 DOF

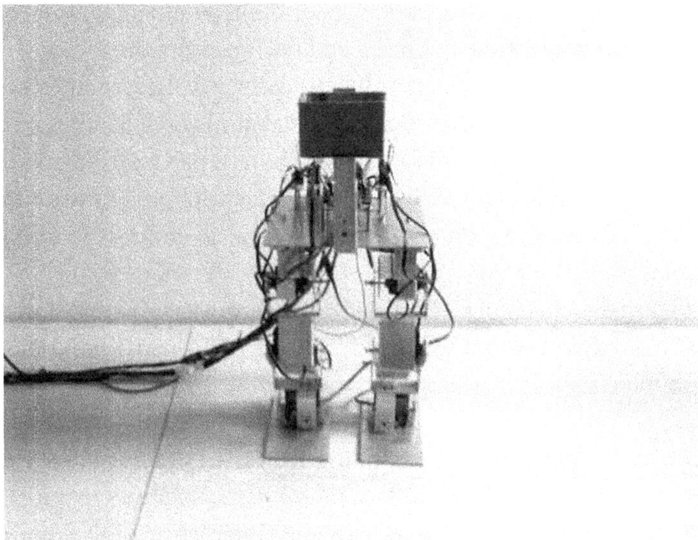

INTERFACE DESIGN ASPECTS

Choice of Programming Language

At present, there is a large number of tools with which graphic user interfaces can be developed. Since these graphic interfaces have been developed for different applications, their complexities and kinds of programming language vary.

For our purpose, MATLAB software was chosen over other programming languages because it is focused on, and directed to, numerical calculation, solving differential equations, simulation, etc. Specifically, thanks to reasons set forth by Smith (2006), namely: 1) development based on high level scripts, 2) developed applications are not dependent on the operating system, 3) ability to work in real time, 4) its language does not require the use of low-level memory, and, 5) the GUIDE environment for the construction of graphic interfaces.

GUIDE vs. Manual Programming

The GUIDE tool, which MATLAB includes, facilitates greatly the task for the creation and implementation of graphic interfaces. GUIDE allows a graphic interface to be designed in a very short time because it facilitates the positioning of the controls of all the components of the interface, as well programming, because it is capable of generating the code associated with the different elements of the graphic interface automatically. Some of the advantages of using GUIDE are mentioned in the work of Smith (2006), namely, automatic code generation, an object browser that allows visualizing the properties of the controls, and design simplicity.

However, GUIDE has little flexibility when dealing with larger and more complex graphic interfaces. In general, this tool has disadvantages whereby, it requires an additional .fig file that contains the interface design, parts of the code generated automatically lack the simplicity to be readily understood, it takes somewhat more time to be executed, and creation and/or elimination of controls in execution time is more complex.

Because of the advantages of GUIDE and considering its flexibility for the creation of interfaces manually, this tool was used here for the design and implementation of the graphic interface.

Integration of Simulink in the Graphic Interface

One of the characteristics of MATLAB is its easy interaction with Simulink, which is a software that, through the use of interconnected block diagrams, allows making simulations of control systems, mechanical systems, electric systems, solving differential equations, neural networks, etc. (Magrab, 2011).

Having a graphic interface allows for easy design and implementation of controllers, such as adaptive, neural, and others, which ensures good performance of the walk of the biped robot. Consequently, the choice of Simulink represents excellent provision for integrating into the graphic interface, to which computational simulations of the robot can also be made in real time.

Specific instructions, such as *get_param*, *set_param*, *add_block*, among others, allow reading and modifying the properties of the dynamic representations implemented in Simulink. Thanks to these instructions, also, it is possible to integrate dynamic representations with the graphic interface. Furthermore, using a type of functions called listeners, it is possible to read, in real time, values of the variables of interest during the execution of those dynamic representations. It is precisely because of this last reason that herein use is made of Simulink tools based on S functions of level 2 to send and receive data from the robot. These functions allow the configuration of parameters such as sampling time, number of inputs and outputs, specifications that the user of the robot enters in the graphic interface, and other critical project variables.

Modular Structure

The design of the proposed graphic interface has a modular character; i.e., every screen of the graphic interface is oriented toward the development of a task in particular, which is an advantage because a screen with a large number of controls associated to different tasks can confuse the user. In general terms, diagram representation permits an understanding of the design and structure of the graphic interface, as evidenced in the next section.

COMPONENTS OF THE GRAPHIC INTERFACE

Welcome Screen

This screen is designed with the purpose of carrying out initialization tasks of the graphic interface, including, downloading other interface modules, downloading Simulink diagrams, verifying the integrity of the interface, and verifying previous program instances. It shows a welcoming message to the user while waiting for the main screen (see Figure 2).

Figure 2. Structure of the graphic interface

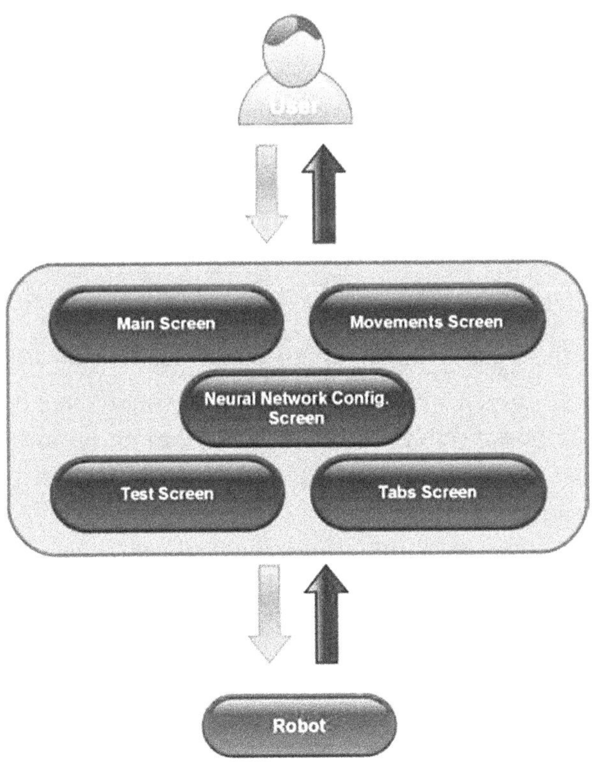

Main Screen

This component has the purpose of carrying out starting and stopping configuration tasks of the communication between the user and the robot, in addition to activating all the other commands associated with the graphic interface.

Testing Screen

This screen allows the user to carry out simple tests of the robot's movements. The movement ranges of the servomotors are established from this, without exceeding mechanical limits of the robot's joints. It also enables verifying the general operation of the biped robot.

This screen gives a graphic representation of the biped robot. It permits entering angular position values between 0 y 180°, through the keyboard. Furthermore, it has sliding controls that allow changing the angular value of the set points imposed to the robot by means of the mouse.

Screen of the Robot's Movement Trajectories

In this screen it is possible to specify movement trajectories that the biped robot must follow. It has a number of buttons that allow carrying out configuration and control actions, and it also has an axle-type control that allows the results of the position reading of the robot to be seen by means of a graph.

The most widely used trajectories in robotics correspond to smooth trajectories derived from a polynomial of order 3, which has the form (Craig, 2006):

$$q\left(t\right) = a_0 + a_1 t + a_2 t^2 + a_3 t^3 \qquad (1)$$

where $q\left(t\right)$ corresponds to the angular position of the links and the constants a_i must be determined from a speed condition of the robot's actuators. If a value of zero is considered for the speed at both the initial and the final instant of the trajectory, smooth movements are ensured in the actuators, and their integrity and that of the robot's mechanical parts is protected.

The movement screen uses the *windowbuttondownfcn* function to determine coordinates of the graph area when the left button of the mouse is pressed. In this way, it is possible to create trajectories given by Equation 1 and graph them directly on the graph area of the screen (see Figure 3).

Figure 3. Screen of the robot's movement trajectories

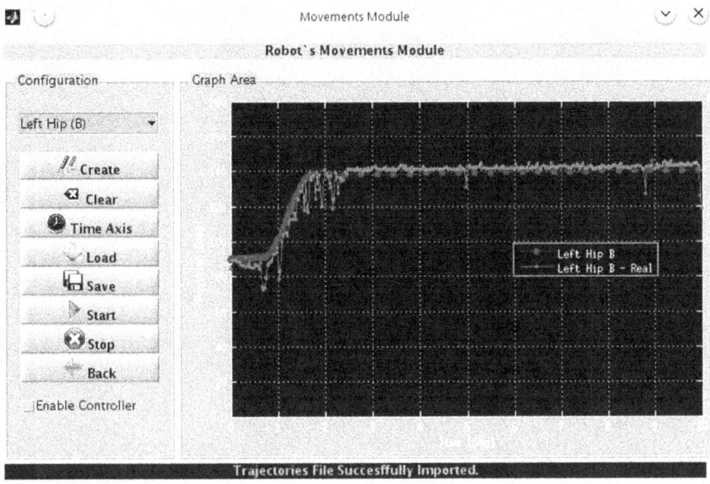

Editing Screen

This screen allows adjustments to be made at the points of the trajectories created by the screen described above. It is also possible to modify a complete trajectory based on Equation 1. If, when using the testing screen to analyze the robot's behavior, verifying its stability during its walk, it is seen that angles involved in the trajectories are not viable, the editing screen permits correcting and making modifications required for the robot to move accordingly (see Figure 4).

Loading Sequence

Upon initializing the operation of the graphic interface, all the system's components are loaded automatically, and that makes the main screen be ready for use. This happens because in this way the user does not have to wait for each element of the interface to be loaded separately every time that it is required. This strategy for loading the application is based on the launching of all the screens of the interface in hidden mode, so that they remain in the memory during their use, and when they are required, all that is needed is to activate the visible mode of the graphic interface (see Figure 5).

Figure 4. Editing screen of the trajectories

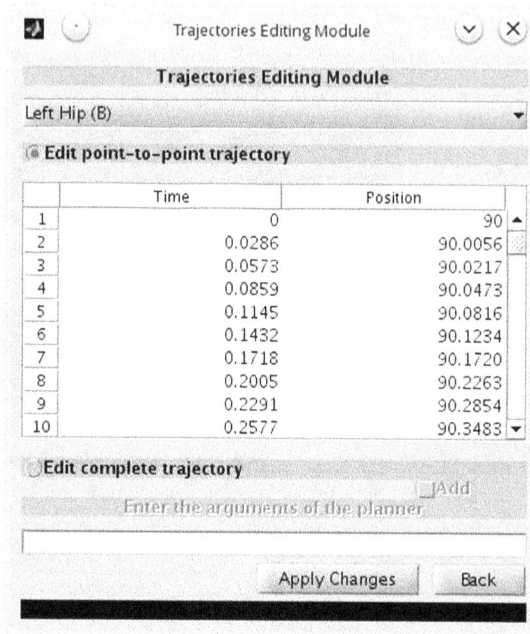

Once the loading of the application's components is finished, the welcome screen closes automatically, leaving the main screen ready to be used.

RESULTS

For simplicity, it was decided not to implement the routines needed for resizing the screens and their contents, so that all the application's components are presented as independent screens, even when their design is modular (see Figure 6).

Once the welcome screen finishes the assigned operations, the main window is placed in the center of the screen; it can be seen that it also contains a panel that allows making the serial communication with the robot and allows the rest of the screens to become visible (see Figure 7).

Figure 5. Loading sequence of the components

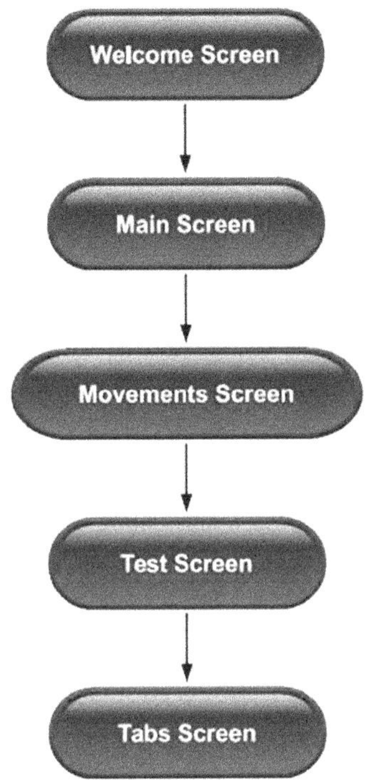

Integration with the Robot

Some movements of the real biped robot are obtained by means of the graphic interface (see Figure 8).

CONCLUSION

The design and implementation of a functional graphic interface that allows operating and controlling a robot by serial communication has been presented. This graphic interface allows the user to carry out tasks operating on the robot without the need to resort to commands that are not user-friendly. The movements screen presents a key functionality in the interface, because since it allows smooth trajectories to be created, it facilitates the fine adjustment of the trajectories with which the biped robot can move around. All this, thanks to the visualization of the trajectories on the graph's area, to the time adjustment functionality of this screen, and to the functionality

Figure 6. Welcome screen and its components

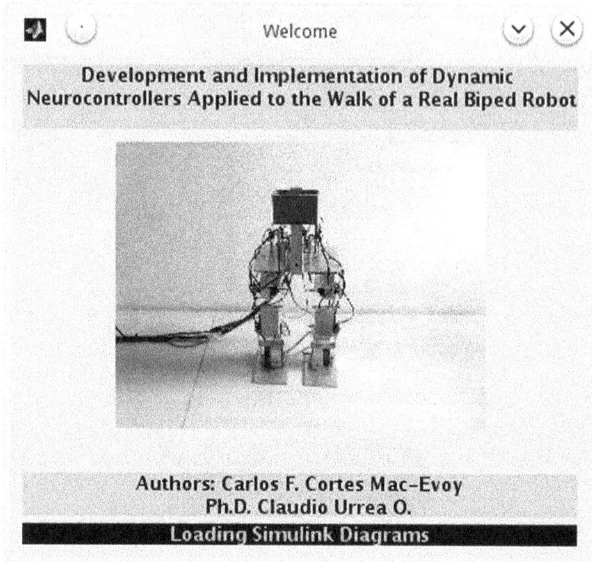

provided by the trajectory editing screen. These attributes of the graphic interface allow the development and testing of the trajectories of the robot's real movements, analyzing their viability from the standpoint of a stable walk.

FUTURE WORK

Although the application has a collection of screens and controls that allow several tasks to be developed with the robot, some characteristics are expected to be implemented in the future, namely:

- **Development:** A graphic interface for simulation that allows visualizing the behavior of essential parts of a robot, such as the actuators.
- **Multilingual Support:** It is expected to make the changes required by the application for the user to select the preferred language, so that the text appearing in the application agrees with this selection.

Figure 7. Main screen

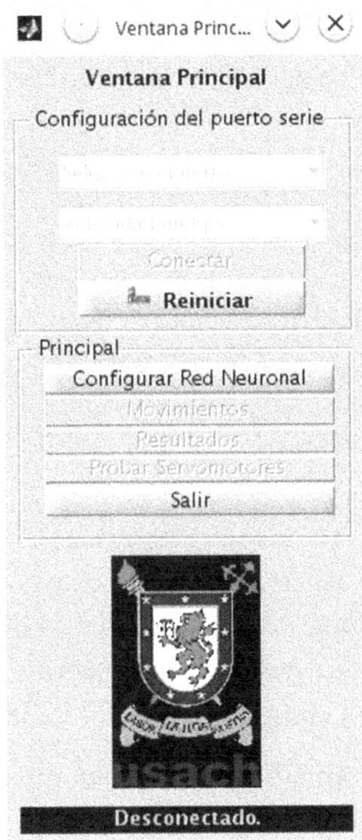

Figure 8. Walking robot

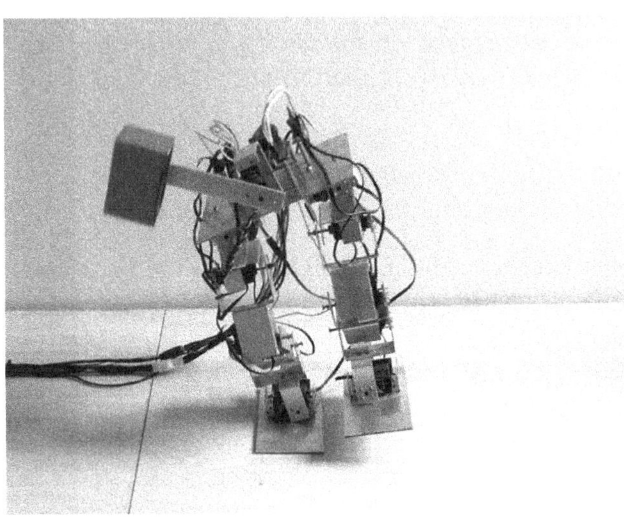

- **Hardware:** It is desired that the application should automatically be able to detect when the robot's hardware is connected to or disconnected from a computer, and in this way correct possible errors.
- **Compilation:** It is expected to generate an application that will be independent of MATLAB. This can be achieved thanks to the compiler that this software has.

ACKNOWLEDGMENT

This work has been supported by Proyectos Basales and the Vicerrectoría de Investigación, Desarrollo e Innovación (VRIDEI), Universidad de Santiago de Chile, Chile.

REFERENCES

Ali, M. L., Islam, A. R., Anom, A. S., & Amziah, N. A. M. (2008). GUI for PC auto-shutdown. *Computers & Electrical Engineering an International Journal*, *34*(3), 257–262. doi:10.1016/j.compeleceng.2007.02.009

Aliane, N. (2010). A Matlab/Simulink-based interactive module for servo systems learning. *Education. IEEE Transactions on Education*, *53*(2), 265–271. doi:10.1109/TE.2009.2014468

Craig, J. (2006). *Robótica tercera edición*. México: Pearson Educación.

Dean-Leon, E., Nair, S., & Knoll, A. (2012). User friendly Matlab-Toolbox for symbolic robot dynamic modeling used for control design. In *Proceedings of the IEEE International Conference on Robotics and Biometrics* (pp. 2181-2188). Guangzhou, China: IEEE. doi:10.1109/ROBIO.2012.6491292

Deperlioglu, O., & Kose, U. (2011). An educational tool for artificial neural networks. *Computers & Electrical Engineering an International Journal*, *37*(3), 392–402. doi:10.1016/j.compeleceng.2011.03.010

Dub, M., & Brotan, L. (2012). MATLAB GUI for DC drive parameter identification. In *Proceedings of a meeting held 15Th International Symposium MECHATRONIKA* (pp. 1-3). Prague, Czech: IEEE.

Espinoza, L. (2010). *Diseño e implementación de un simulador gráfico de un robot bípedo de 14 grados de libertad*. Universidad de Santiago de Chile, Chile: Magister These.

Goel, L., & Shrestha, G. B. (2002). A versatile tool for electric substation reliability assessment. *Computers & Electrical Engineering an International Journal, 28*(6), 561–576. doi:10.1016/S0045-7906(00)00064-1

Lii, N. Y., Leidner, D., Schiele, A., Birkenkampf, P., Pleintinger, B., & Bayer, R. (2015). Command robots from orbit with supervised autonomy: An introduction to the meteron supvis-justin experiment. In *Proceedings of the ACM/IEEE International Conference on Human-Robots Interaction* (HRI). doi:10.1145/2701973.2702022

Magrab, E. B. (2011). *An engineer's guide to MATLAB: With applications from mechanical, aerospace, electrical, civil, and biological systems engineering*. Upper Saddle River, N. J.: Prentice Hall.

Robot Operating System. (n. d.). Retrieved March 24, 2014 from http://www.ros.org/about-ros/

Rodriguez, G., Aromataris, L., Donolo, M., Hernández, J., & Moitre, D. (2007). Software for calculation of electric power systems parameters. *Computers & Electrical Engineering an International Journal, 29*(5), 643–651. doi:10.1016/S0045-7906(01)00053-2

Smith, S. T. (2006). *MATLAB: Advanced GUI development*. USA: Dog Ear Publishing.

Zhiguo, L., Chong, L., Jun, P., Habin, Z., & Hong, W. (2014, June 29–July 4). Motion simulation platform for a humanoid robot base on MATLAB. In *Proceedings of the 2014 11th World Congress on Intelligent Control and Automation (WCICA)* (pp. 13–18).

Related References

To continue our tradition of advancing academic research, we have compiled a list of recommended IGI Global readings. These references will provide additional information and guidance to further enrich your knowledge and assist you with your own research and future publications.

Adeyemo, O. (2013). The nationwide health information network: A biometric approach to prevent medical identity theft. In *User-driven healthcare: Concepts, methodologies, tools, and applications* (pp. 1636–1649). Hershey, PA: IGI Global. doi:10.4018/978-1-4666-2770-3.ch081

Adler, M., & Henman, P. (2009). Justice beyond the courts: The implications of computerisation for procedural justice in social security. In A. Martínez & P. Abat (Eds.), *E-justice: Using information communication technologies in the court system* (pp. 65–86). Hershey, PA: IGI Global. doi:10.4018/978-1-59904-998-4.ch005

Aflalo, E., & Gabay, E. (2013). An information system for coping with student dropout. In L. Tomei (Ed.), *Learning tools and teaching approaches through ICT advancements* (pp. 176–187). Hershey, PA: IGI Global. doi:10.4018/978-1-4666-2017-9.ch016

Ahmed, M. A., Janssen, M., & van den Hoven, J. (2012). Value sensitive transfer (VST) of systems among countries: Towards a framework. *International Journal of Electronic Government Research*, *8*(1), 26–42. doi:10.4018/jegr.2012010102

Aikins, S. K. (2008). Issues and trends in internet-based citizen participation. In G. Garson & M. Khosrow-Pour (Eds.), *Handbook of research on public information technology* (pp. 31–40). Hershey, PA: IGI Global. doi:10.4018/978-1-59904-857-4.ch004

Aikins, S. K. (2009). A comparative study of municipal adoption of internet-based citizen participation. In C. Reddick (Ed.), *Handbook of research on strategies for local e-government adoption and implementation: Comparative studies* (pp. 206–230). Hershey, PA: IGI Global. doi:10.4018/978-1-60566-282-4.ch011

Aikins, S. K. (2012). Improving e-government project management: Best practices and critical success factors. In *Digital democracy: Concepts, methodologies, tools, and applications* (pp. 1314–1332). Hershey, PA: IGI Global. doi:10.4018/978-1-4666-1740-7.ch065

Akabawi, M. S. (2011). Ghabbour group ERP deployment: Learning from past technology failures. In E. Business Research and Case Center (Ed.), Cases on business and management in the MENA region: New trends and opportunities (pp. 177-203). Hershey, PA: IGI Global. doi:10.4018/978-1-60960-583-4.ch012

Akabawi, M. S. (2013). Ghabbour group ERP deployment: Learning from past technology failures. In *Industrial engineering: Concepts, methodologies, tools, and applications* (pp. 933–958). Hershey, PA: IGI Global. doi:10.4018/978-1-4666-1945-6.ch051

Akbulut, A. Y., & Motwani, J. (2008). Integration and information sharing in e-government. In G. Putnik & M. Cruz-Cunha (Eds.), *Encyclopedia of networked and virtual organizations* (pp. 729–734). Hershey, PA: IGI Global. doi:10.4018/978-1-59904-885-7.ch096

Akers, E. J. (2008). Technology diffusion in public administration. In G. Garson & M. Khosrow-Pour (Eds.), *Handbook of research on public information technology* (pp. 339–348). Hershey, PA: IGI Global. doi:10.4018/978-1-59904-857-4.ch033

Al-Shafi, S. (2008). Free wireless internet park services: An investigation of technology adoption in Qatar from a citizens perspective. *Journal of Cases on Information Technology, 10*(3), 21–34. doi:10.4018/jcit.2008070103

Al-Shafi, S., & Weerakkody, V. (2009). Implementing free wi-fi in public parks: An empirical study in Qatar. *International Journal of Electronic Government Research, 5*(3), 21–35. doi:10.4018/jegr.2009070102

Aladwani, A. M. (2002). Organizational actions, computer attitudes and end-user satisfaction in public organizations: An empirical study. In C. Snodgrass & E. Szewczak (Eds.), *Human factors in information systems* (pp. 153–168). Hershey, PA: IRM Press. doi:10.4018/978-1-931777-10-0.ch012

Aladwani, A. M. (2002). Organizational actions, computer attitudes, and end-user satisfaction in public organizations: An empirical study. *Journal of Organizational and End User Computing, 14*(1), 42–49. doi:10.4018/joeuc.2002010104

Allen, B., Juillet, L., Paquet, G., & Roy, J. (2005). E-government and private-public partnerships: Relational challenges and strategic directions. In M. Khosrow-Pour (Ed.), *Practicing e-government: A global perspective* (pp. 364–382). Hershey, PA: IGI Global. doi:10.4018/978-1-59140-637-2.ch016

Alshawaf, A., & Knalil, O. E. (2008). IS success factors and IS organizational impact: Does ownership type matter in Kuwait? *International Journal of Enterprise Information Systems, 4*(2), 13–33. doi:10.4018/jeis.2008040102

Ambali, A. R. (2009). Digital divide and its implication on Malaysian e-government: Policy initiatives. In H. Rahman (Ed.), *Social and political implications of data mining: Knowledge management in e-government* (pp. 267–287). Hershey, PA: IGI Global. doi:10.4018/978-1-60566-230-5.ch016

Amoretti, F. (2007). Digital international governance. In A. Anttiroiko & M. Malkia (Eds.), *Encyclopedia of digital government* (pp. 365–370). Hershey, PA: IGI Global. doi:10.4018/978-1-59140-789-8.ch056

Amoretti, F. (2008). Digital international governance. In A. Anttiroiko (Ed.), *Electronic government: Concepts, methodologies, tools, and applications* (pp. 688–696). Hershey, PA: IGI Global. doi:10.4018/978-1-59904-947-2.ch058

Amoretti, F. (2008). E-government at supranational level in the European Union. In A. Anttiroiko (Ed.), *Electronic government: Concepts, methodologies, tools, and applications* (pp. 1047–1055). Hershey, PA: IGI Global. doi:10.4018/978-1-59904-947-2.ch079

Amoretti, F. (2008). E-government regimes. In A. Anttiroiko (Ed.), *Electronic government: Concepts, methodologies, tools, and applications* (pp. 3846–3856). Hershey, PA: IGI Global. doi:10.4018/978-1-59904-947-2.ch280

Amoretti, F. (2009). Electronic constitution: A Braudelian perspective. In F. Amoretti (Ed.), *Electronic constitution: Social, cultural, and political implications* (pp. 1–19). Hershey, PA: IGI Global. doi:10.4018/978-1-60566-254-1.ch001

Amoretti, F., & Musella, F. (2009). Institutional isomorphism and new technologies. In M. Khosrow-Pour (Ed.), *Encyclopedia of information science and technology* (2nd ed., pp. 2066–2071). Hershey, PA: IGI Global. doi:10.4018/978-1-60566-026-4.ch325

Andersen, K. V., & Henriksen, H. Z. (2007). E-government research: Capabilities, interaction, orientation, and values. In D. Norris (Ed.), *Current issues and trends in e-government research* (pp. 269–288). Hershey, PA: CyberTech Publishing. doi:10.4018/978-1-59904-283-1.ch013

Anderson, K. V., & Henriksen, H. Z. (2005). The first leg of e-government research: Domains and application areas 19982003. *International Journal of Electronic Government Research, 1*(4), 26–44. doi:10.4018/jegr.2005100102

Anttiroiko, A. (2009). Democratic e-governance. In M. Khosrow-Pour (Ed.), *Encyclopedia of information science and technology* (2nd ed., pp. 990–995). Hershey, PA: IGI Global. doi:10.4018/978-1-60566-026-4.ch158

Association, I. R. (2010). Networking and telecommunications: Concepts, methodologies, tools and applications (Vols. 1–3). Hershey, PA: IGI Global. doi:10.4018/978-1-60566-986-1

Association, I. R. (2010). Web-based education: Concepts, methodologies, tools and applications (Vols. 1–3). Hershey, PA: IGI Global. doi:10.4018/978-1-61520-963-7

Baker, P. M., Bell, A., & Moon, N. W. (2009). Accessibility issues in municipal wireless networks. In C. Reddick (Ed.), *Handbook of research on strategies for local e-government adoption and implementation: Comparative studies* (pp. 569–588). Hershey, PA: IGI Global. doi:10.4018/978-1-60566-282-4.ch030

Becker, S. A., Keimer, R., & Muth, T. (2010). A case on university and community collaboration: The sci-tech entrepreneurial training services (ETS) program. In S. Becker & R. Niebuhr (Eds.), *Cases on technology innovation: Entrepreneurial successes and pitfalls* (pp. 68–90). Hershey, PA: IGI Global. doi:10.4018/978-1-61520-609-4.ch003

Becker, S. A., Keimer, R., & Muth, T. (2012). A case on university and community collaboration: The sci-tech entrepreneurial training services (ETS) program. In Regional development: Concepts, methodologies, tools, and applications (pp. 947-969). Hershey, PA: IGI Global. doi:10.4018/978-1-4666-0882-5.ch507

Bernardi, R. (2012). Information technology and resistance to public sector reforms: A case study in Kenya. In T. Papadopoulos & P. Kanellis (Eds.), *Public sector reform using information technologies: Transforming policy into practice* (pp. 59–78). Hershey, PA: IGI Global. doi:10.4018/978-1-60960-839-2.ch004

Bernardi, R. (2013). Information technology and resistance to public sector reforms: A case study in Kenya. In *User-driven healthcare: Concepts, methodologies, tools, and applications* (pp. 14–33). Hershey, PA: IGI Global. doi:10.4018/978-1-4666-2770-3.ch002

Bolívar, M. P., Pérez, M. D., & Hernández, A. M. (2012). Municipal e-government services in emerging economies: The Latin-American and Caribbean experiences. In Y. Chen & P. Chu (Eds.), *Electronic governance and cross-boundary collaboration: Innovations and advancing tools* (pp. 198–226). Hershey, PA: IGI Global. doi:10.4018/978-1-60960-753-1.ch011

Borycki, E. M., & Kushniruk, A. W. (2010). Use of clinical simulations to evaluate the impact of health information systems and ubiquitous computing devices upon health professional work. In S. Mohammed & J. Fiaidhi (Eds.), *Ubiquitous health and medical informatics: The ubiquity 2.0 trend and beyond* (pp. 552–573). Hershey, PA: IGI Global. doi:10.4018/978-1-61520-777-0.ch026

Borycki, E. M., & Kushniruk, A. W. (2011). Use of clinical simulations to evaluate the impact of health information systems and ubiquitous computing devices upon health professional work. In *Clinical technologies: Concepts, methodologies, tools and applications* (pp. 532–553). Hershey, PA: IGI Global. doi:10.4018/978-1-60960-561-2.ch220

Buchan, J. (2011). Developing a dynamic and responsive online learning environment: A case study of a large Australian university. In B. Czerkawski (Ed.), *Free and open source software for e-learning: Issues, successes and challenges* (pp. 92–109). Hershey, PA: IGI Global. doi:10.4018/978-1-61520-917-0.ch006

Buenger, A. W. (2008). Digital convergence and cybersecurity policy. In G. Garson & M. Khosrow-Pour (Eds.), *Handbook of research on public information technology* (pp. 395–405). Hershey, PA: IGI Global. doi:10.4018/978-1-59904-857-4.ch038

Burn, J. M., & Loch, K. D. (2002). The societal impact of world wide web - Key challenges for the 21st century. In A. Salehnia (Ed.), *Ethical issues of information systems* (pp. 88–106). Hershey, PA: IRM Press. doi:10.4018/978-1-931777-15-5.ch007

Burn, J. M., & Loch, K. D. (2003). The societal impact of the world wide web-Key challenges for the 21st century. In M. Khosrow-Pour (Ed.), *Advanced topics in information resources management* (Vol. 2, pp. 32–51). Hershey, PA: IGI Global. doi:10.4018/978-1-59140-062-2.ch002

Bwalya, K. J., Du Plessis, T., & Rensleigh, C. (2012). The "quicksilver initiatives" as a framework for e-government strategy design in developing economies. In K. Bwalya & S. Zulu (Eds.), *Handbook of research on e-government in emerging economies: Adoption, e-participation, and legal frameworks* (pp. 605–623). Hershey, PA: IGI Global. doi:10.4018/978-1-4666-0324-0.ch031

Cabotaje, C. E., & Alampay, E. A. (2013). Social media and citizen engagement: Two cases from the Philippines. In S. Saeed & C. Reddick (Eds.), *Human-centered system design for electronic governance* (pp. 225–238). Hershey, PA: IGI Global. doi:10.4018/978-1-4666-3640-8.ch013

Camillo, A., Di Pietro, L., Di Virgilio, F., & Franco, M. (2013). Work-groups conflict at PetroTech-Italy, S.R.L.: The influence of culture on conflict dynamics. In B. Christiansen, E. Turkina, & N. Williams (Eds.), *Cultural and technological influences on global business* (pp. 272–289). Hershey, PA: IGI Global. doi:10.4018/978-1-4666-3966-9.ch015

Capra, E., Francalanci, C., & Marinoni, C. (2008). Soft success factors for m-government. In A. Anttiroiko (Ed.), *Electronic government: Concepts, methodologies, tools, and applications* (pp. 1213–1233). Hershey, PA: IGI Global. doi:10.4018/978-1-59904-947-2.ch089

Cartelli, A. (2009). The implementation of practices with ICT as a new teaching-learning paradigm. In A. Cartelli & M. Palma (Eds.), *Encyclopedia of information communication technology* (pp. 413–417). Hershey, PA: IGI Global. doi:10.4018/978-1-59904-845-1.ch055

Charalabidis, Y., Lampathaki, F., & Askounis, D. (2010). Investigating the landscape in national interoperability frameworks. *International Journal of E-Services and Mobile Applications*, *2*(4), 28–41. doi:10.4018/jesma.2010100103

Charalabidis, Y., Lampathaki, F., & Askounis, D. (2012). Investigating the landscape in national interoperability frameworks. In A. Scupola (Ed.), *Innovative mobile platform developments for electronic services design and delivery* (pp. 218–231). Hershey, PA: IGI Global. doi:10.4018/978-1-4666-1568-7.ch013

Chen, I. (2005). Distance education associations. In C. Howard, J. Boettcher, L. Justice, K. Schenk, P. Rogers, & G. Berg (Eds.), *Encyclopedia of distance learning* (pp. 599–612). Hershey, PA: IGI Global. doi:10.4018/978-1-59140-555-9.ch087

Chen, I. (2008). Distance education associations. In L. Tomei (Ed.), *Online and distance learning: Concepts, methodologies, tools, and applications* (pp. 562–579). Hershey, PA: IGI Global. doi:10.4018/978-1-59904-935-9.ch048

Chen, Y. (2008). Managing IT outsourcing for digital government. In A. Anttiroiko (Ed.), *Electronic government: Concepts, methodologies, tools, and applications* (pp. 3107–3114). Hershey, PA: IGI Global. doi:10.4018/978-1-59904-947-2.ch229

Chen, Y., & Dimitrova, D. V. (2006). Electronic government and online engagement: Citizen interaction with government via web portals. *International Journal of Electronic Government Research, 2*(1), 54–76. doi:10.4018/jegr.2006010104

Chen, Y., & Knepper, R. (2005). Digital government development strategies: Lessons for policy makers from a comparative perspective. In W. Huang, K. Siau, & K. Wei (Eds.), *Electronic government strategies and implementation* (pp. 394–420). Hershey, PA: IGI Global. doi:10.4018/978-1-59140-348-7.ch017

Chen, Y., & Knepper, R. (2008). Digital government development strategies: Lessons for policy makers from a comparative perspective. In H. Rahman (Ed.), *Developing successful ICT strategies: Competitive advantages in a global knowledge-driven society* (pp. 334–356). Hershey, PA: IGI Global. doi:10.4018/978-1-59904-654-9.ch017

Cherian, E. J., & Ryan, T. W. (2014). Incongruent needs: Why differences in the iron-triangle of priorities make health information technology adoption and use difficult. In C. El Morr (Ed.), *Research perspectives on the role of informatics in health policy and management* (pp. 209–221). Hershey, PA: IGI Global. doi:10.4018/978-1-4666-4321-5.ch012

Cho, H. J., & Hwang, S. (2010). Government 2.0 in Korea: Focusing on e-participation services. In C. Reddick (Ed.), *Politics, democracy and e-government: Participation and service delivery* (pp. 94–114). Hershey, PA: IGI Global. doi:10.4018/978-1-61520-933-0.ch006

Chorus, C., & Timmermans, H. (2010). Ubiquitous travel environments and travel control strategies: Prospects and challenges. In M. Wachowicz (Ed.), *Movement-aware applications for sustainable mobility: Technologies and approaches* (pp. 30–51). Hershey, PA: IGI Global. doi:10.4018/978-1-61520-769-5.ch003

Chuanshen, R. (2007). E-government construction and China's administrative litigation act. In A. Anttiroiko & M. Malkia (Eds.), *Encyclopedia of digital government* (pp. 507–510). Hershey, PA: IGI Global. doi:10.4018/978-1-59140-789-8.ch077

Ciaghi, A., & Villafiorita, A. (2012). Law modeling and BPR for public administration improvement. In K. Bwalya & S. Zulu (Eds.), *Handbook of research on e-government in emerging economies: Adoption, e-participation, and legal frameworks* (pp. 391–410). Hershey, PA: IGI Global. doi:10.4018/978-1-4666-0324-0.ch019

Ciaramitaro, B. L., & Skrocki, M. (2012). mHealth: Mobile healthcare. In B. Ciaramitaro (Ed.), Mobile technology consumption: Opportunities and challenges (pp. 99-109). Hershey, PA: IGI Global. doi:10.4018/978-1-61350-150-4.ch007

Comite, U. (2012). Innovative processes and managerial effectiveness of e-procurement in healthcare. In A. Manoharan & M. Holzer (Eds.), *Active citizen participation in e-government: A global perspective* (pp. 206–229). Hershey, PA: IGI Global. doi:10.4018/978-1-4666-0116-1.ch011

Cordella, A. (2013). E-government success: How to account for ICT, administrative rationalization, and institutional change. In J. Gil-Garcia (Ed.), *E-government success factors and measures: Theories, concepts, and methodologies* (pp. 40–51). Hershey, PA: IGI Global. doi:10.4018/978-1-4666-4058-0.ch003

Cropf, R. A. (2009). ICT and e-democracy. In M. Khosrow-Pour (Ed.), *Encyclopedia of information science and technology* (2nd ed., pp. 1789–1793). Hershey, PA: IGI Global. doi:10.4018/978-1-60566-026-4.ch281

Cropf, R. A. (2009). The virtual public sphere. In M. Pagani (Ed.), *Encyclopedia of multimedia technology and networking* (2nd ed., pp. 1525–1530). Hershey, PA: IGI Global. doi:10.4018/978-1-60566-014-1.ch206

D'Abundo, M. L. (2013). Electronic health record implementation in the United States healthcare industry: Making the process of change manageable. In V. Wang (Ed.), *Handbook of research on technologies for improving the 21st century workforce: Tools for lifelong learning* (pp. 272–286). Hershey, PA: IGI Global. doi:10.4018/978-1-4666-2181-7.ch018

Damurski, L. (2012). E-participation in urban planning: Online tools for citizen engagement in Poland and in Germany. *International Journal of E-Planning Research*, *1*(3), 40–67. doi:10.4018/ijepr.2012070103

de Almeida, M. O. (2007). E-government strategy in Brazil: Increasing transparency and efficiency through e-government procurement. In M. Gascó-Hernandez (Ed.), *Latin America online: Cases, successes and pitfalls* (pp. 34–82). Hershey, PA: IRM Press. doi:10.4018/978-1-59140-974-8.ch002

de Juana Espinosa, S. (2008). Empirical study of the municipalitites' motivations for adopting online presence. In A. Anttiroiko (Ed.), *Electronic government: Concepts, methodologies, tools, and applications* (pp. 3593–3608). Hershey, PA: IGI Global. doi:10.4018/978-1-59904-947-2.ch262 de Souza Dias, D. (2002). Motivation for using information technology. In C. Snodgrass & E. Szewczak (Eds.), *Human factors in information systems* (pp. 55–60). Hershey, PA: IRM Press. doi:10.4018/978-1-931777-10-0.ch005

Demediuk, P. (2006). Government procurement ICT's impact on the sustainability of SMEs and regional communities. In S. Marshall, W. Taylor, & X. Yu (Eds.), *Encyclopedia of developing regional communities with information and communication technology* (pp. 321–324). Hershey, PA: IGI Global. doi:10.4018/978-1-59140-575-7.ch056

Devonshire, E., Forsyth, H., Reid, S., & Simpson, J. M. (2013). The challenges and opportunities of online postgraduate coursework programs in a traditional university context. In B. Tynan, J. Willems, & R. James (Eds.), *Outlooks and opportunities in blended and distance learning* (pp. 353–368). Hershey, PA: IGI Global. doi:10.4018/978-1-4666-4205-8.ch026

Di Cerbo, F., Scotto, M., Sillitti, A., Succi, G., & Vernazza, T. (2007). Toward a GNU/Linux distribution for corporate environments. In S. Sowe, I. Stamelos, & I. Samoladas (Eds.), *Emerging free and open source software practices* (pp. 215–236). Hershey, PA: IGI Global. doi:10.4018/978-1-59904-210-7.ch010

Diesner, J., & Carley, K. M. (2005). Revealing social structure from texts: Meta-matrix text analysis as a novel method for network text analysis. In V. Narayanan & D. Armstrong (Eds.), *Causal mapping for research in information technology* (pp. 81–108). Hershey, PA: IGI Global. doi:10.4018/978-1-59140-396-8.ch004

Dologite, D. G., Mockler, R. J., Bai, Q., & Viszhanyo, P. F. (2006). IS change agents in practice in a US-Chinese joint venture. In M. Hunter & F. Tan (Eds.), *Advanced topics in global information management* (Vol. 5, pp. 331–352). Hershey, PA: IGI Global. doi:10.4018/978-1-59140-923-6.ch015

Drnevich, P., Brush, T. H., & Luckock, G. T. (2011). Process and structural implications for IT-enabled outsourcing. *International Journal of Strategic Information Technology and Applications*, *2*(4), 30–43. doi:10.4018/jsita.2011100103

Dwivedi, A. N. (2009). Handbook of research on information technology management and clinical data administration in healthcare (Vols. 1–2). Hershey, PA: IGI Global. doi:10.4018/978-1-60566-356-2

Elbeltagi, I., McBride, N., & Hardaker, G. (2006). Evaluating the factors affecting DSS usage by senior managers in local authorities in Egypt. In M. Hunter & F. Tan (Eds.), *Advanced topics in global information management* (Vol. 5, pp. 283–307). Hershey, PA: IGI Global. doi:10.4018/978-1-59140-923-6.ch013

Eom, S., & Fountain, J. E. (2013). Enhancing information services through public-private partnerships: Information technology knowledge transfer underlying structures to develop shared services in the U.S. and Korea. In J. Gil-Garcia (Ed.), *E-government success around the world: Cases, empirical studies, and practical recommendations* (pp. 15–40). Hershey, PA: IGI Global. doi:10.4018/978-1-4666-4173-0.ch002

Esteves, T., Leuenberger, D., & Van Leuven, N. (2012). Reaching citizen 2.0: How government uses social media to send public messages during times of calm and times of crisis. In K. Kloby & M. D'Agostino (Eds.), *Citizen 2.0: Public and governmental interaction through web 2.0 technologies* (pp. 250–268). Hershey, PA: IGI Global. doi:10.4018/978-1-4666-0318-9.ch013

Estevez, E., Fillottrani, P., Janowski, T., & Ojo, A. (2012). Government information sharing: A framework for policy formulation. In Y. Chen & P. Chu (Eds.), *Electronic governance and cross-boundary collaboration: Innovations and advancing tools* (pp. 23–55). Hershey, PA: IGI Global. doi:10.4018/978-1-60960-753-1.ch002

Ezz, I. E. (2008). E-governement emerging trends: Organizational challenges. In A. Anttiroiko (Ed.), *Electronic government: Concepts, methodologies, tools, and applications* (pp. 3721–3737). Hershey, PA: IGI Global. doi:10.4018/978-1-59904-947-2.ch269

Fabri, M. (2009). The Italian style of e-justice in a comparative perspective. In A. Martínez & P. Abat (Eds.), *E-justice: Using information communication technologies in the court system* (pp. 1–19). Hershey, PA: IGI Global. doi:10.4018/978-1-59904-998-4.ch001

Fagbe, T., & Adekola, O. D. (2010). Workplace safety and personnel well-being: The impact of information technology. *International Journal of Green Computing, 1*(1), 28–33. doi:10.4018/jgc.2010010103

Fagbe, T., & Adekola, O. D. (2011). Workplace safety and personnel well-being: The impact of information technology. In *Global business: Concepts, methodologies, tools and applications* (pp. 1438–1444). Hershey, PA: IGI Global. doi:10.4018/978-1-60960-587-2.ch509

Farmer, L. (2008). Affective collaborative instruction with librarians. In S. Kelsey & K. St.Amant (Eds.), *Handbook of research on computer mediated communication* (pp. 15–24). Hershey, PA: IGI Global. doi:10.4018/978-1-59904-863-5.ch002

Favier, L., & Mekhantar, J. (2007). Use of OSS by local e-administration: The French situation. In K. St.Amant & B. Still (Eds.), *Handbook of research on open source software: Technological, economic, and social perspectives* (pp. 428–444). Hershey, PA: IGI Global. doi:10.4018/978-1-59140-999-1.ch033

Fernando, S. (2009). Issues of e-learning in third world countries. In M. Khosrow-Pour (Ed.), *Encyclopedia of information science and technology* (2nd ed., pp. 2273–2277). Hershey, PA: IGI Global. doi:10.4018/978-1-60566-026-4.ch360

Filho, J. R., & dos Santos Junior, J. R. (2009). Local e-government in Brazil: Poor interaction and local politics as usual. In C. Reddick (Ed.), *Handbook of research on strategies for local e-government adoption and implementation: Comparative studies* (pp. 863–878). Hershey, PA: IGI Global. doi:10.4018/978-1-60566-282-4.ch045

Fletcher, P. D. (2004). Portals and policy: Implications of electronic access to U.S. federal government information services. In A. Pavlichev & G. Garson (Eds.), *Digital government: Principles and best practices* (pp. 52–62). Hershey, PA: IGI Global. doi:10.4018/978-1-59140-122-3.ch004

Fletcher, P. D. (2008). Portals and policy: Implications of electronic access to U.S. federal government information services. In A. Anttiroiko (Ed.), *Electronic government: Concepts, methodologies, tools, and applications* (pp. 3970–3979). Hershey, PA: IGI Global. doi:10.4018/978-1-59904-947-2.ch289

Forlano, L. (2004). The emergence of digital government: International perspectives. In A. Pavlichev & G. Garson (Eds.), *Digital government: Principles and best practices* (pp. 34–51). Hershey, PA: IGI Global. doi:10.4018/978-1-59140-122-3.ch003

Franzel, J. M., & Coursey, D. H. (2004). Government web portals: Management issues and the approaches of five states. In A. Pavlichev & G. Garson (Eds.), *Digital government: Principles and best practices* (pp. 63–77). Hershey, PA: IGI Global. doi:10.4018/978-1-59140-122-3.ch005

Gaivéo, J. M. (2013). Security of ICTs supporting healthcare activities. In M. Cruz-Cunha, I. Miranda, & P. Gonçalves (Eds.), *Handbook of research on ICTs for human-centered healthcare and social care services* (pp. 208–228). Hershey, PA: IGI Global. doi:10.4018/978-1-4666-3986-7.ch011

Garson, G. D. (1999). *Information technology and computer applications in public administration: Issues and trends*. Hershey, PA: IGI Global. doi:10.4018/978-1-87828-952-0

Garson, G. D. (2003). Toward an information technology research agenda for public administration. In G. Garson (Ed.), *Public information technology: Policy and management issues* (pp. 331–357). Hershey, PA: IGI Global. doi:10.4018/978-1-59140-060-8.ch014

Garson, G. D. (2004). The promise of digital government. In A. Pavlichev & G. Garson (Eds.), *Digital government: Principles and best practices* (pp. 2–15). Hershey, PA: IGI Global. doi:10.4018/978-1-59140-122-3.ch001

Garson, G. D. (2007). An information technology research agenda for public administration. In G. Garson (Ed.), *Modern public information technology systems: Issues and challenges* (pp. 365–392). Hershey, PA: IGI Global. doi:10.4018/978-1-59904-051-6.ch018

Gasco, M. (2007). Civil servants' resistance towards e-government development. In A. Anttiroiko & M. Malkia (Eds.), *Encyclopedia of digital government* (pp. 190–195). Hershey, PA: IGI Global. doi:10.4018/978-1-59140-789-8.ch028

Gasco, M. (2008). Civil servants' resistance towards e-government development. In A. Anttiroiko (Ed.), *Electronic government: Concepts, methodologies, tools, and applications* (pp. 2580–2588). Hershey, PA: IGI Global. doi:10.4018/978-1-59904-947-2.ch190

Ghere, R. K. (2010). Accountability and information technology enactment: Implications for social empowerment. In E. Ferro, Y. Dwivedi, J. Gil-Garcia, & M. Williams (Eds.), *Handbook of research on overcoming digital divides: Constructing an equitable and competitive information society* (pp. 515–532). Hershey, PA: IGI Global. doi:10.4018/978-1-60566-699-0.ch028

Gibson, I. W. (2012). Simulation modeling of healthcare delivery. In A. Kolker & P. Story (Eds.), *Management engineering for effective healthcare delivery: Principles and applications* (pp. 69–89). Hershey, PA: IGI Global. doi:10.4018/978-1-60960-872-9.ch003

Gil-Garcia, J. R. (2007). Exploring e-government benefits and success factors. In A. Anttiroiko & M. Malkia (Eds.), *Encyclopedia of digital government* (pp. 803–811). Hershey, PA: IGI Global. doi:10.4018/978-1-59140-789-8.ch122

Gil-Garcia, J. R., & González Miranda, F. (2010). E-government and opportunities for participation: The case of the Mexican state web portals. In C. Reddick (Ed.), *Politics, democracy and e-government: Participation and service delivery* (pp. 56–74). Hershey, PA: IGI Global. doi:10.4018/978-1-61520-933-0.ch004

Goldfinch, S. (2012). Public trust in government, trust in e-government, and use of e-government. In Z. Yan (Ed.), *Encyclopedia of cyber behavior* (pp. 987–995). Hershey, PA: IGI Global. doi:10.4018/978-1-4666-0315-8.ch081

Goodyear, M. (2012). Organizational change contributions to e-government project transitions. In S. Aikins (Ed.), *Managing e-government projects: Concepts, issues, and best practices* (pp. 1–21). Hershey, PA: IGI Global. doi:10.4018/978-1-4666-0086-7.ch001

Gordon, S., & Mulligan, P. (2003). Strategic models for the delivery of personal financial services: The role of infocracy. In S. Gordon (Ed.), *Computing information technology: The human side* (pp. 220–232). Hershey, PA: IRM Press. doi:10.4018/978-1-93177-752-0.ch014

Gordon, T. F. (2007). Legal knowledge systems. In A. Anttiroiko & M. Malkia (Eds.), *Encyclopedia of digital government* (pp. 1161–1166). Hershey, PA: IGI Global. doi:10.4018/978-1-59140-789-8.ch175

Graham, J. E., & Semich, G. W. (2008). Integrating technology to transform pedagogy: Revisiting the progress of the three phase TUI model for faculty development. In L. Tomei (Ed.), *Adapting information and communication technologies for effective education* (pp. 1–12). Hershey, PA: IGI Global. doi:10.4018/978-1-59904-922-9.ch001

Grandinetti, L., & Pisacane, O. (2012). Web services for healthcare management. In D. Prakash Vidyarthi (Ed.), *Technologies and protocols for the future of internet design: Reinventing the web* (pp. 60–94). Hershey, PA: IGI Global. doi:10.4018/978-1-4666-0203-8.ch004

Groenewegen, P., & Wagenaar, F. P. (2008). VO as an alternative to hierarchy in the Dutch police sector. In G. Putnik & M. Cruz-Cunha (Eds.), *Encyclopedia of networked and virtual organizations* (pp. 1851–1857). Hershey, PA: IGI Global. doi:10.4018/978-1-59904-885-7.ch245

Gronlund, A. (2001). Building an infrastructure to manage electronic services. In S. Dasgupta (Ed.), *Managing internet and intranet technologies in organizations: Challenges and opportunities* (pp. 71–103). Hershey, PA: IGI Global. doi:10.4018/978-1-878289-95-7.ch006

Gronlund, A. (2002). Introduction to electronic government: Design, applications and management. In Å. Grönlund (Ed.), *Electronic government: Design, applications and management* (pp. 1–21). Hershey, PA: IGI Global. doi:10.4018/978-1-930708-19-8.ch001

Gupta, A., Woosley, R., Crk, I., & Sarnikar, S. (2009). An information technology architecture for drug effectiveness reporting and post-marketing surveillance. In J. Tan (Ed.), *Medical informatics: Concepts, methodologies, tools, and applications* (pp. 631–646). Hershey, PA: IGI Global. doi:10.4018/978-1-60566-050-9.ch047

Hallin, A., & Lundevall, K. (2007). mCity: User focused development of mobile services within the city of Stockholm. In I. Kushchu (Ed.), Mobile government: An emerging direction in e-government (pp. 12-29). Hershey, PA: IGI Global. doi:10.4018/978-1-59140-884-0.ch002

Hallin, A., & Lundevall, K. (2009). mCity: User focused development of mobile services within the city of Stockholm. In S. Clarke (Ed.), Evolutionary concepts in end user productivity and performance: Applications for organizational progress (pp. 268-280). Hershey, PA: IGI Global. doi:10.4018/978-1-60566-136-0.ch017

Hallin, A., & Lundevall, K. (2009). mCity: User focused development of mobile services within the city of Stockholm. In D. Taniar (Ed.), Mobile computing: Concepts, methodologies, tools, and applications (pp. 3455-3467). Hershey, PA: IGI Global. doi:10.4018/978-1-60566-054-7.ch253

Hanson, A. (2005). Overcoming barriers in the planning of a virtual library. In M. Khosrow-Pour (Ed.), *Encyclopedia of information science and technology* (pp. 2255–2259). Hershey, PA: IGI Global. doi:10.4018/978-1-59140-553-5.ch397

Haque, A. (2008). Information technology and surveillance: Implications for public administration in a new word order. In T. Loendorf & G. Garson (Eds.), *Patriotic information systems* (pp. 177–185). Hershey, PA: IGI Global. doi:10.4018/978-1-59904-594-8.ch008

Hauck, R. V., Thatcher, S. M., & Weisband, S. P. (2012). Temporal aspects of information technology use: Increasing shift work effectiveness. In J. Wang (Ed.), *Advancing the service sector with evolving technologies: Techniques and principles* (pp. 87–104). Hershey, PA: IGI Global. doi:10.4018/978-1-4666-0044-7.ch006

Hawk, S., & Witt, T. (2006). Telecommunications courses in information systems programs. *International Journal of Information and Communication Technology Education*, *2*(1), 79–92. doi:10.4018/jicte.2006010107

Helms, M. M., Moore, R., & Ahmadi, M. (2009). Information technology (IT) and the healthcare industry: A SWOT analysis. In J. Tan (Ed.), *Medical informatics: Concepts, methodologies, tools, and applications* (pp. 134–152). Hershey, PA: IGI Global. doi:10.4018/978-1-60566-050-9.ch012

Hendrickson, S. M., & Young, M. E. (2014). Electronic records management at a federally funded research and development center. In J. Krueger (Ed.), *Cases on electronic records and resource management implementation in diverse environments* (pp. 334–350). Hershey, PA: IGI Global. doi:10.4018/978-1-4666-4466-3.ch020

Henman, P. (2010). Social policy and information communication technologies. In J. Martin & L. Hawkins (Eds.), *Information communication technologies for human services education and delivery: Concepts and cases* (pp. 215–229). Hershey, PA: IGI Global. doi:10.4018/978-1-60566-735-5.ch014

Hismanoglu, M. (2011). Important issues in online education: E-pedagogy and marketing. In U. Demiray & S. Sever (Eds.), *Marketing online education programs: Frameworks for promotion and communication* (pp. 184–209). Hershey, PA: IGI Global. doi:10.4018/978-1-60960-074-7.ch012

Ho, K. K. (2008). The e-government development, IT strategies, and portals of the Hong Kong SAR government. In A. Anttiroiko (Ed.), *Electronic government: Concepts, methodologies, tools, and applications* (pp. 715–733). Hershey, PA: IGI Global. doi:10.4018/978-1-59904-947-2.ch060

Holden, S. H. (2003). The evolution of information technology management at the federal level: Implications for public administration. In G. Garson (Ed.), *Public information technology: Policy and management issues* (pp. 53–73). Hershey, PA: IGI Global. doi:10.4018/978-1-59140-060-8.ch003

Holden, S. H. (2007). The evolution of federal information technology management literature: Does IT finally matter? In G. Garson (Ed.), *Modern public information technology systems: Issues and challenges* (pp. 17–34). Hershey, PA: IGI Global. doi:10.4018/978-1-59904-051-6.ch002

Holland, J. W. (2009). Automation of American criminal justice. In M. Khosrow-Pour (Ed.), *Encyclopedia of information science and technology* (2nd ed., pp. 300–302). Hershey, PA: IGI Global. doi:10.4018/978-1-60566-026-4.ch051

Holloway, K. (2013). Fair use, copyright, and academic integrity in an online academic environment. In *Digital rights management: Concepts, methodologies, tools, and applications* (pp. 917–928). Hershey, PA: IGI Global. doi:10.4018/978-1-4666-2136-7.ch044

Horiuchi, C. (2005). E-government databases. In L. Rivero, J. Doorn, & V. Ferraggine (Eds.), *Encyclopedia of database technologies and applications* (pp. 206–210). Hershey, PA: IGI Global. doi:10.4018/978-1-59140-560-3.ch035

Horiuchi, C. (2006). Creating IS quality in government settings. In E. Duggan & J. Reichgelt (Eds.), *Measuring information systems delivery quality* (pp. 311–327). Hershey, PA: IGI Global. doi:10.4018/978-1-59140-857-4.ch014

Hsiao, N., Chu, P., & Lee, C. (2012). Impact of e-governance on businesses: Model development and case study. In *Digital democracy: Concepts, methodologies, tools, and applications* (pp. 1407–1425). Hershey, PA: IGI Global. doi:10.4018/978-1-4666-1740-7.ch070

Huang, T., & Lee, C. (2010). Evaluating the impact of e-government on citizens: Cost-benefit analysis. In C. Reddick (Ed.), *Citizens and e-government: Evaluating policy and management* (pp. 37–52). Hershey, PA: IGI Global. doi:10.4018/978-1-61520-931-6.ch003

Hunter, M. G., Diochon, M., Pugsley, D., & Wright, B. (2002). Unique challenges for small business adoption of information technology: The case of the Nova Scotia ten. In S. Burgess (Ed.), *Managing information technology in small business: Challenges and solutions* (pp. 98–117). Hershey, PA: IGI Global. doi:10.4018/978-1-930708-35-8.ch006

Hurskainen, J. (2003). Integration of business systems and applications in merger and alliance: Case metso automation. In T. Reponen (Ed.), *Information technology enabled global customer service* (pp. 207–225). Hershey, PA: IGI Global. doi:10.4018/978-1-59140-048-6.ch012

Iazzolino, G., & Pietrantonio, R. (2011). The soveria.it project: A best practice of e-government in southern Italy. In D. Piaggesi, K. Sund, & W. Castelnovo (Eds.), *Global strategy and practice of e-governance: Examples from around the world* (pp. 34–56). Hershey, PA: IGI Global. doi:10.4018/978-1-60960-489-9.ch003

Imran, A., & Gregor, S. (2012). A process model for successful e-government adoption in the least developed countries: A case of Bangladesh. In F. Tan (Ed.), *International comparisons of information communication technologies: Advancing applications* (pp. 321–350). Hershey, PA: IGI Global. doi:10.4018/978-1-61350-480-2.ch014

Inoue, Y., & Bell, S. T. (2005). Electronic/digital government innovation, and publishing trends with IT. In M. Khosrow-Pour (Ed.), *Encyclopedia of information science and technology* (pp. 1018–1023). Hershey, PA: IGI Global. doi:10.4018/978-1-59140-553-5.ch180

IRMA. (2010). Electronic services: Concepts, methodologies, tools and applications (3 Volumes). Hershey, PA: IGI Global. doi:10.4018/978-1-61520-967-5

IRMA. (2010). Information resources management: Concepts, methodologies, tools and applications (4 Volumes). Hershey, PA: IGI Global. doi:10.4018/978-1-61520-965-1

Islam, M. M., & Ehsan, M. (2013). Understanding e-governance: A theoretical approach. In M. Islam & M. Ehsan (Eds.), *From government to e-governance: Public administration in the digital age* (pp. 38–49). Hershey, PA: IGI Global. doi:10.4018/978-1-4666-1909-8.ch003

Jaeger, B. (2009). E-government and e-democracy in the making. In M. Khosrow-Pour (Ed.), *Encyclopedia of information science and technology* (2nd ed., pp. 1318–1322). Hershey, PA: IGI Global. doi:10.4018/978-1-60566-026-4.ch208

Jain, R. B. (2007). Revamping the administrative structure and processes in India for online diplomacy. In A. Anttiroiko & M. Malkia (Eds.), *Encyclopedia of digital government* (pp. 1418–1423). Hershey, PA: IGI Global. doi:10.4018/978-1-59140-789-8.ch217

Jain, R. B. (2008). Revamping the administrative structure and processes in India for online diplomacy. In A. Anttiroiko (Ed.), *Electronic government: Concepts, methodologies, tools, and applications* (pp. 3142–3149). Hershey, PA: IGI Global. doi:10.4018/978-1-59904-947-2.ch233

Jauhiainen, J. S., & Inkinen, T. (2009). E-governance and the information society in periphery. In C. Reddick (Ed.), *Handbook of research on strategies for local e-government adoption and implementation: Comparative studies* (pp. 497–514). Hershey, PA: IGI Global. doi:10.4018/978-1-60566-282-4.ch026

Jensen, M. J. (2009). Electronic democracy and citizen influence in government. In C. Reddick (Ed.), *Handbook of research on strategies for local e-government adoption and implementation: Comparative studies* (pp. 288–305). Hershey, PA: IGI Global. doi:10.4018/978-1-60566-282-4.ch015

Jiao, Y., Hurson, A. R., Potok, T. E., & Beckerman, B. G. (2009). Integrating mobile-based systems with healthcare databases. In J. Erickson (Ed.), *Database technologies: Concepts, methodologies, tools, and applications* (pp. 484–504). Hershey, PA: IGI Global. doi:10.4018/978-1-60566-058-5.ch031

Joia, L. A. (2002). A systematic model to integrate information technology into metabusinesses: A case study in the engineering realms. In F. Tan (Ed.), *Advanced topics in global information management* (Vol. 1, pp. 250–267). Hershey, PA: IGI Global. doi:10.4018/978-1-930708-43-3.ch016

Jones, T. H., & Song, I. (2000). Binary equivalents of ternary relationships in entity-relationship modeling: A logical decomposition approach. *Journal of Database Management*, *11*(2), 12–19. doi:10.4018/jdm.2000040102

Juana-Espinosa, S. D. (2007). Empirical study of the municipalitites' motivations for adopting online presence. In L. Al-Hakim (Ed.), *Global e-government: Theory, applications and benchmarking* (pp. 261–279). Hershey, PA: IGI Global. doi:10.4018/978-1-59904-027-1.ch015

Jun, K., & Weare, C. (2012). Bridging from e-government practice to e-government research: Past trends and future directions. In K. Bwalya & S. Zulu (Eds.), *Handbook of research on e-government in emerging economies: Adoption, e-participation, and legal frameworks* (pp. 263–289). Hershey, PA: IGI Global. doi:10.4018/978-1-4666-0324-0.ch013

Junqueira, A., Diniz, E. H., & Fernandez, M. (2010). Electronic government implementation projects with multiple agencies: Analysis of the electronic invoice project under PMBOK framework. In J. Cordoba-Pachon & A. Ochoa-Arias (Eds.), *Systems thinking and e-participation: ICT in the governance of society* (pp. 135–153). Hershey, PA: IGI Global. doi:10.4018/978-1-60566-860-4.ch009

Juntunen, A. (2009). Joint service development with the local authorities. In C. Reddick (Ed.), *Handbook of research on strategies for local e-government adoption and implementation: Comparative studies* (pp. 902–920). Hershey, PA: IGI Global. doi:10.4018/978-1-60566-282-4.ch047

Kamel, S. (2001). *Using DSS for crisis management*. Hershey, PA: IGI Global. doi:10.4018/978-1-87828-961-2.ch020

Kamel, S. (2006). DSS for strategic decision making. In M. Khosrow-Pour (Ed.), *Cases on information technology and organizational politics & culture* (pp. 230–246). Hershey, PA: IGI Global. doi:10.4018/978-1-59904-411-8.ch015

Kamel, S. (2009). The software industry in Egypt as a potential contributor to economic growth. In M. Khosrow-Pour (Ed.), *Encyclopedia of information science and technology* (2nd ed., pp. 3531–3537). Hershey, PA: IGI Global. doi:10.4018/978-1-60566-026-4.ch562

Kamel, S., & Hussein, M. (2008). Xceed: Pioneering the contact center industry in Egypt. *Journal of Cases on Information Technology, 10*(1), 67–91. doi:10.4018/jcit.2008010105

Kamel, S., & Wahba, K. (2003). The use of a hybrid model in web-based education: "The Global campus project. In A. Aggarwal (Ed.), *Web-based education: Learning from experience* (pp. 331–346). Hershey, PA: IGI Global. doi:10.4018/978-1-59140-102-5.ch020

Kardaras, D. K., & Papathanassiou, E. A. (2008). An exploratory study of the e-government services in Greece. In G. Garson & M. Khosrow-Pour (Eds.), *Handbook of research on public information technology* (pp. 162–174). Hershey, PA: IGI Global. doi:10.4018/978-1-59904-857-4.ch016

Kassahun, A. E., Molla, A., & Sarkar, P. (2012). Government process reengineering: What we know and what we need to know. In *Digital democracy: Concepts, methodologies, tools, and applications* (pp. 1730–1752). Hershey, PA: IGI Global. doi:10.4018/978-1-4666-1740-7.ch086

Khan, B. (2005). Technological issues. In B. Khan (Ed.), *Managing e-learning strategies: Design, delivery, implementation and evaluation* (pp. 154–180). Hershey, PA: IGI Global. doi:10.4018/978-1-59140-634-1.ch004

Khasawneh, A., Bsoul, M., Obeidat, I., & Al Azzam, I. (2012). Technology fears: A study of e-commerce loyalty perception by Jordanian customers. In J. Wang (Ed.), *Advancing the service sector with evolving technologies: Techniques and principles* (pp. 158–165). Hershey, PA: IGI Global. doi:10.4018/978-1-4666-0044-7.ch010

Khatibi, V., & Montazer, G. A. (2012). E-research methodology. In A. Juan, T. Daradoumis, M. Roca, S. Grasman, & J. Faulin (Eds.), *Collaborative and distributed e-research: Innovations in technologies, strategies and applications* (pp. 62–81). Hershey, PA: IGI Global. doi:10.4018/978-1-4666-0125-3.ch003

Kidd, T. (2011). The dragon in the school's backyard: A review of literature on the uses of technology in urban schools. In L. Tomei (Ed.), *Online courses and ICT in education: Emerging practices and applications* (pp. 242–257). Hershey, PA: IGI Global. doi:10.4018/978-1-60960-150-8.ch019

Kidd, T. T. (2010). My experience tells the story: Exploring technology adoption from a qualitative perspective - A pilot study. In H. Song & T. Kidd (Eds.), *Handbook of research on human performance and instructional technology* (pp. 247–262). Hershey, PA: IGI Global. doi:10.4018/978-1-60566-782-9.ch015

Kieley, B., Lane, G., Paquet, G., & Roy, J. (2002). e-Government in Canada: Services online or public service renewal? In Å. Grönlund (Ed.), Electronic government: Design, applications and management (pp. 340-355). Hershey, PA: IGI Global. doi:10.4018/978-1-930708-19-8.ch016

Kim, P. (2012). "Stay out of the way! My kid is video blogging through a phone!": A lesson learned from math tutoring social media for children in underserved communities. In *Wireless technologies: Concepts, methodologies, tools and applications* (pp. 1415–1428). Hershey, PA: IGI Global. doi:10.4018/978-1-61350-101-6.ch517

Kirlidog, M. (2010). Financial aspects of national ICT strategies. In S. Kamel (Ed.), *E-strategies for technological diffusion and adoption: National ICT approaches for socioeconomic development* (pp. 277–292). Hershey, PA: IGI Global. doi:10.4018/978-1-60566-388-3.ch016

Kisielnicki, J. (2006). Transfer of information and knowledge in the project management. In E. Coakes & S. Clarke (Eds.), *Encyclopedia of communities of practice in information and knowledge management* (pp. 544–551). Hershey, PA: IGI Global. doi:10.4018/978-1-59140-556-6.ch091

Kittner, M., & Van Slyke, C. (2006). Reorganizing information technology services in an academic environment. In M. Khosrow-Pour (Ed.), *Cases on the human side of information technology* (pp. 49–66). Hershey, PA: IGI Global. doi:10.4018/978-1-59904-405-7.ch004

Knoell, H. D. (2008). Semi virtual workplaces in German financial service enterprises. In P. Zemliansky & K. St.Amant (Eds.), *Handbook of research on virtual workplaces and the new nature of business practices* (pp. 570–581). Hershey, PA: IGI Global. doi:10.4018/978-1-59904-893-2.ch041

Koh, S. L., & Maguire, S. (2009). Competing in the age of information technology in a developing economy: Experiences of an Indian bank. In S. Koh & S. Maguire (Eds.), *Information and communication technologies management in turbulent business environments* (pp. 326–350). Hershey, PA: IGI Global. doi:10.4018/978-1-60566-424-8.ch018

Kollmann, T., & Häsel, M. (2009). Competence of information technology professionals in internet-based ventures. In I. Lee (Ed.), *Electronic business: Concepts, methodologies, tools, and applications* (pp. 1905–1919). Hershey, PA: IGI Global. doi:10.4018/978-1-60566-056-1.ch118

Kollmann, T., & Häsel, M. (2009). Competence of information technology professionals in internet-based ventures. In A. Cater-Steel (Ed.), *Information technology governance and service management: Frameworks and adaptations* (pp. 239–253). Hershey, PA: IGI Global. doi:10.4018/978-1-60566-008-0.ch013

Kollmann, T., & Häsel, M. (2010). Competence of information technology professionals in internet-based ventures. In *Electronic services: Concepts, methodologies, tools and applications* (pp. 1551–1565). Hershey, PA: IGI Global. doi:10.4018/978-1-61520-967-5.ch094

Kraemer, K., & King, J. L. (2006). Information technology and administrative reform: Will e-government be different? *International Journal of Electronic Government Research*, *2*(1), 1–20. doi:10.4018/jegr.2006010101

Kraemer, K., & King, J. L. (2008). Information technology and administrative reform: Will e-government be different? In D. Norris (Ed.), *E-government research: Policy and management* (pp. 1–20). Hershey, PA: IGI Global. doi:10.4018/978-1-59904-913-7.ch001

Lampathaki, F., Tsiakaliaris, C., Stasis, A., & Charalabidis, Y. (2011). National interoperability frameworks: The way forward. In Y. Charalabidis (Ed.), *Interoperability in digital public services and administration: Bridging e-government and e-business* (pp. 1–24). Hershey, PA: IGI Global. doi:10.4018/978-1-61520-887-6.ch001

Lan, Z., & Scott, C. R. (1996). The relative importance of computer-mediated information versus conventional non-computer-mediated information in public managerial decision making. *Information Resources Management Journal*, *9*(1), 27–0. doi:10.4018/irmj.1996010103

Law, W. (2004). *Public sector data management in a developing economy*. Hershey, PA: IGI Global. doi:10.4018/978-1-59140-259-6.ch034

Law, W. K. (2005). Information resources development challenges in a cross-cultural environment. In M. Khosrow-Pour (Ed.), *Encyclopedia of information science and technology* (pp. 1476–1481). Hershey, PA: IGI Global. doi:10.4018/978-1-59140-553-5.ch259

Law, W. K. (2009). Cross-cultural challenges for information resources management. In M. Khosrow-Pour (Ed.), *Encyclopedia of information science and technology* (2nd ed., pp. 840–846). Hershey, PA: IGI Global. doi:10.4018/978-1-60566-026-4.ch136

Law, W. K. (2011). Cross-cultural challenges for information resources management. In *Global business: Concepts, methodologies, tools and applications* (pp. 1924–1932). Hershey, PA: IGI Global. doi:10.4018/978-1-60960-587-2.ch704

Malkia, M., & Savolainen, R. (2004). eTransformation in government, politics and society: Conceptual framework and introduction. In M. Malkia, A. Anttiroiko, & R. Savolainen (Eds.), eTransformation in governance: New directions in government and politics (pp. 1-21). Hershey, PA: IGI Global. doi:10.4018/978-1-59140-130-8.ch001

Mandujano, S. (2011). Network manageability security. In D. Kar & M. Syed (Eds.), *Network security, administration and management: Advancing technology and practice* (pp. 158–181). Hershey, PA: IGI Global. doi:10.4018/978-1-60960-777-7.ch009

Marich, M. J., Schooley, B. L., & Horan, T. A. (2012). A normative enterprise architecture for guiding end-to-end emergency response decision support. In M. Jennex (Ed.), *Managing crises and disasters with emerging technologies: Advancements* (pp. 71–87). Hershey, PA: IGI Global. doi:10.4018/978-1-4666-0167-3.ch006

Markov, R., & Okujava, S. (2008). Costs, benefits, and risks of e-government portals. In G. Putnik & M. Cruz-Cunha (Eds.), *Encyclopedia of networked and virtual organizations* (pp. 354–363). Hershey, PA: IGI Global. doi:10.4018/978-1-59904-885-7.ch047

Martin, N., & Rice, J. (2013). Evaluating and designing electronic government for the future: Observations and insights from Australia. In V. Weerakkody (Ed.), *E-government services design, adoption, and evaluation* (pp. 238–258). Hershey, PA: IGI Global. doi:10.4018/978-1-4666-2458-0.ch014

i. Martinez, A. C. (2008). Accessing administration's information via internet in Spain. In F. Tan (Ed.), *Global information technologies: Concepts, methodologies, tools, and applications* (pp. 2558–2573). Hershey, PA: IGI Global. doi:10.4018/978-1-59904-939-7.ch186

Mbarika, V. W., Meso, P. N., & Musa, P. F. (2006). A disconnect in stakeholders' perceptions from emerging realities of teledensity growth in Africa's least developed countries. In M. Hunter & F. Tan (Eds.), *Advanced topics in global information management* (Vol. 5, pp. 263–282). Hershey, PA: IGI Global. doi:10.4018/978-1-59140-923-6.ch012

Mbarika, V. W., Meso, P. N., & Musa, P. F. (2008). A disconnect in stakeholders' perceptions from emerging realities of teledensity growth in Africa's least developed countries. In F. Tan (Ed.), *Global information technologies: Concepts, methodologies, tools, and applications* (pp. 2948–2962). Hershey, PA: IGI Global. doi:10.4018/978-1-59904-939-7.ch209

Means, T., Olson, E., & Spooner, J. (2013). Discovering ways that don't work on the road to success: Strengths and weaknesses revealed by an active learning studio classroom project. In A. Benson, J. Moore, & S. Williams van Rooij (Eds.), *Cases on educational technology planning, design, and implementation: A project management perspective* (pp. 94–113). Hershey, PA: IGI Global. doi:10.4018/978-1-4666-4237-9.ch006

Melitski, J., Holzer, M., Kim, S., Kim, C., & Rho, S. (2008). Digital government worldwide: An e-government assessment of municipal web sites. In G. Garson & M. Khosrow-Pour (Eds.), *Handbook of research on public information technology* (pp. 790–804). Hershey, PA: IGI Global. doi:10.4018/978-1-59904-857-4.ch069

Memmola, M., Palumbo, G., & Rossini, M. (2009). Web & RFID technology: New frontiers in costing and process management for rehabilitation medicine. In L. Al-Hakim & M. Memmola (Eds.), *Business web strategy: Design, alignment, and application* (pp. 145–169). Hershey, PA: IGI Global. doi:10.4018/978-1-60566-024-0.ch008

Meng, Z., Fahong, Z., & Lei, L. (2008). Information technology and environment. In Y. Kurihara, S. Takaya, H. Harui, & H. Kamae (Eds.), *Information technology and economic development* (pp. 201–212). Hershey, PA: IGI Global. doi:10.4018/978-1-59904-579-5.ch014

Mentzingen de Moraes, A. J., Ferneda, E., Costa, I., & Spinola, M. D. (2011). Practical approach for implementation of governance process in IT: Information technology areas. In N. Shi & G. Silvius (Eds.), *Enterprise IT governance, business value and performance measurement* (pp. 19–40). Hershey, PA: IGI Global. doi:10.4018/978-1-60566-346-3.ch002

Merwin, G. A. Jr, McDonald, J. S., & Odera, L. C. (2008). Economic development: Government's cutting edge in IT. In M. Raisinghani (Ed.), *Handbook of research on global information technology management in the digital economy* (pp. 1–37). Hershey, PA: IGI Global. doi:10.4018/978-1-59904-875-8.ch001

Meso, P., & Duncan, N. (2002). Can national information infrastructures enhance social development in the least developed countries? An empirical investigation. In M. Dadashzadeh (Ed.), *Information technology management in developing countries* (pp. 23–51). Hershey, PA: IRM Press. doi:10.4018/978-1-931777-03-2.ch002

Meso, P. N., & Duncan, N. B. (2002). Can national information infrastructures enhance social development in the least developed countries? In F. Tan (Ed.), *Advanced topics in global information management* (Vol. 1, pp. 207–226). Hershey, PA: IGI Global. doi:10.4018/978-1-930708-43-3.ch014

Middleton, M. (2008). Evaluation of e-government web sites. In G. Garson & M. Khosrow-Pour (Eds.), *Handbook of research on public information technology* (pp. 699–710). Hershey, PA: IGI Global. doi:10.4018/978-1-59904-857-4.ch063

Mingers, J. (2010). Pluralism, realism, and truth: The keys to knowledge in information systems research. In D. Paradice (Ed.), *Emerging systems approaches in information technologies: Concepts, theories, and applications* (pp. 86–98). Hershey, PA: IGI Global. doi:10.4018/978-1-60566-976-2.ch006

Mital, K. M. (2012). ICT, unique identity and inclusive growth: An Indian perspective. In A. Manoharan & M. Holzer (Eds.), *E-governance and civic engagement: Factors and determinants of e-democracy* (pp. 584–612). Hershey, PA: IGI Global. doi:10.4018/978-1-61350-083-5.ch029

Mizell, A. P. (2008). Helping close the digital divide for financially disadvantaged seniors. In F. Tan (Ed.), *Global information technologies: Concepts, methodologies, tools, and applications* (pp. 2396–2402). Hershey, PA: IGI Global. doi:10.4018/978-1-59904-939-7.ch173

Molinari, F., Wills, C., Koumpis, A., & Moumtzi, V. (2011). A citizen-centric platform to support networking in the area of e-democracy. In H. Rahman (Ed.), *Cases on adoption, diffusion and evaluation of global e-governance systems: Impact at the grass roots* (pp. 282–302). Hershey, PA: IGI Global. doi:10.4018/978-1-61692-814-8.ch014

Molinari, F., Wills, C., Koumpis, A., & Moumtzi, V. (2013). A citizen-centric platform to support networking in the area of e-democracy. In H. Rahman (Ed.), *Cases on progressions and challenges in ICT utilization for citizen-centric governance* (pp. 265–297). Hershey, PA: IGI Global. doi:10.4018/978-1-4666-2071-1.ch013

Monteverde, F. (2010). The process of e-government public policy inclusion in the governmental agenda: A framework for assessment and case study. In J. Cordoba-Pachon & A. Ochoa-Arias (Eds.), *Systems thinking and e-participation: ICT in the governance of society* (pp. 233–245). Hershey, PA: IGI Global. doi:10.4018/978-1-60566-860-4.ch015

Moodley, S. (2008). Deconstructing the South African government's ICT for development discourse. In A. Anttiroiko (Ed.), *Electronic government: Concepts, methodologies, tools, and applications* (pp. 622–631). Hershey, PA: IGI Global. doi:10.4018/978-1-59904-947-2.ch053

Moodley, S. (2008). Deconstructing the South African government's ICT for development discourse. In C. Van Slyke (Ed.), *Information communication technologies: Concepts, methodologies, tools, and applications* (pp. 816–825). Hershey, PA: IGI Global. doi:10.4018/978-1-59904-949-6.ch052

Mora, M., Cervantes-Perez, F., Gelman-Muravchik, O., Forgionne, G. A., & Mejia-Olvera, M. (2003). DMSS implementation research: A conceptual analysis of the contributions and limitations of the factor-based and stage-based streams. In G. Forgionne, J. Gupta, & M. Mora (Eds.), *Decision-making support systems: Achievements and challenges for the new decade* (pp. 331–356). Hershey, PA: IGI Global. doi:10.4018/978-1-59140-045-5.ch020

Mörtberg, C., & Elovaara, P. (2010). Attaching people and technology: Between e and government. In S. Booth, S. Goodman, & G. Kirkup (Eds.), *Gender issues in learning and working with information technology: Social constructs and cultural contexts* (pp. 83–98). Hershey, PA: IGI Global. doi:10.4018/978-1-61520-813-5.ch005

Murphy, J., Harper, E., Devine, E. C., Burke, L. J., & Hook, M. L. (2011). Case study: Lessons learned when embedding evidence-based knowledge in a nurse care planning and documentation system. In A. Cashin & R. Cook (Eds.), *Evidence-based practice in nursing informatics: Concepts and applications* (pp. 174–190). Hershey, PA: IGI Global. doi:10.4018/978-1-60960-034-1.ch014

Mutula, S. M. (2013). E-government's role in poverty alleviation: Case study of South Africa. In H. Rahman (Ed.), *Cases on progressions and challenges in ICT utilization for citizen-centric governance* (pp. 44–68). Hershey, PA: IGI Global. doi:10.4018/978-1-4666-2071-1.ch003

Nath, R., & Angeles, R. (2005). Relationships between supply characteristics and buyer-supplier coupling in e-procurement: An empirical analysis. [IJEBR]. *International Journal of E-Business Research, 1*(2), 40–55. doi:10.4018/jebr.2005040103

Nissen, M. E. (2006). Application cases in government. In M. Nissen (Ed.), *Harnessing knowledge dynamics: Principled organizational knowing & learning* (pp. 152–181). Hershey, PA: IRM Press. doi:10.4018/978-1-59140-773-7.ch008

Norris, D. F. (2003). Leading-edge information technologies and American local governments. In G. Garson (Ed.), *Public information technology: Policy and management issues* (pp. 139–169). Hershey, PA: IGI Global. doi:10.4018/978-1-59140-060-8.ch007

Norris, D. F. (2008). Information technology among U.S. local governments. In G. Garson & M. Khosrow-Pour (Eds.), *Handbook of research on public information technology* (pp. 132–144). Hershey, PA: IGI Global. doi:10.4018/978-1-59904-857-4.ch013

Northrop, A. (1999). The challenge of teaching information technology in public administration graduate programs. In G. Garson (Ed.), *Information technology and computer applications in public administration: Issues and trends* (pp. 1–22). Hershey, PA: IGI Global. doi:10.4018/978-1-87828-952-0.ch001

Northrop, A. (2003). Information technology and public administration: The view from the profession. In G. Garson (Ed.), *Public information technology: Policy and management issues* (pp. 1–19). Hershey, PA: IGI Global. doi:10.4018/978-1-59140-060-8.ch001

Northrop, A. (2007). Lip service? How PA journals and textbooks view information technology. In G. Garson (Ed.), *Modern public information technology systems: Issues and challenges* (pp. 1–16). Hershey, PA: IGI Global. doi:10.4018/978-1-59904-051-6.ch001

Null, E. (2013). Legal and political barriers to municipal networks in the United States. In A. Abdelaal (Ed.), *Social and economic effects of community wireless networks and infrastructures* (pp. 27–56). Hershey, PA: IGI Global. doi:10.4018/978-1-4666-2997-4.ch003

Okunoye, A., Frolick, M., & Crable, E. (2006). ERP implementation in higher education: An account of pre-implementation and implementation phases. *Journal of Cases on Information Technology*, *8*(2), 110–132. doi:10.4018/jcit.2006040106

Olasina, G. (2012). A review of egovernment services in Nigeria. In A. Tella & A. Issa (Eds.), *Library and information science in developing countries: Contemporary issues* (pp. 205–221). Hershey, PA: IGI Global. doi:10.4018/978-1-61350-335-5.ch015

Orgeron, C. P. (2008). A model for reengineering IT job classes in state government. In G. Garson & M. Khosrow-Pour (Eds.), *Handbook of research on public information technology* (pp. 735–746). Hershey, PA: IGI Global. doi:10.4018/978-1-59904-857-4.ch066

Owsinski, J. W., & Pielak, A. M. (2011). Local authority websites in rural areas: Measuring quality and functionality, and assessing the role. In Z. Andreopoulou, B. Manos, N. Polman, & D. Viaggi (Eds.), *Agricultural and environmental informatics, governance and management: Emerging research applications* (pp. 39–60). Hershey, PA: IGI Global. doi:10.4018/978-1-60960-621-3.ch003

Owsiński, J. W., Pielak, A. M., Sęp, K., & Stańczak, J. (2014). Local web-based networks in rural municipalities: Extension, density, and meaning. In Z. Andreopoulou, V. Samathrakis, S. Louca, & M. Vlachopoulou (Eds.), *E-innovation for sustainable development of rural resources during global economic crisis* (pp. 126–151). Hershey, PA: IGI Global. doi:10.4018/978-1-4666-4550-9.ch011

Pagani, M., & Pasinetti, C. (2008). Technical and functional quality in the development of t-government services. In A. Anttiroiko (Ed.), *Electronic government: Concepts, methodologies, tools, and applications* (pp. 2943–2965). Hershey, PA: IGI Global. doi:10.4018/978-1-59904-947-2.ch220

Pani, A. K., & Agrahari, A. (2005). On e-markets in emerging economy: An Indian experience. In M. Khosrow-Pour (Ed.), *Advanced topics in electronic commerce* (Vol. 1, pp. 287–299). Hershey, PA: IGI Global. doi:10.4018/978-1-59140-819-2.ch015

Papadopoulos, T., Angelopoulos, S., & Kitsios, F. (2011). A strategic approach to e-health interoperability using e-government frameworks. In A. Lazakidou, K. Siassiakos, & K. Ioannou (Eds.), *Wireless technologies for ambient assisted living and healthcare: Systems and applications* (pp. 213–229). Hershey, PA: IGI Global. doi:10.4018/978-1-61520-805-0.ch012

Papadopoulos, T., Angelopoulos, S., & Kitsios, F. (2013). A strategic approach to e-health interoperability using e-government frameworks. In *User-driven healthcare: Concepts, methodologies, tools, and applications* (pp. 791–807). Hershey, PA: IGI Global. doi:10.4018/978-1-4666-2770-3.ch039

Papaleo, G., Chiarella, D., Aiello, M., & Caviglione, L. (2012). Analysis, development and deployment of statistical anomaly detection techniques for real e-mail traffic. In T. Chou (Ed.), *Information assurance and security technologies for risk assessment and threat management: Advances* (pp. 47–71). Hershey, PA: IGI Global. doi:10.4018/978-1-61350-507-6.ch003

Papp, R. (2003). Information technology & FDA compliance in the pharmaceutical industry. In M. Khosrow-Pour (Ed.), *Annals of cases on information technology* (Vol. 5, pp. 262–273). Hershey, PA: IGI Global. doi:10.4018/978-1-59140-061-5.ch017

Parsons, T. W. (2007). Developing a knowledge management portal. In A. Tatnall (Ed.), *Encyclopedia of portal technologies and applications* (pp. 223–227). Hershey, PA: IGI Global. doi:10.4018/978-1-59140-989-2.ch039

Passaris, C. E. (2007). Immigration and digital government. In A. Anttiroiko & M. Malkia (Eds.), *Encyclopedia of digital government* (pp. 988–994). Hershey, PA: IGI Global. doi:10.4018/978-1-59140-789-8.ch148

Pavlichev, A. (2004). The e-government challenge for public administration. In A. Pavlichev & G. Garson (Eds.), *Digital government: Principles and best practices* (pp. 276–290). Hershey, PA: IGI Global. doi:10.4018/978-1-59140-122-3.ch018

Penrod, J. I., & Harbor, A. F. (2000). Designing and implementing a learning organization-oriented information technology planning and management process. In L. Petrides (Ed.), *Case studies on information technology in higher education: Implications for policy and practice* (pp. 7–19). Hershey, PA: IGI Global. doi:10.4018/978-1-878289-74-2.ch001

Planas-Silva, M. D., & Joseph, R. C. (2011). Perspectives on the adoption of electronic resources for use in clinical trials. In M. Guah (Ed.), *Healthcare delivery reform and new technologies: Organizational initiatives* (pp. 19–28). Hershey, PA: IGI Global. doi:10.4018/978-1-60960-183-6.ch002

Pomazalová, N., & Rejman, S. (2013). The rationale behind implementation of new electronic tools for electronic public procurement. In N. Pomazalová (Ed.), *Public sector transformation processes and internet public procurement: Decision support systems* (pp. 85–117). Hershey, PA: IGI Global. doi:10.4018/978-1-4666-2665-2.ch006

Postorino, M. N. (2012). City competitiveness and airport: Information science perspective. In M. Bulu (Ed.), *City competitiveness and improving urban subsystems: Technologies and applications* (pp. 61–83). Hershey, PA: IGI Global. doi:10.4018/978-1-61350-174-0.ch004

Poupa, C. (2002). Electronic government in Switzerland: Priorities for 2001-2005 - Electronic voting and federal portal. In Å. Grönlund (Ed.), *Electronic government: Design, applications and management* (pp. 356–369). Hershey, PA: IGI Global. doi:10.4018/978-1-930708-19-8.ch017

Powell, S. R. (2010). Interdisciplinarity in telecommunications and networking. In *Networking and telecommunications: Concepts, methodologies, tools and applications* (pp. 33–40). Hershey, PA: IGI Global. doi:10.4018/978-1-60566-986-1.ch004

Priya, P. S., & Mathiyalagan, N. (2011). A study of the implementation status of two e-governance projects in land revenue administration in India. In M. Shareef, V. Kumar, U. Kumar, & Y. Dwivedi (Eds.), *Stakeholder adoption of e-government services: Driving and resisting factors* (pp. 214–230). Hershey, PA: IGI Global. doi:10.4018/978-1-60960-601-5.ch011

Prysby, C., & Prysby, N. (2000). Electronic mail, employee privacy and the workplace. In L. Janczewski (Ed.), *Internet and intranet security management: Risks and solutions* (pp. 251–270). Hershey, PA: IGI Global. doi:10.4018/978-1-878289-71-1.ch009

Prysby, C. L., & Prysby, N. D. (2003). Electronic mail in the public workplace: Issues of privacy and public disclosure. In G. Garson (Ed.), *Public information technology: Policy and management issues* (pp. 271–298). Hershey, PA: IGI Global. doi:10.4018/978-1-59140-060-8.ch012

Prysby, C. L., & Prysby, N. D. (2007). You have mail, but who is reading it? Issues of e-mail in the public workplace. In G. Garson (Ed.), *Modern public information technology systems: Issues and challenges* (pp. 312–336). Hershey, PA: IGI Global. doi:10.4018/978-1-59904-051-6.ch016

Radl, A., & Chen, Y. (2005). Computer security in electronic government: A state-local education information system. *International Journal of Electronic Government Research, 1*(1), 79–99. doi:10.4018/jegr.2005010105

Rahman, H. (2008). Information dynamics in developing countries. In C. Van Slyke (Ed.), *Information communication technologies: Concepts, methodologies, tools, and applications* (pp. 104–114). Hershey, PA: IGI Global. doi:10.4018/978-1-59904-949-6.ch008

Ramanathan, J. (2009). Adaptive IT architecture as a catalyst for network capability in government. In P. Saha (Ed.), *Advances in government enterprise architecture* (pp. 149–172). Hershey, PA: IGI Global. doi:10.4018/978-1-60566-068-4.ch007

Ramos, I., & Berry, D. M. (2006). Social construction of information technology supporting work. In M. Khosrow-Pour (Ed.), *Cases on information technology: Lessons learned* (Vol. 7, pp. 36–52). Hershey, PA: IGI Global. doi:10.4018/978-1-59140-673-0.ch003

Ray, D., Gulla, U., Gupta, M. P., & Dash, S. S. (2009). Interoperability and constituents of interoperable systems in public sector. In V. Weerakkody, M. Janssen, & Y. Dwivedi (Eds.), *Handbook of research on ICT-enabled transformational government: A global perspective* (pp. 175–195). Hershey, PA: IGI Global. doi:10.4018/978-1-60566-390-6.ch010

Reddick, C. G. (2007). E-government and creating a citizen-centric government: A study of federal government CIOs. In G. Garson (Ed.), *Modern public information technology systems: Issues and challenges* (pp. 143–165). Hershey, PA: IGI Global. doi:10.4018/978-1-59904-051-6.ch008

Reddick, C. G. (2010). Citizen-centric e-government. In C. Reddick (Ed.), *Homeland security preparedness and information systems: Strategies for managing public policy* (pp. 45–75). Hershey, PA: IGI Global. doi:10.4018/978-1-60566-834-5.ch002

Reddick, C. G. (2010). E-government and creating a citizen-centric government: A study of federal government CIOs. In C. Reddick (Ed.), *Homeland security preparedness and information systems: Strategies for managing public policy* (pp. 230–250). Hershey, PA: IGI Global. doi:10.4018/978-1-60566-834-5.ch012

Reddick, C. G. (2010). Perceived effectiveness of e-government and its usage in city governments: Survey evidence from information technology directors. In C. Reddick (Ed.), *Homeland security preparedness and information systems: Strategies for managing public policy* (pp. 213–229). Hershey, PA: IGI Global. doi:10.4018/978-1-60566-834-5.ch011

Reddick, C. G. (2012). Customer relationship management adoption in local governments in the United States. In S. Chhabra & M. Kumar (Eds.), *Strategic enterprise resource planning models for e-government: Applications and methodologies* (pp. 111–124). Hershey, PA: IGI Global. doi:10.4018/978-1-60960-863-7.ch008

Reeder, F. S., & Pandy, S. M. (2008). Identifying effective funding models for e-government. In A. Anttiroiko (Ed.), *Electronic government: Concepts, methodologies, tools, and applications* (pp. 1108–1138). Hershey, PA: IGI Global. doi:10.4018/978-1-59904-947-2.ch083

Riesco, D., Acosta, E., & Montejano, G. (2003). An extension to a UML activity graph from workflow. In L. Favre (Ed.), *UML and the unified process* (pp. 294–314). Hershey, PA: IRM Press. doi:10.4018/978-1-93177-744-5.ch015

Ritzhaupt, A. D., & Gill, T. G. (2008). A hybrid and novel approach to teaching computer programming in MIS curriculum. In S. Negash, M. Whitman, A. Woszczynski, K. Hoganson, & H. Mattord (Eds.), *Handbook of distance learning for real-time and asynchronous information technology education* (pp. 259–281). Hershey, PA: IGI Global. doi:10.4018/978-1-59904-964-9.ch014

Roche, E. M. (1993). International computing and the international regime. *Journal of Global Information Management, 1*(2), 33–44. doi:10.4018/jgim.1993040103

Rocheleau, B. (2007). Politics, accountability, and information management. In G. Garson (Ed.), *Modern public information technology systems: Issues and challenges* (pp. 35–71). Hershey, PA: IGI Global. doi:10.4018/978-1-59904-051-6.ch003

Rodrigues Filho, J. (2010). E-government in Brazil: Reinforcing dominant institutions or reducing citizenship? In C. Reddick (Ed.), *Politics, democracy and e-government: Participation and service delivery* (pp. 347–362). Hershey, PA: IGI Global. doi:10.4018/978-1-61520-933-0.ch021

Rodriguez, S. R., & Thorp, D. A. (2013). eLearning for industry: A case study of the project management process. In A. Benson, J. Moore, & S. Williams van Rooij (Eds.), Cases on educational technology planning, design, and implementation: A project management perspective (pp. 319-342). Hershey, PA: IGI Global. doi:10.4018/978-1-4666-4237-9.ch017

Roman, A. V. (2013). Delineating three dimensions of e-government success: Security, functionality, and transformation. In J. Gil-Garcia (Ed.), *E-government success factors and measures: Theories, concepts, and methodologies* (pp. 171–192). Hershey, PA: IGI Global. doi:10.4018/978-1-4666-4058-0.ch010

Ross, S. C., Tyran, C. K., & Auer, D. J. (2008). Up in smoke: Rebuilding after an IT disaster. In H. Nemati (Ed.), *Information security and ethics: Concepts, methodologies, tools, and applications* (pp. 3659–3675). Hershey, PA: IGI Global. doi:10.4018/978-1-59904-937-3.ch248

Ross, S. C., Tyran, C. K., Auer, D. J., Junell, J. M., & Williams, T. G. (2005). Up in smoke: Rebuilding after an IT disaster. *Journal of Cases on Information Technology*, *7*(2), 31–49. doi:10.4018/jcit.2005040103

Roy, J. (2008). Security, sovereignty, and continental interoperability: Canada's elusive balance. In T. Loendorf & G. Garson (Eds.), *Patriotic information systems* (pp. 153–176). Hershey, PA: IGI Global. doi:10.4018/978-1-59904-594-8.ch007

Rubeck, R. F., & Miller, G. A. (2009). vGOV: Remote video access to government services. In A. Scupola (Ed.), Cases on managing e-services (pp. 253-268). Hershey, PA: IGI Global. doi:10.4018/978-1-60566-064-6.ch017

Saekow, A., & Boonmee, C. (2011). The challenges of implementing e-government interoperability in Thailand: Case of official electronic correspondence letters exchange across government departments. In Y. Charalabidis (Ed.), *Interoperability in digital public services and administration: Bridging e-government and e-business* (pp. 40–61). Hershey, PA: IGI Global. doi:10.4018/978-1-61520-887-6.ch003

Saekow, A., & Boonmee, C. (2012). The challenges of implementing e-government interoperability in Thailand: Case of official electronic correspondence letters exchange across government departments. In *Digital democracy: Concepts, methodologies, tools, and applications* (pp. 1883–1905). Hershey, PA: IGI Global. doi:10.4018/978-1-4666-1740-7.ch094

Sagsan, M., & Medeni, T. (2012). Understanding "knowledge management (KM) paradigms" from social media perspective: An empirical study on discussion group for KM at professional networking site. In M. Cruz-Cunha, P. Gonçalves, N. Lopes, E. Miranda, & G. Putnik (Eds.), *Handbook of research on business social networking: Organizational, managerial, and technological dimensions* (pp. 738–755). Hershey, PA: IGI Global. doi:10.4018/978-1-61350-168-9.ch039

Sahi, G., & Madan, S. (2013). Information security threats in ERP enabled e-governance: Challenges and solutions. In *Enterprise resource planning: Concepts, methodologies, tools, and applications* (pp. 825–837). Hershey, PA: IGI Global. doi:10.4018/978-1-4666-4153-2.ch048

Sanford, C., & Bhattacherjee, A. (2008). IT implementation in a developing country municipality: A sociocognitive analysis. *International Journal of Technology and Human Interaction*, *4*(3), 68–93. doi:10.4018/jthi.2008070104

Schelin, S. H. (2003). E-government: An overview. In G. Garson (Ed.), *Public information technology: Policy and management issues* (pp. 120–138). Hershey, PA: IGI Global. doi:10.4018/978-1-59140-060-8.ch006

Schelin, S. H. (2004). Training for digital government. In A. Pavlichev & G. Garson (Eds.), *Digital government: Principles and best practices* (pp. 263–275). Hershey, PA: IGI Global. doi:10.4018/978-1-59140-122-3.ch017

Schelin, S. H. (2007). E-government: An overview. In G. Garson (Ed.), *Modern public information technology systems: Issues and challenges* (pp. 110–126). Hershey, PA: IGI Global. doi:10.4018/978-1-59904-051-6.ch006

Schelin, S. H., & Garson, G. (2004). Theoretical justification of critical success factors. In G. Garson & S. Schelin (Eds.), *IT solutions series: Humanizing information technology: Advice from experts* (pp. 4–15). Hershey, PA: CyberTech Publishing. doi:10.4018/978-1-59140-245-9.ch002

Scime, A. (2002). Information systems and computer science model curricula: A comparative look. In M. Dadashzadeh, A. Saber, & S. Saber (Eds.), *Information technology education in the new millennium* (pp. 146–158). Hershey, PA: IRM Press. doi:10.4018/978-1-931777-05-6.ch018

Scime, A. (2009). Computing curriculum analysis and development. In M. Khosrow-Pour (Ed.), *Encyclopedia of information science and technology* (2nd ed., pp. 667–671). Hershey, PA: IGI Global. doi:10.4018/978-1-60566-026-4.ch108

Scime, A., & Wania, C. (2008). Computing curricula: A comparison of models. In C. Van Slyke (Ed.), *Information communication technologies: Concepts, methodologies, tools, and applications* (pp. 1270–1283). Hershey, PA: IGI Global. doi:10.4018/978-1-59904-949-6.ch088

Seidman, S. B. (2009). An international perspective on professional software engineering credentials. In H. Ellis, S. Demurjian, & J. Naveda (Eds.), *Software engineering: Effective teaching and learning approaches and practices* (pp. 351–361). Hershey, PA: IGI Global. doi:10.4018/978-1-60566-102-5.ch018

Seifert, J. W. (2007). E-government act of 2002 in the United States. In A. Anttiroiko & M. Malkia (Eds.), *Encyclopedia of digital government* (pp. 476–481). Hershey, PA: IGI Global. doi:10.4018/978-1-59140-789-8.ch072

Seifert, J. W., & Relyea, H. C. (2008). E-government act of 2002 in the United States. In A. Anttiroiko (Ed.), *Electronic government: Concepts, methodologies, tools, and applications* (pp. 154–161). Hershey, PA: IGI Global. doi:10.4018/978-1-59904-947-2.ch013

Seufert, S. (2002). E-learning business models: Framework and best practice examples. In M. Raisinghani (Ed.), *Cases on worldwide e-commerce: Theory in action* (pp. 70–94). Hershey, PA: IGI Global. doi:10.4018/978-1-930708-27-3.ch004

Shareef, M. A., & Archer, N. (2012). E-government service development. In M. Shareef, N. Archer, & S. Dutta (Eds.), *E-government service maturity and development: Cultural, organizational and technological perspectives* (pp. 1–14). Hershey, PA: IGI Global. doi:10.4018/978-1-60960-848-4.ch001

Shareef, M. A., & Archer, N. (2012). E-government initiatives: Review studies on different countries. In M. Shareef, N. Archer, & S. Dutta (Eds.), *E-government service maturity and development: Cultural, organizational and technological perspectives* (pp. 40–76). Hershey, PA: IGI Global. doi:10.4018/978-1-60960-848-4.ch003

Shareef, M. A., Kumar, U., & Kumar, V. (2011). E-government development: Performance evaluation parameters. In M. Shareef, V. Kumar, U. Kumar, & Y. Dwivedi (Eds.), *Stakeholder adoption of e-government services: Driving and resisting factors* (pp. 197–213). Hershey, PA: IGI Global. doi:10.4018/978-1-60960-601-5.ch010

Shareef, M. A., Kumar, U., Kumar, V., & Niktash, M. (2012). Electronic-government vision: Case studies for objectives, strategies, and initiatives. In M. Shareef, N. Archer, & S. Dutta (Eds.), *E-government service maturity and development: Cultural, organizational and technological perspectives* (pp. 15–39). Hershey, PA: IGI Global. doi:10.4018/978-1-60960-848-4.ch002

Shukla, P., Kumar, A., & Anu Kumar, P. B.Anu Kumar P.B. (2013). Impact of national culture on business continuity management system implementation. *International Journal of Risk and Contingency Management*, *2*(3), 23–36. doi:10.4018/ijrcm.2013070102

Shulman, S. W. (2007). The federal docket management system and the prospect for digital democracy in U S rulemaking. In G. Garson (Ed.), *Modern public information technology systems: Issues and challenges* (pp. 166–184). Hershey, PA: IGI Global. doi:10.4018/978-1-59904-051-6.ch009

Simonovic, S. (2007). Problems of offline government in e-Serbia. In A. Anttiroiko & M. Malkia (Eds.), *Encyclopedia of digital government* (pp. 1342–1351). Hershey, PA: IGI Global. doi:10.4018/978-1-59140-789-8.ch205

Simonovic, S. (2008). Problems of offline government in e-Serbia. In A. Anttiroiko (Ed.), *Electronic government: Concepts, methodologies, tools, and applications* (pp. 2929–2942). Hershey, PA: IGI Global. doi:10.4018/978-1-59904-947-2.ch219

Singh, A. M. (2005). Information systems and technology in South Africa. In M. Khosrow-Pour (Ed.), *Encyclopedia of information science and technology* (pp. 1497–1502). Hershey, PA: IGI Global. doi:10.4018/978-1-59140-553-5.ch263

Singh, S., & Naidoo, G. (2005). Towards an e-government solution: A South African perspective. In W. Huang, K. Siau, & K. Wei (Eds.), *Electronic government strategies and implementation* (pp. 325–353). Hershey, PA: IGI Global. doi:10.4018/978-1-59140-348-7.ch014

Snoke, R., & Underwood, A. (2002). Generic attributes of IS graduates: An analysis of Australian views. In F. Tan (Ed.), *Advanced topics in global information management* (Vol. 1, pp. 370–384). Hershey, PA: IGI Global. doi:10.4018/978-1-930708-43-3.ch023

Sommer, L. (2006). Revealing unseen organizations in higher education: A study framework and application example. In A. Metcalfe (Ed.), *Knowledge management and higher education: A critical analysis* (pp. 115–146). Hershey, PA: IGI Global. doi:10.4018/978-1-59140-509-2.ch007

Song, H., Kidd, T., & Owens, E. (2011). Examining technological disparities and instructional practices in English language arts classroom: Implications for school leadership and teacher training. In L. Tomei (Ed.), *Online courses and ICT in education: Emerging practices and applications* (pp. 258–274). Hershey, PA: IGI Global. doi:10.4018/978-1-60960-150-8.ch020

Speaker, P. J., & Kleist, V. F. (2003). Using information technology to meet electronic commerce and MIS education demands. In A. Aggarwal (Ed.), *Web-based education: Learning from experience* (pp. 280–291). Hershey, PA: IGI Global. doi:10.4018/978-1-59140-102-5.ch017

Spitler, V. K. (2007). Learning to use IT in the workplace: Mechanisms and masters. In M. Mahmood (Ed.), *Contemporary issues in end user computing* (pp. 292–323). Hershey, PA: IGI Global. doi:10.4018/978-1-59140-926-7.ch013

Stellefson, M. (2011). Considerations for marketing distance education courses in health education: Five important questions to examine before development. In U. Demiray & S. Sever (Eds.), *Marketing online education programs: Frameworks for promotion and communication* (pp. 222–234). Hershey, PA: IGI Global. doi:10.4018/978-1-60960-074-7.ch014

Straub, D. W., & Loch, K. D. (2006). Creating and developing a program of global research. *Journal of Global Information Management*, *14*(2), 1–28. doi:10.4018/jgim.2006040101

Straub, D. W., Loch, K. D., & Hill, C. E. (2002). Transfer of information technology to the Arab world: A test of cultural influence modeling. In M. Dadashzadeh (Ed.), *Information technology management in developing countries* (pp. 92–134). Hershey, PA: IRM Press. doi:10.4018/978-1-931777-03-2.ch005

Straub, D. W., Loch, K. D., & Hill, C. E. (2003). Transfer of information technology to the Arab world: A test of cultural influence modeling. In F. Tan (Ed.), *Advanced topics in global information management* (Vol. 2, pp. 141–172). Hershey, PA: IGI Global. doi:10.4018/978-1-59140-064-6.ch009

Suki, N. M., Ramayah, T., Ming, M. K., & Suki, N. M. (2013). Factors enhancing employed job seekers intentions to use social networking sites as a job search tool. In A. Mesquita (Ed.), *User perception and influencing factors of technology in everyday life* (pp. 265–281). Hershey, PA: IGI Global. doi:10.4018/978-1-4666-1954-8.ch018

Suomi, R. (2006). Introducing electronic patient records to hospitals: Innovation adoption paths. In T. Spil & R. Schuring (Eds.), *E-health systems diffusion and use: The innovation, the user and the use IT model* (pp. 128–146). Hershey, PA: IGI Global. doi:10.4018/978-1-59140-423-1.ch008

Swim, J., & Barker, L. (2012). Pathways into a gendered occupation: Brazilian women in IT. *International Journal of Social and Organizational Dynamics in IT*, *2*(4), 34–51. doi:10.4018/ijsodit.2012100103

Tarafdar, M., & Vaidya, S. D. (2006). Adoption and implementation of IT in developing nations: Experiences from two public sector enterprises in India. In M. Khosrow-Pour (Ed.), *Cases on information technology planning, design and implementation* (pp. 208–233). Hershey, PA: IGI Global. doi:10.4018/978-1-59904-408-8.ch013

Tarafdar, M., & Vaidya, S. D. (2008). Adoption and implementation of IT in developing nations: Experiences from two public sector enterprises in India. In G. Garson & M. Khosrow-Pour (Eds.), *Handbook of research on public information technology* (pp. 905–924). Hershey, PA: IGI Global. doi:10.4018/978-1-59904-857-4.ch076

Thesing, Z. (2007). Zarina thesing, pumpkin patch. In M. Hunter (Ed.), *Contemporary chief information officers: Management experiences* (pp. 83–94). Hershey, PA: IGI Global. doi:10.4018/978-1-59904-078-3.ch007

Thomas, J. C. (2004). Public involvement in public administration in the information age: Speculations on the effects of technology. In M. Malkia, A. Anttiroiko, & R. Savolainen (Eds.), *eTransformation in governance: New directions in government and politics* (pp. 67–84). Hershey, PA: IGI Global. doi:10.4018/978-1-59140-130-8.ch004

Treiblmaier, H., & Chong, S. (2013). Trust and perceived risk of personal information as antecedents of online information disclosure: Results from three countries. In F. Tan (Ed.), *Global diffusion and adoption of technologies for knowledge and information sharing* (pp. 341–361). Hershey, PA: IGI Global. doi:10.4018/978-1-4666-2142-8.ch015

van Grembergen, W., & de Haes, S. (2008). IT governance in practice: Six case studies. In W. van Grembergen & S. De Haes (Eds.), *Implementing information technology governance: Models, practices and cases* (pp. 125–237). Hershey, PA: IGI Global. doi:10.4018/978-1-59904-924-3.ch004

van Os, G., Homburg, V., & Bekkers, V. (2013). Contingencies and convergence in European social security: ICT coordination in the back office of the welfare state. In M. Cruz-Cunha, I. Miranda, & P. Gonçalves (Eds.), *Handbook of research on ICTs and management systems for improving efficiency in healthcare and social care* (pp. 268–287). Hershey, PA: IGI Global. doi:10.4018/978-1-4666-3990-4.ch013

Velloso, A. B., Gassenferth, W., & Machado, M. A. (2012). Evaluating IBMEC-RJ's intranet usability using fuzzy logic. In M. Cruz-Cunha, P. Gonçalves, N. Lopes, E. Miranda, & G. Putnik (Eds.), *Handbook of research on business social networking: Organizational, managerial, and technological dimensions* (pp. 185–205). Hershey, PA: IGI Global. doi:10.4018/978-1-61350-168-9.ch010

Villablanca, A. C., Baxi, H., & Anderson, K. (2009). Novel data interface for evaluating cardiovascular outcomes in women. In A. Dwivedi (Ed.), *Handbook of research on information technology management and clinical data administration in healthcare* (pp. 34–53). Hershey, PA: IGI Global. doi:10.4018/978-1-60566-356-2.ch003

Villablanca, A. C., Baxi, H., & Anderson, K. (2011). Novel data interface for evaluating cardiovascular outcomes in women. In *Clinical technologies: Concepts, methodologies, tools and applications* (pp. 2094–2113). Hershey, PA: IGI Global. doi:10.4018/978-1-60960-561-2.ch806

Virkar, S. (2011). Information and communication technologies in administrative reform for development: Exploring the case of property tax systems in Karnataka, India. In J. Steyn, J. Van Belle, & E. Mansilla (Eds.), *ICTs for global development and sustainability: Practice and applications* (pp. 127–149). Hershey, PA: IGI Global. doi:10.4018/978-1-61520-997-2.ch006

Virkar, S. (2013). Designing and implementing e-government projects: Actors, influences, and fields of play. In S. Saeed & C. Reddick (Eds.), *Human-centered system design for electronic governance* (pp. 88–110). Hershey, PA: IGI Global. doi:10.4018/978-1-4666-3640-8.ch007

Wallace, A. (2009). E-justice: An Australian perspective. In A. Martínez & P. Abat (Eds.), *E-justice: Using information communication technologies in the court system* (pp. 204–228). Hershey, PA: IGI Global. doi:10.4018/978-1-59904-998-4.ch014

Wang, G. (2012). E-democratic administration and bureaucratic responsiveness: A primary study of bureaucrats' perceptions of the civil service e-mail box in Taiwan. In K. Kloby & M. D'Agostino (Eds.), *Citizen 2.0: Public and governmental interaction through web 2.0 technologies* (pp. 146–173). Hershey, PA: IGI Global. doi:10.4018/978-1-4666-0318-9.ch009

Wangpipatwong, S., Chutimaskul, W., & Papasratorn, B. (2011). Quality enhancing the continued use of e-government web sites: Evidence from e-citizens of Thailand. In V. Weerakkody (Ed.), *Applied technology integration in governmental organizations: New e-government research* (pp. 20–36). Hershey, PA: IGI Global. doi:10.4018/978-1-60960-162-1.ch002

Wedemeijer, L. (2006). Long-term evolution of a conceptual schema at a life insurance company. In M. Khosrow-Pour (Ed.), *Cases on database technologies and applications* (pp. 202–226). Hershey, PA: IGI Global. doi:10.4018/978-1-59904-399-9.ch012

Whybrow, E. (2008). Digital access, ICT fluency, and the economically disadvantages: Approaches to minimize the digital divide. In F. Tan (Ed.), *Global information technologies: Concepts, methodologies, tools, and applications* (pp. 1409–1422). Hershey, PA: IGI Global. doi:10.4018/978-1-59904-939-7.ch102

Whybrow, E. (2008). Digital access, ICT fluency, and the economically disadvantages: Approaches to minimize the digital divide. In C. Van Slyke (Ed.), *Information communication technologies: Concepts, methodologies, tools, and applications* (pp. 764–777). Hershey, PA: IGI Global. doi:10.4018/978-1-59904-949-6.ch049

Wickramasinghe, N., & Geisler, E. (2010). Key considerations for the adoption and implementation of knowledge management in healthcare operations. In M. Saito, N. Wickramasinghe, M. Fuji, & E. Geisler (Eds.), *Redesigning innovative healthcare operation and the role of knowledge management* (pp. 125–142). Hershey, PA: IGI Global. doi:10.4018/978-1-60566-284-8.ch009

Wickramasinghe, N., & Geisler, E. (2012). Key considerations for the adoption and implementation of knowledge management in healthcare operations. In *Organizational learning and knowledge: Concepts, methodologies, tools and applications* (pp. 1316–1328). Hershey, PA: IGI Global. doi:10.4018/978-1-60960-783-8.ch405

Wickramasinghe, N., & Goldberg, S. (2007). A framework for delivering m-health excellence. In L. Al-Hakim (Ed.), *Web mobile-based applications for healthcare management* (pp. 36–61). Hershey, PA: IRM Press. doi:10.4018/978-1-59140-658-7.ch002

Wickramasinghe, N., & Goldberg, S. (2008). Critical success factors for delivering m-health excellence. In N. Wickramasinghe & E. Geisler (Eds.), *Encyclopedia of healthcare information systems* (pp. 339–351). Hershey, PA: IGI Global. doi:10.4018/978-1-59904-889-5.ch045

Wyld, D. (2009). Radio frequency identification (RFID) technology. In J. Symonds, J. Ayoade, & D. Parry (Eds.), *Auto-identification and ubiquitous computing applications* (pp. 279–293). Hershey, PA: IGI Global. doi:10.4018/978-1-60566-298-5.ch017

Yaghmaei, F. (2010). Understanding computerised information systems usage in community health. In J. Rodrigues (Ed.), *Health information systems: Concepts, methodologies, tools, and applications* (pp. 1388–1399). Hershey, PA: IGI Global. doi:10.4018/978-1-60566-988-5.ch088

Yee, G., El-Khatib, K., Korba, L., Patrick, A. S., Song, R., & Xu, Y. (2005). Privacy and trust in e-government. In W. Huang, K. Siau, & K. Wei (Eds.), *Electronic government strategies and implementation* (pp. 145–190). Hershey, PA: IGI Global. doi:10.4018/978-1-59140-348-7.ch007

Yeh, S., & Chu, P. (2010). Evaluation of e-government services: A citizen-centric approach to citizen e-complaint services. In C. Reddick (Ed.), *Citizens and e-government: Evaluating policy and management* (pp. 400–417). Hershey, PA: IGI Global. doi:10.4018/978-1-61520-931-6.ch022

Young-Jin, S., & Seang-tae, K. (2008). E-government concepts, measures, and best practices. In A. Anttiroiko (Ed.), *Electronic government: Concepts, methodologies, tools, and applications* (pp. 32–57). Hershey, PA: IGI Global. doi:10.4018/978-1-59904-947-2.ch004

Yun, H. J., & Opheim, C. (2012). New technology communication in American state governments: The impact on citizen participation. In K. Bwalya & S. Zulu (Eds.), *Handbook of research on e-government in emerging economies: Adoption, e-participation, and legal frameworks* (pp. 573–590). Hershey, PA: IGI Global. doi:10.4018/978-1-4666-0324-0.ch029

Zhang, N., Guo, X., Chen, G., & Chau, P. Y. (2011). User evaluation of e-government systems: A Chinese cultural perspective. In F. Tan (Ed.), *International enterprises and global information technologies: Advancing management practices* (pp. 63–84). Hershey, PA: IGI Global. doi:10.4018/978-1-60960-605-3.ch004

Zuo, Y., & Hu, W. (2011). Trust-based information risk management in a supply chain network. In J. Wang (Ed.), *Supply chain optimization, management and integration: Emerging applications* (pp. 181–196). Hershey, PA: IGI Global. doi:10.4018/978-1-60960-135-5.ch013

Compilation of References

Adams, F., & Steadman, A. (2004). Intentional Action in Ordinary Language: Core Concept or Pragmatic Understanding? *Analysis*, *64*(2), 173–181. doi:10.1093/analys/64.2.173

Adeyemo, O. (2013). The nationwide health information network: A biometric approach to prevent medical identity theft. In *User-driven healthcare: Concepts, methodologies, tools, and applications* (pp. 1636–1649). Hershey, PA: IGI Global. doi:10.4018/978-1-4666-2770-3.ch081

Adler, M., & Henman, P. (2009). Justice beyond the courts: The implications of computerisation for procedural justice in social security. In A. Martínez & P. Abat (Eds.), *E-justice: Using information communication technologies in the court system* (pp. 65–86). Hershey, PA: IGI Global. doi:10.4018/978-1-59904-998-4.ch005

Aflalo, E., & Gabay, E. (2013). An information system for coping with student dropout. In L. Tomei (Ed.), *Learning tools and teaching approaches through ICT advancements* (pp. 176–187). Hershey, PA: IGI Global. doi:10.4018/978-1-4666-2017-9.ch016

Ahmed, M. A., Janssen, M., & van den Hoven, J. (2012). Value sensitive transfer (VST) of systems among countries: Towards a framework. *International Journal of Electronic Government Research*, *8*(1), 26–42. doi:10.4018/jegr.2012010102

Aida, Y. (2012). *Gunslinger Girl* (Vols. 1-15). Tokyo: ASCII Media Works.

Aikins, S. K. (2008). Issues and trends in internet-based citizen participation. In G. Garson & M. Khosrow-Pour (Eds.), *Handbook of research on public information technology* (pp. 31–40). Hershey, PA: IGI Global. doi:10.4018/978-1-59904-857-4.ch004

Aikins, S. K. (2009). A comparative study of municipal adoption of internet-based citizen participation. In C. Reddick (Ed.), *Handbook of research on strategies for local e-government adoption and implementation: Comparative studies* (pp. 206–230). Hershey, PA: IGI Global. doi:10.4018/978-1-60566-282-4.ch011

Aikins, S. K. (2012). Improving e-government project management: Best practices and critical success factors. In *Digital democracy: Concepts, methodologies, tools, and applications* (pp. 1314–1332). Hershey, PA: IGI Global. doi:10.4018/978-1-4666-1740-7.ch065

Akabawi, M. S. (2011). Ghabbour group ERP deployment: Learning from past technology failures. In E. Business Research and Case Center (Ed.), Cases on business and management in the MENA region: New trends and opportunities (pp. 177-203). Hershey, PA: IGI Global. doi:10.4018/978-1-60960-583-4.ch012

Akabawi, M. S. (2013). Ghabbour group ERP deployment: Learning from past technology failures. In *Industrial engineering: Concepts, methodologies, tools, and applications* (pp. 933–958). Hershey, PA: IGI Global. doi:10.4018/978-1-4666-1945-6.ch051

Akbulut, A. Y., & Motwani, J. (2008). Integration and information sharing in e-government. In G. Putnik & M. Cruz-Cunha (Eds.), *Encyclopedia of networked and virtual organizations* (pp. 729–734). Hershey, PA: IGI Global. doi:10.4018/978-1-59904-885-7.ch096

Akers, E. J. (2008). Technology diffusion in public administration. In G. Garson & M. Khosrow-Pour (Eds.), *Handbook of research on public information technology* (pp. 339–348). Hershey, PA: IGI Global. doi:10.4018/978-1-59904-857-4.ch033

Aladwani, A. M. (2002). Organizational actions, computer attitudes and end-user satisfaction in public organizations: An empirical study. In C. Snodgrass & E. Szewczak (Eds.), *Human factors in information systems* (pp. 153–168). Hershey, PA: IRM Press. doi:10.4018/978-1-931777-10-0.ch012

Aladwani, A. M. (2002). Organizational actions, computer attitudes, and end-user satisfaction in public organizations: An empirical study. *Journal of Organizational and End User Computing*, *14*(1), 42–49. doi:10.4018/joeuc.2002010104

Alekseeva, I. Y., Arshinov, V. I., & Chekletsov, V. (2013). "Techno-People" against "Post-People": NBIKS — revolution and the future of man. *Philosophical Issues*, *3*, 12–21.

Alemi, M., Meghdari, A., & Ghazisaedy, M. (2014). Employing Humanoid Robots for Teaching English Language in Iranian Junior High-Schools. *International Journal of Humanoid Robotics*.

Aliane, N. (2010). A Matlab/Simulink-based interactive module for servo systems learning. *Education. IEEE Transactions on Education*, *53*(2), 265–271. doi:10.1109/TE.2009.2014468

Ali, M. L., Islam, A. R., Anom, A. S., & Amziah, N. A. M. (2008). GUI for PC auto-shutdown. *Computers & Electrical Engineering an International Journal*, *34*(3), 257–262. doi:10.1016/j.compeleceng.2007.02.009

Allen, B., Juillet, L., Paquet, G., & Roy, J. (2005). E-government and private-public partnerships: Relational challenges and strategic directions. In M. Khosrow-Pour (Ed.), *Practicing e-government: A global perspective* (pp. 364–382). Hershey, PA: IGI Global. doi:10.4018/978-1-59140-637-2.ch016

Al-Shafi, S. (2008). Free wireless internet park services: An investigation of technology adoption in Qatar from a citizens perspective. *Journal of Cases on Information Technology*, *10*(3), 21–34. doi:10.4018/jcit.2008070103

Al-Shafi, S., & Weerakkody, V. (2009). Implementing free wi-fi in public parks: An empirical study in Qatar. *International Journal of Electronic Government Research*, *5*(3), 21–35. doi:10.4018/jegr.2009070102

Alshawaf, A., & Knalil, O. E. (2008). IS success factors and IS organizational impact: Does ownership type matter in Kuwait? *International Journal of Enterprise Information Systems*, *4*(2), 13–33. doi:10.4018/jeis.2008040102

amazon.com. (2017). *Joy For All Silver Cat With White Mitts*. Retrieved from https://www.amazon.com/Joy-All-Silver-White-Mitts/dp/B017JQQ00Q

Ambali, A. R. (2009). Digital divide and its implication on Malaysian e-government: Policy initiatives. In H. Rahman (Ed.), *Social and political implications of data mining: Knowledge management in e-government* (pp. 267–287). Hershey, PA: IGI Global. doi:10.4018/978-1-60566-230-5.ch016

Amoretti, F. (2007). Digital international governance. In A. Anttiroiko & M. Malkia (Eds.), *Encyclopedia of digital government* (pp. 365–370). Hershey, PA: IGI Global. doi:10.4018/978-1-59140-789-8.ch056

Amoretti, F. (2008). Digital international governance. In A. Anttiroiko (Ed.), *Electronic government: Concepts, methodologies, tools, and applications* (pp. 688–696). Hershey, PA: IGI Global. doi:10.4018/978-1-59904-947-2.ch058

Amoretti, F. (2008). E-government at supranational level in the European Union. In A. Anttiroiko (Ed.), *Electronic government: Concepts, methodologies, tools, and applications* (pp. 1047–1055). Hershey, PA: IGI Global. doi:10.4018/978-1-59904-947-2.ch079

Amoretti, F. (2008). E-government regimes. In A. Anttiroiko (Ed.), *Electronic government: Concepts, methodologies, tools, and applications* (pp. 3846–3856). Hershey, PA: IGI Global. doi:10.4018/978-1-59904-947-2.ch280

Amoretti, F. (2009). Electronic constitution: A Braudelian perspective. In F. Amoretti (Ed.), *Electronic constitution: Social, cultural, and political implications* (pp. 1–19). Hershey, PA: IGI Global. doi:10.4018/978-1-60566-254-1.ch001

Amoretti, F., & Musella, F. (2009). Institutional isomorphism and new technologies. In M. Khosrow-Pour (Ed.), *Encyclopedia of information science and technology* (2nd ed., pp. 2066–2071). Hershey, PA: IGI Global. doi:10.4018/978-1-60566-026-4.ch325

Andersen, K. V., & Henriksen, H. Z. (2007). E-government research: Capabilities, interaction, orientation, and values. In D. Norris (Ed.), *Current issues and trends in e-government research* (pp. 269–288). Hershey, PA: CyberTech Publishing. doi:10.4018/978-1-59904-283-1.ch013

Anderson, K. V., & Henriksen, H. Z. (2005). The first leg of e-government research: Domains and application areas 19982003. *International Journal of Electronic Government Research*, *1*(4), 26–44. doi:10.4018/jegr.2005100102

Anttiroiko, A. (2009). Democratic e-governance. In M. Khosrow-Pour (Ed.), *Encyclopedia of information science and technology* (2nd ed., pp. 990–995). Hershey, PA: IGI Global. doi:10.4018/978-1-60566-026-4.ch158

Asaka, M. (Director). (2003). Gunslinger Girl [Television series]. Japan: Madhouse Studios.

Asimov, I. (2013). Ja, robot. (Z. A. Królicki, Trans.). Poznań: Dom Wydawniczy Rebis. (Original work published from: 1950).

Asimov, I., & Silverberg, R. (1998). The Positronic Man (P. Hermanowski, Trans.). Poznań: Dom Wydawniczy Rebis. (Original work published from 1992).

Association, I. R. (2010). Networking and telecommunications: Concepts, methodologies, tools and applications (Vols. 1–3). Hershey, PA: IGI Global. doi:10.4018/978-1-60566-986-1

Association, I. R. (2010). Web-based education: Concepts, methodologies, tools and applications (Vols. 1–3). Hershey, PA: IGI Global. doi:10.4018/978-1-61520-963-7

Atmanspacher, H. (2012). Dual-aspect monism à la Pauli and Jung. *Journal of Consciousness Studies*, *19*, 96–120.

Aubrey, J. S., & Harrison, K. (2004). The gender-role content of childrens favorite television programs and its links to their gender-related perceptions. *Media Psychology*, *6*(2), 111–146. doi:10.1207/s1532785xmep0602_1

Avner, T. (2015). Nå åpner hotellet som drives av menneskeroboter. *Aftenposten*. Retrieved from http://www.aftenposten.no/article/ap-7899510.html

Ayer, A. J. (1952). *Language, truth and logic* (2nd ed.). New York: Dover.

Baeva, L. V. (2014). *E-culture*. In M. Khosrow-Pour (Ed.), *Encyclopedia of Information Science and Technology* (3rd ed., pp. 6847–6854). USA: Information Resources Management Association.

Bailey, A., & Richards, B. (2013). Horgan and Tienson on phenomenology and intentionality. *Philosophical Studies*, *167*(2), 313–326. doi:10.1007/s11098-013-0089-7

Baker, K., & Raney, A. A. (2007). Equally super? Gender-role stereotyping of superheroes in childrens animated programs. *Mass Communication & Society*, *10*(1), 25–41. doi:10.1080/15205430709337003

Baker, P. M., Bell, A., & Moon, N. W. (2009). Accessibility issues in municipal wireless networks. In C. Reddick (Ed.), *Handbook of research on strategies for local e-government adoption and implementation: Comparative studies* (pp. 569–588). Hershey, PA: IGI Global. doi:10.4018/978-1-60566-282-4.ch030

Baryshnikov, P. N. (2014). The typology of immortality in the theoretical field of French transhumanism. *Philosophical problems of IT and cyberspace 1*(7), 98-127.

Becker, S. A., Keimer, R., & Muth, T. (2012). A case on university and community collaboration: The sci-tech entrepreneurial training services (ETS) program. In Regional development: Concepts, methodologies, tools, and applications (pp. 947-969). Hershey, PA: IGI Global. doi:10.4018/978-1-4666-0882-5.ch507

Becker, S. A., Keimer, R., & Muth, T. (2010). A case on university and community collaboration: The sci-tech entrepreneurial training services (ETS) program. In S. Becker & R. Niebuhr (Eds.), *Cases on technology innovation: Entrepreneurial successes and pitfalls* (pp. 68–90). Hershey, PA: IGI Global. doi:10.4018/978-1-61520-609-4.ch003

Benjamin, W. (2008). Attested author. In M. Jennings, W. B. Doherty, & T. Y. Levin (Eds.), *The work of art in the age of its technological reproducibility and other writings on media* (pp. 171–172). Cambridge: Harvard University Press.

Bernardi, R. (2012). Information technology and resistance to public sector reforms: A case study in Kenya. In T. Papadopoulos & P. Kanellis (Eds.), *Public sector reform using information technologies: Transforming policy into practice* (pp. 59–78). Hershey, PA: IGI Global. doi:10.4018/978-1-60960-839-2.ch004

Bernardi, R. (2013). Information technology and resistance to public sector reforms: A case study in Kenya. In *User-driven healthcare: Concepts, methodologies, tools, and applications* (pp. 14–33). Hershey, PA: IGI Global. doi:10.4018/978-1-4666-2770-3.ch002

Bethel, C. (2011). Sharing a Secret: Interactions between a Child, Robot, and Adult. *Paper presented at the 6th ACM/IEEE International Conference on Human-Robot Interaction*, Lausanne, Switzerland. doi:10.1109/ICSMC.2011.6084051

Bethel, C., Eakin, D., Anreddy, S., Stuart, J., & Carruth, D. (2013). Eyewitnesses are misled by human but not Robot Interviewers. In *Proceedings of the 2013 8Th ACM/IEEE International Conference on Human-Robot Interaction (HRI)* (pp. 25-32).

Bickmore, T., Caruso, L., Clough-Gorr, K., & Heeren, T. (2005). Its just like you talk to a friend relational agents for older adults. *Interacting with Computers*, *17*(6), 711–735. doi:10.1016/j.intcom.2005.09.002

Bijker, W. E. (2001). Understanding technological culture through a constructivist view of science, technology, and society. *Visions of STS: Counterpoints in science, technology and society studies*.

Block, N. (1980). *Readings in philosophy of psychology*. London: Methuen.

Block, N. (1995). On a confusion about a function of consciousness. *Behavioral and Brain Sciences*, *18*(02), 265–266. doi:10.1017/S0140525X00038188

Blomkamp, N. (Producer). & Blomkamp, N. (Director). (2015). *Chappie* [Motion picture]. US: Media Rights Capital.

Bogost, I. (2007). *Persuasive games: The expressive power of videogames*. Cambridge, MA: MIT Press.

Bolívar, M. P., Pérez, M. D., & Hernández, A. M. (2012). Municipal e-government services in emerging economies: The Latin-American and Caribbean experiences. In Y. Chen & P. Chu (Eds.), *Electronic governance and cross-boundary collaboration: Innovations and advancing tools* (pp. 198–226). Hershey, PA: IGI Global. doi:10.4018/978-1-60960-753-1.ch011

Bolter, J. D. (1984). *Turing's man: Western culture in the computer age*. North Carolina: The University of North Carolina Press.

Borycki, E. M., & Kushniruk, A. W. (2010). Use of clinical simulations to evaluate the impact of health information systems and ubiquitous computing devices upon health professional work. In S. Mohammed & J. Fiaidhi (Eds.), *Ubiquitous health and medical informatics: The ubiquity 2.0 trend and beyond* (pp. 552–573). Hershey, PA: IGI Global. doi:10.4018/978-1-61520-777-0.ch026

Borycki, E. M., & Kushniruk, A. W. (2011). Use of clinical simulations to evaluate the impact of health information systems and ubiquitous computing devices upon health professional work. In *Clinical technologies: Concepts, methodologies, tools and applications* (pp. 532–553). Hershey, PA: IGI Global. doi:10.4018/978-1-60960-561-2.ch220

Boyd, R. (1993). Metaphor and theory change: What is 'metaphor' a metaphor for? In A. Ortony (Ed.), *Metaphor and thought* (2nd ed., pp. 481–532). Cambridge, UK: Cambridge University Press. doi:10.1017/CBO9781139173865.023

Breazeal, C. (2011). Social Robots for Health Applications. In *Proceedings of the 2011 Annual International Conference of The IEEE Engineering In Medicine And Biology Society*. doi:10.1109/IEMBS.2011.6091328

Breazeal, C. (2003). Towards Sociable Robots. *Robotics and Autonomous Systems*, *42*(3-4), 167–175. doi:10.1016/S0921-8890(02)00373-1

Breazeal, C., Brooks, A., Gray, J., Hoffman, G., Kidd, C., Lee, H., & Chilongo, D. et al. (2004). Tutelage and collaboration for humanoid robots. *International Journal of Humanoid Robots*, *1*(2), 315–348. doi:10.1142/S0219843604000150

Brefczynski-Lewis, J. A., Lutz, A., Schaefer, H. S., Levinson, D. B., & Davidson, R. J. (2007). Neural correlates of attentional expertise in long-term meditation practitioners. *Proceedings of the National Academy of Sciences of the United States of America*, *104*(27), 11483–11488. doi:10.1073/pnas.0606552104 PMID:17596341

Brooks, R. A. (1991, August). Intelligence without reason. In *Proceedings of the 12th International Joint Conference on Artificial Intelligence IJCAI-91* (*Vol. 1*, pp. 569-595).

Brooks, R. A. (1999). *Cambrian intelligence: The early history of the new AI*. Cambridge, MA: MIT Press.

Brooks, R., & Flynn, A. M. (1989). Fast, cheap and out of control: A robot invasion of the solar system. *Journal of the British Interplanetary Society*, *42*, 478–485. Retrieved from http://people.csail.mit.edu/brooks/papers/fast-cheap.pdf

Brunner, C., & Fritsch, J. (2011). FCJ-124 Interactive environments as fields of transduction. *The Fibreculture Journal, Issue 18,* 2011. Retrieved from http://eighteen.fibreculturejournal.org/2011/10/09/fcj-124-interactive-environments-as-fields-of-transduction

Brzechwa, J. (1990). *Akademia Pana Kleksa.* Wrocław: Siedmioróg.

Buchan, J. (2011). Developing a dynamic and responsive online learning environment: A case study of a large Australian university. In B. Czerkawski (Ed.), *Free and open source software for e-learning: Issues, successes and challenges* (pp. 92–109). Hershey, PA: IGI Global. doi:10.4018/978-1-61520-917-0.ch006

Buckingham, D. (1996). *Moving images: Understanding children's emotional responses to television.* Manchester: Manchester University Press.

Buckingham, D. (2013). *Media education: Literacy, learning, and contemporary culture.* Cambridge: Polity Press.

Buckingham, D. (Ed.). (2005). *The media literacy of children and young people.* London: University of London.

Buckley, D. (2013). *Cycler / ReCycler Presentation Robot. Davidbuckley.net.* Retrieved from http://davidbuckley.net/FR/Cycler/CyclerPresentationRobot.htm

Buenger, A. W. (2008). Digital convergence and cybersecurity policy. In G. Garson & M. Khosrow-Pour (Eds.), *Handbook of research on public information technology* (pp. 395–405). Hershey, PA: IGI Global. doi:10.4018/978-1-59904-857-4.ch038

Burge, T. (1998). Computer Proof, Apriori Knowledge, and Other Minds The Sixth Philosophical Perspectives Lecture. *Noûs (Detroit, Mich.), 32*(S12), 1–37. doi:10.1111/0029-4624.32.s12.1

Burn, J. M., & Loch, K. D. (2002). The societal impact of world wide web - Key challenges for the 21st century. In A. Salehnia (Ed.), *Ethical issues of information systems* (pp. 88–106). Hershey, PA: IRM Press. doi:10.4018/978-1-931777-15-5.ch007

Burn, J. M., & Loch, K. D. (2003). The societal impact of the world wide web-Key challenges for the 21st century. In M. Khosrow-Pour (Ed.), *Advanced topics in information resources management* (Vol. 2, pp. 32–51). Hershey, PA: IGI Global. doi:10.4018/978-1-59140-062-2.ch002

Butler, J. (1990). *Gender trouble: Feminism and the subversion of identity.* New York, NY: Routledge.

Bwalya, K. J., Du Plessis, T., & Rensleigh, C. (2012). The "quicksilver initiatives" as a framework for e-government strategy design in developing economies. In K. Bwalya & S. Zulu (Eds.), *Handbook of research on e-government in emerging economies: Adoption, e-participation, and legal frameworks* (pp. 605–623). Hershey, PA: IGI Global. doi:10.4018/978-1-4666-0324-0.ch031

Byerly, K. M., & Ross, K. (2008). *Women and media: A critical introduction.* Malden, MA: Blackwell.

Cabibihan, J., Javed, H., Ang, M., & Aljunied, S. (2013). Why Robots? A Survey on the Roles and Benefits of Social Robots in the Therapy of Children with Autism. *International Journal Of Social Robotics*, *5*(4), 593–618. doi:10.1007/s12369-013-0202-2

Cabotaje, C. E., & Alampay, E. A. (2013). Social media and citizen engagement: Two cases from the Philippines. In S. Saeed & C. Reddick (Eds.), *Human-centered system design for electronic governance* (pp. 225–238). Hershey, PA: IGI Global. doi:10.4018/978-1-4666-3640-8.ch013

Calo, J. (2016, March). Robots in American Law. *Paper presented to the Robert Strauss Centre*, University of Texas, Austin.

Cameron, J. (1984). *The Terminator script*. Retrieved from www.script-o-rama.com/movie_scripts/t/terminator-script-transcript-arnold-schwartzenegger.html

Camillo, A., Di Pietro, L., Di Virgilio, F., & Franco, M. (2013). Work-groups conflict at PetroTech-Italy, S.R.L.: The influence of culture on conflict dynamics. In B. Christiansen, E. Turkina, & N. Williams (Eds.), *Cultural and technological influences on global business* (pp. 272–289). Hershey, PA: IGI Global. doi:10.4018/978-1-4666-3966-9.ch015

Campbell, K. K. (1970). *Body and mind*. New York, NY: Doubleday. doi:10.1007/978-1-349-00678-6

Capra, E., Francalanci, C., & Marinoni, C. (2008). Soft success factors for m-government. In A. Anttiroiko (Ed.), *Electronic government: Concepts, methodologies, tools, and applications* (pp. 1213–1233). Hershey, PA: IGI Global. doi:10.4018/978-1-59904-947-2.ch089

Cartelli, A. (2009). The implementation of practices with ICT as a new teaching-learning paradigm. In A. Cartelli & M. Palma (Eds.), *Encyclopedia of information communication technology* (pp. 413–417). Hershey, PA: IGI Global. doi:10.4018/978-1-59904-845-1.ch055

Chabot, P. (2003). *The philosophy of Simondon between technology and individuation* [Kindle version].

Chalmers, D. (2003). The Nature of Narrow Content. *Philosophical Issues*, *13*(1), 46–66. doi:10.1111/1533-6077.00004

Chalmers, D. J. (1996). *The conscious mind*. New York, NY: Oxford University Press.

Chalmers, D. J. (2002). Consciousness and its place in nature. In D. J. Chalmers (Ed.), *Philosophy of mind: Classical and contemporary readings* (pp. 247–272). Oxford, UK: Oxford University Press.

Charalabidis, Y., Lampathaki, F., & Askounis, D. (2010). Investigating the landscape in national interoperability frameworks. *International Journal of E-Services and Mobile Applications*, *2*(4), 28–41. doi:10.4018/jesma.2010100103

Charalabidis, Y., Lampathaki, F., & Askounis, D. (2012). Investigating the landscape in national interoperability frameworks. In A. Scupola (Ed.), *Innovative mobile platform developments for electronic services design and delivery* (pp. 218–231). Hershey, PA: IGI Global. doi:10.4018/978-1-4666-1568-7.ch013

Charmaz, K., & Smith, J. (2003). Grounded theory. *Qualitative psychology: A practical guide to research methods*, 81-110.

Chen, I. (2005). Distance education associations. In C. Howard, J. Boettcher, L. Justice, K. Schenk, P. Rogers, & G. Berg (Eds.), *Encyclopedia of distance learning* (pp. 599–612). Hershey, PA: IGI Global. doi:10.4018/978-1-59140-555-9.ch087

Chen, I. (2008). Distance education associations. In L. Tomei (Ed.), *Online and distance learning: Concepts, methodologies, tools, and applications* (pp. 562–579). Hershey, PA: IGI Global. doi:10.4018/978-1-59904-935-9.ch048

Chen, Y. (2008). Managing IT outsourcing for digital government. In A. Anttiroiko (Ed.), *Electronic government: Concepts, methodologies, tools, and applications* (pp. 3107–3114). Hershey, PA: IGI Global. doi:10.4018/978-1-59904-947-2.ch229

Chen, Y., & Dimitrova, D. V. (2006). Electronic government and online engagement: Citizen interaction with government via web portals. *International Journal of Electronic Government Research*, *2*(1), 54–76. doi:10.4018/jegr.2006010104

Chen, Y., & Knepper, R. (2005). Digital government development strategies: Lessons for policy makers from a comparative perspective. In W. Huang, K. Siau, & K. Wei (Eds.), *Electronic government strategies and implementation* (pp. 394–420). Hershey, PA: IGI Global. doi:10.4018/978-1-59140-348-7.ch017

Chen, Y., & Knepper, R. (2008). Digital government development strategies: Lessons for policy makers from a comparative perspective. In H. Rahman (Ed.), *Developing successful ICT strategies: Competitive advantages in a global knowledge-driven society* (pp. 334–356). Hershey, PA: IGI Global. doi:10.4018/978-1-59904-654-9.ch017

Cherian, E. J., & Ryan, T. W. (2014). Incongruent needs: Why differences in the iron-triangle of priorities make health information technology adoption and use difficult. In C. El Morr (Ed.), *Research perspectives on the role of informatics in health policy and management* (pp. 209–221). Hershey, PA: IGI Global. doi:10.4018/978-1-4666-4321-5.ch012

Chesher, C. (2017). Toy robots on YouTube: Consumption and peer production at the robotic moment. *Convergence*.

Chimba, M., & Kitzinger, J. (2010). Bimbo or boffin? Women in science: An analysis of media representations and how female scientists negotiate cultural contradictions. *Public Understanding of Science (Bristol, England)*, *19*(5), 609–624. PMID:21553601

Cho, H. J., & Hwang, S. (2010). Government 2.0 in Korea: Focusing on e-participation services. In C. Reddick (Ed.), *Politics, democracy and e-government: Participation and service delivery* (pp. 94–114). Hershey, PA: IGI Global. doi:10.4018/978-1-61520-933-0.ch006

Chorus, C., & Timmermans, H. (2010). Ubiquitous travel environments and travel control strategies: Prospects and challenges. In M. Wachowicz (Ed.), *Movement-aware applications for sustainable mobility: Technologies and approaches* (pp. 30–51). Hershey, PA: IGI Global. doi:10.4018/978-1-61520-769-5.ch003

Chuanshen, R. (2007). E-government construction and China's administrative litigation act. In A. Anttiroiko & M. Malkia (Eds.), *Encyclopedia of digital government* (pp. 507–510). Hershey, PA: IGI Global. doi:10.4018/978-1-59140-789-8.ch077

Chun, W. H. (2008). *Control and freedom: Power and paranoia in the age of fiber optics.* Cambridge, MA: MIT Press.

Ciaghi, A., & Villafiorita, A. (2012). Law modeling and BPR for public administration improvement. In K. Bwalya & S. Zulu (Eds.), *Handbook of research on e-government in emerging economies: Adoption, e-participation, and legal frameworks* (pp. 391–410). Hershey, PA: IGI Global. doi:10.4018/978-1-4666-0324-0.ch019

Ciaramitaro, B. L., & Skrocki, M. (2012). mHealth: Mobile healthcare. In B. Ciaramitaro (Ed.), Mobile technology consumption: Opportunities and challenges (pp. 99-109). Hershey, PA: IGI Global. doi:10.4018/978-1-61350-150-4.ch007

Clynes, M. E., & Kline, N. S. (1960). *Cyborgs and space.* In *Astronautics*, 26-27, 74-75. Retrieved from http://web.mit.edu/digitalapollo/Documents/Chapter1/cyborgs.pdf

Clynes, M., & Kline, N. (1960). Cyborgs and Space. *Astronautics,* (September): 26–27, 74–76.

Coady, C. (1992). *Testimony.* Oxford: Clarendon Press.

Columbus, C., Petersen, W., Katz, G., Mark, L., Miller, N., Radcliffe, M., & Bernathan, M. (Producers) & Columbus C. (Director). (1999). *Bicentennial Man* [Motion picture]. United States & Canada: Buena Vista Pictures, Columbia TriStar, Sony Pictures Entertainment Inc.

Comite, U. (2012). Innovative processes and managerial effectiveness of e-procurement in healthcare. In A. Manoharan & M. Holzer (Eds.), *Active citizen participation in e-government: A global perspective* (pp. 206–229). Hershey, PA: IGI Global. doi:10.4018/978-1-4666-0116-1.ch011

Cook, N. D. (2000). On defining awareness and consciousness: The importance of the neuronal membrane. In *Proceedings of the Tokyo-99 Conference on Consciousness.* Singapore: World Scientific.

Cook, N. D. (2002a). Bihemispheric language: How the two hemispheres collaborate in the processing of language. In *Proceedings of the British Academy* (Vol. 106).

Cook, N. D. (2002b). *Tone of voice and mind: The connections between intonation, emotion, cognition and consciousness.* Amsterdam, Netherlands: John Benjamins. doi:10.1075/aicr.47

Cook, N. D. (2008). The neuron-level phenomena underlying cognition and consciousness: Synaptic activity and the action potential. *Neuroscience, 153*(3), 556–570. doi:10.1016/j.neuroscience.2008.02.042 PMID:18406536

Cordella, A. (2013). E-government success: How to account for ICT, administrative rationalization, and institutional change. In J. Gil-Garcia (Ed.), *E-government success factors and measures: Theories, concepts, and methodologies* (pp. 40–51). Hershey, PA: IGI Global. doi:10.4018/978-1-4666-4058-0.ch003

Cortazzi, M. (2014). *Narrative analysis* (Vol. 12). Routledge.

Craig, A. D. (2003). Interoception: The sense of the physiological condition of the body. *Current Opinion in Neurobiology*, *13*(4), 500–505. doi:10.1016/S0959-4388(03)00090-4 PMID:12965300

Craig, J. (2006). *Robótica tercera edición*. México: Pearson Educación.

Cropf, R. A. (2009). ICT and e-democracy. In M. Khosrow-Pour (Ed.), *Encyclopedia of information science and technology* (2nd ed., pp. 1789–1793). Hershey, PA: IGI Global. doi:10.4018/978-1-60566-026-4.ch281

Cropf, R. A. (2009). The virtual public sphere. In M. Pagani (Ed.), *Encyclopedia of multimedia technology and networking* (2nd ed., pp. 1525–1530). Hershey, PA: IGI Global. doi:10.4018/978-1-60566-014-1.ch206

Cummings, e. (1994). *Next to of course god America. 100 selected poems* (p. 31). New York: Grove.

Custom Robotic Wildlife. (2017). Retrieved from http://customroboticwildlife.com

Cvencek, D., Meltzoff, A., & Greenwald, A. (2011). Math-Gender Stereotypes in Elementary School Children. *Child Development*, *82*(3), 766–779. doi:10.1111/j.1467-8624.2010.01529.x PMID:21410915

D'Abundo, M. L. (2013). Electronic health record implementation in the United States healthcare industry: Making the process of change manageable. In V. Wang (Ed.), *Handbook of research on technologies for improving the 21st century workforce: Tools for lifelong learning* (pp. 272–286). Hershey, PA: IGI Global. doi:10.4018/978-1-4666-2181-7.ch018

Damasio, A. R. (1999). *The feeling of what happens: Body and emotion in the making of consciousness*. New York: Harcourt, Brace & Co.

Damurski, L. (2012). E-participation in urban planning: Online tools for citizen engagement in Poland and in Germany. *International Journal of E-Planning Research*, *1*(3), 40–67. doi:10.4018/ijepr.2012070103

Daugman, J. G. (1993). An information-theoretic view of analog representation in striate cortex. In E. L. Schwartz (Ed.), *Computational Neuroscience* (pp. 403–423). Cambridge: MIT Press.

Dawson, M., Dupuis, B., & Wilson, M. (2010). *From bricks to brains: The embodied cognitive science of LEGO robots*. Edmonton: Athabasca University Press.

de Almeida, M. O. (2007). E-government strategy in Brazil: Increasing transparency and efficiency through e-government procurement. In M. Gascó-Hernandez (Ed.), *Latin America online: Cases, successes and pitfalls* (pp. 34–82). Hershey, PA: IRM Press. doi:10.4018/978-1-59140-974-8.ch002

De Beauvoir, S. (1953). *The second sex. New York, NY.* Alfred: Knopf.

de Juana Espinosa, S. (2008). Empirical study of the municipalitites' motivations for adopting online presence. In A. Anttiroiko (Ed.), *Electronic government: Concepts, methodologies, tools, and applications* (pp. 3593–3608). Hershey, PA: IGI Global. doi:10.4018/978-1-59904-947-2.ch262

de Souza Dias, D. (2002). Motivation for using information technology. In C. Snodgrass & E. Szewczak (Eds.), *Human factors in information systems* (pp. 55–60). Hershey, PA: IRM Press. doi:10.4018/978-1-931777-10-0.ch005

De Valois, R. L., & De Valois, K. K. (1993). A multi-stage color model. *Color Vision*, *33*, 1053–1065. PMID:8506645

Dean-Leon, E., Nair, S., & Knoll, A. (2012). User friendly Matlab-Toolbox for symbolic robot dynamic modeling used for control design. In *Proceedings of the IEEE International Conference on Robotics and Biometrics* (pp. 2181-2188). Guangzhou, China: IEEE. doi:10.1109/ROBIO.2012.6491292

DeLuce, D. (2014). Can Robots help stop the Ebola outbreak? *Medicalxpress.com*. Retrieved from http://medicalxpress.com/news/2014-11-robots-ebola-outbreak.html

Demediuk, P. (2006). Government procurement ICT's impact on the sustainability of SMEs and regional communities. In S. Marshall, W. Taylor, & X. Yu (Eds.), *Encyclopedia of developing regional communities with information and communication technology* (pp. 321–324). Hershey, PA: IGI Global. doi:10.4018/978-1-59140-575-7.ch056

Dennett, D. (1994). Original Intentionality. In E. Dietrich (Ed.), *Thinking computers and virtual persons* (1st ed.). San Diego: Academic Press. doi:10.1016/B978-0-12-215495-9.50009-0

Denton, D. A. (2006). *The primordial emotions: The dawning of consciousness*. Oxford, UK: Oxford University Press. doi:10.1093/acprof:oso/9780199203147.001.0001

Deperlioglu, O., & Kose, U. (2011). An educational tool for artificial neural networks. *Computers & Electrical Engineering an International Journal*, *37*(3), 392–402. doi:10.1016/j.compeleceng.2011.03.010

Devonshire, E., Forsyth, H., Reid, S., & Simpson, J. M. (2013). The challenges and opportunities of online postgraduate coursework programs in a traditional university context. In B. Tynan, J. Willems, & R. James (Eds.), *Outlooks and opportunities in blended and distance learning* (pp. 353–368). Hershey, PA: IGI Global. doi:10.4018/978-1-4666-4205-8.ch026

Di Cerbo, F., Scotto, M., Sillitti, A., Succi, G., & Vernazza, T. (2007). Toward a GNU/Linux distribution for corporate environments. In S. Sowe, I. Stamelos, & I. Samoladas (Eds.), *Emerging free and open source software practices* (pp. 215–236). Hershey, PA: IGI Global. doi:10.4018/978-1-59904-210-7.ch010

Dick, P. K. (2002). Impostor. In Impostor. Test na człowieczeństwo. (M. Gawlik-Małkowska, Trans.). Warszawa: Prószyński i S-ka. (Original work published from 1959), 375-392.

Dick, P. K. (2003). Elektryczna mrówka, In Opowiadania najlepsze. (M. Begiert & J. Karłowski & T. Oljasz, Trans.). Poznań: Dom Wydawniczy Rebis. (Original work published from 1969).

Dick, P. K. (2011). Blade Runner: Czy androidy marzą o elektrycznych owcach? (Kędzierski S.). Warszawa: Dom Wydawniczy Rebis. (Original work published from 1968).

Diesner, J., & Carley, K. M. (2005). Revealing social structure from texts: Meta-matrix text analysis as a novel method for network text analysis. In V. Narayanan & D. Armstrong (Eds.), *Causal mapping for research in information technology* (pp. 81–108). Hershey, PA: IGI Global. doi:10.4018/978-1-59140-396-8.ch004

Dietrich, E. (1994). *Thinking computers and virtual persons*. San Diego: Academic Press.

Dinh, L. (2013). Talyst's InSite remote dispensing technology - compliant with short-cycle dispensing. *GlobeNewswire News Room*. Retrieved from http://globenewswire.com/news-release/2013/03/26/533801/10026396/en/Talyst-s-InSite-Remote-Dispensing-Technology-Compliant-With-Short-Cycle-Dispensing.html

Dmitruk, N. (2013). *Tożsamość cyborga w komiksie i animacji japońskiej (na wybranych przykładach)*. Unpublished Thesis, University of Wrocław, Wrocław.

DMTLS. (2006). cyborg-ken [Blog post]. Retrieved from http://our-modern-playground.blogspot.ru/2006/09/cyborg-ken.html

Dockterman, E. (2015, July 1). Robot Kills Man at Volkswagen Plant. *TIME*. Retrieved from http://time.com/3944181/robot-kills-man-volkswagen-plant

Dockterman, E. (2015a). Robot Kills Man at Volkswagen Plant. *TIME*. Retrieved from http://time.com/3944181/robot-kills-man-volkswagen-plant

Dockterman, E. (2015b). Meet NASA's 'Superhero' Valkyrie Robot [Video]. *TIME*. from http://time.com/3944181/robot-kills-man-volkswagen-plant

Dologite, D. G., Mockler, R. J., Bai, Q., & Viszhanyo, P. F. (2006). IS change agents in practice in a US-Chinese joint venture. In M. Hunter & F. Tan (Eds.), *Advanced topics in global information management* (Vol. 5, pp. 331–352). Hershey, PA: IGI Global. doi:10.4018/978-1-59140-923-6.ch015

Drnevich, P., Brush, T. H., & Luckock, G. T. (2011). Process and structural implications for IT-enabled outsourcing. *International Journal of Strategic Information Technology and Applications*, *2*(4), 30–43. doi:10.4018/jsita.2011100103

Dub, M., & Brotan, L. (2012). MATLAB GUI for DC drive parameter identification. In *Proceedings of a meeting held 15Th International Symposium MECHATRONIKA* (pp. 1-3). Prague, Czech: IEEE.

Dudo, A., Brossard, D., Shanahan, J., Scheufele, D., Morgan, M., & Signorielli, N. (2011). Science on television in the 21st century: Recent trends in portrayals and their contributions to public attitudes towards science. *Communication Research*, *38*(6), 754–777. doi:10.1177/0093650210384988

Dwivedi, A. N. (2009). Handbook of research on information technology management and clinical data administration in healthcare (Vols. 1–2). Hershey, PA: IGI Global. doi:10.4018/978-1-60566-356-2

Edmond, G. (2016). Expert Evidence. In P. Cane & J. Conaghan (Eds.), *The new Oxford companion to law* (pp. 437–440). Oxford: Oxford University Press.

Elbeltagi, I., McBride, N., & Hardaker, G. (2006). Evaluating the factors affecting DSS usage by senior managers in local authorities in Egypt. In M. Hunter & F. Tan (Eds.), *Advanced topics in global information management* (Vol. 5, pp. 283–307). Hershey, PA: IGI Global. doi:10.4018/978-1-59140-923-6.ch013

Elgin, C. (2002). Take it from Me: The Epistemological Status of Testimony. *Philosophy and Phenomenological Research*, *65*(2), 291–308. doi:10.1111/j.1933-1592.2002.tb00203.x

Emelin, V. A. (2013). Cyborgization and disability of technologically extended human. *National Psychological Journal*, *1*(9), 62–70.

Engelhardt, H. T., & Caplan, A. L. (1987). *Scientific controversies: Case studies in the resolution and closure of disputes in science and technology*. Cambridge University Press. doi:10.1017/CBO9780511628719

Eom, S., & Fountain, J. E. (2013). Enhancing information services through public-private partnerships: Information technology knowledge transfer underlying structures to develop shared services in the U.S. and Korea. In J. Gil-Garcia (Ed.), *E-government success around the world: Cases, empirical studies, and practical recommendations* (pp. 15–40). Hershey, PA: IGI Global. doi:10.4018/978-1-4666-4173-0.ch002

Espinoza, L. (2010). *Diseño e implementación de un simulador gráfico de un robot bípedo de 14 grados de libertad.* Universidad de Santiago de Chile, Chile: Magister These.

Esteves, T., Leuenberger, D., & Van Leuven, N. (2012). Reaching citizen 2.0: How government uses social media to send public messages during times of calm and times of crisis. In K. Kloby & M. D'Agostino (Eds.), *Citizen 2.0: Public and governmental interaction through web 2.0 technologies* (pp. 250–268). Hershey, PA: IGI Global. doi:10.4018/978-1-4666-0318-9.ch013

Estevez, E., Fillottrani, P., Janowski, T., & Ojo, A. (2012). Government information sharing: A framework for policy formulation. In Y. Chen & P. Chu (Eds.), *Electronic governance and cross-boundary collaboration: Innovations and advancing tools* (pp. 23–55). Hershey, PA: IGI Global. doi:10.4018/978-1-60960-753-1.ch002

European Commission. (2016). CORDIS: Projects & results service: spatio-temporal representations and activities for cognitive control in long-term scenarios. *Cordis.europa.eu.* Retrieved from http://cordis.europa.eu/project/rcn/107156_en.html

Ezz, I. E. (2008). E-governement emerging trends: Organizational challenges. In A. Anttiroiko (Ed.), *Electronic government: Concepts, methodologies, tools, and applications* (pp. 3721–3737). Hershey, PA: IGI Global. doi:10.4018/978-1-59904-947-2.ch269

Fabri, M. (2009). The Italian style of e-justice in a comparative perspective. In A. Martínez & P. Abat (Eds.), *E-justice: Using information communication technologies in the court system* (pp. 1–19). Hershey, PA: IGI Global. doi:10.4018/978-1-59904-998-4.ch001

Fagbe, T., & Adekola, O. D. (2010). Workplace safety and personnel well-being: The impact of information technology. *International Journal of Green Computing*, *1*(1), 28–33. doi:10.4018/jgc.2010010103

Fagbe, T., & Adekola, O. D. (2011). Workplace safety and personnel well-being: The impact of information technology. In *Global business: Concepts, methodologies, tools and applications* (pp. 1438–1444). Hershey, PA: IGI Global. doi:10.4018/978-1-60960-587-2.ch509

Falconer, D. (2016). The Panasonic Hospi robot help in the delivery of medicines in Changi hospital. *Digitalavmagazine.com*. Retrieved from http://www. digitalavmagazine.com/en/2015/08/25/the-panasonic-hospi-robot-help-in-the-delivery-of-medicines-in-the-hospital-of-cgh

Farmer, L. (2008). Affective collaborative instruction with librarians. In S. Kelsey & K. St.Amant (Eds.), *Handbook of research on computer mediated communication* (pp. 15–24). Hershey, PA: IGI Global. doi:10.4018/978-1-59904-863-5.ch002

Favier, L., & Mekhantar, J. (2007). Use of OSS by local e-administration: The French situation. In K. St.Amant & B. Still (Eds.), *Handbook of research on open source software: Technological, economic, and social perspectives* (pp. 428–444). Hershey, PA: IGI Global. doi:10.4018/978-1-59140-999-1.ch033

Fenlon, W. (2013). 10 years of Roomba hacks: from magic marker art to combat. *Adam Savage's Tested*. Retrieved from http://www.tested.com/tech/robots/453965-magic-marker-art-kinect-10-years-roomba-hacks

Fernandez, C. (2015, July 02). Robot kills man at Volkswagen plant in Germany. *Daily Mail*. Retrieved from http://www.dailymail.co.uk/news/article-3146547/Robot-kills-man-Volkswagen-plant-Germany.html

Fernando, S. (2009). Issues of e-learning in third world countries. In M. Khosrow-Pour (Ed.), *Encyclopedia of information science and technology* (2nd ed., pp. 2273–2277). Hershey, PA: IGI Global. doi:10.4018/978-1-60566-026-4.ch360

Filho, J. R., & dos Santos Junior, J. R. (2009). Local e-government in Brazil: Poor interaction and local politics as usual. In C. Reddick (Ed.), *Handbook of research on strategies for local e-government adoption and implementation: Comparative studies* (pp. 863–878). Hershey, PA: IGI Global. doi:10.4018/978-1-60566-282-4.ch045

Fink, E., Segall, L., Farkas, J., Quart, J., Hunt, R., Hottman, A. K., . . . Daly, C. (Producers). (2017). I love you, Bot. In *Mostly Human*. Retrieved from http://money.cnn.com/mostly-human/i-love-you-bot

Fisch, S. M. (2005). Children's learning from television. It's not just "violence." *TelevIZIon*, 18(E), 10-14.

Fisch, S. M., & Truglio, R. T. (Eds.). (2001). *G" is for growing. Thirty years of research on children and Sesame Street*. Mahwah, N.J.: Erlbaum.

Fisher, C. (2011). Pluto and Doubling. *The Comics Journal*. Retrieved from http://www.tcj.com/pluto-and-doubling

Fletcher, P. D. (2004). Portals and policy: Implications of electronic access to U.S. federal government information services. In A. Pavlichev & G. Garson (Eds.), *Digital government: Principles and best practices* (pp. 52–62). Hershey, PA: IGI Global. doi:10.4018/978-1-59140-122-3.ch004

Fletcher, P. D. (2008). Portals and policy: Implications of electronic access to U.S. federal government information services. In A. Anttiroiko (Ed.), *Electronic government: Concepts, methodologies, tools, and applications* (pp. 3970–3979). Hershey, PA: IGI Global. doi:10.4018/978-1-59904-947-2.ch289

Forlano, L. (2004). The emergence of digital government: International perspectives. In A. Pavlichev & G. Garson (Eds.), *Digital government: Principles and best practices* (pp. 34–51). Hershey, PA: IGI Global. doi:10.4018/978-1-59140-122-3.ch003

Franzel, J. M., & Coursey, D. H. (2004). Government web portals: Management issues and the approaches of five states. In A. Pavlichev & G. Garson (Eds.), *Digital government: Principles and best practices* (pp. 63–77). Hershey, PA: IGI Global. doi:10.4018/978-1-59140-122-3.ch005

Franzosi, R. (1998). Narrative analysis-or why (and how) sociologists should be interested in narrative. *Annual Review of Sociology*, *24*(1), 517–554. doi:10.1146/annurev.soc.24.1.517

Fricker, E. (1995). Critical Notice. *Mind*, *104*(414), 393–411. doi:10.1093/mind/104.414.393

Fujigaki, Y. (2009). STS in Japan and East Asia: Governance of science and technology and public engagement. *East Asian Science. Technology and Society*, *3*(4), 511–518.

Fukuyama, F. (2003). *Our posthuman future: Consequences of the biotechnology revolution*. London: Profile Books.

Gaivéo, J. M. (2013). Security of ICTs supporting healthcare activities. In M. Cruz-Cunha, I. Miranda, & P. Gonçalves (Eds.), *Handbook of research on ICTs for human-centered healthcare and social care services* (pp. 208–228). Hershey, PA: IGI Global. doi:10.4018/978-1-4666-3986-7.ch011

Garson, G. D. (1999). *Information technology and computer applications in public administration: Issues and trends*. Hershey, PA: IGI Global. doi:10.4018/978-1-87828-952-0

Garson, G. D. (2003). Toward an information technology research agenda for public administration. In G. Garson (Ed.), *Public information technology: Policy and management issues* (pp. 331–357). Hershey, PA: IGI Global. doi:10.4018/978-1-59140-060-8.ch014

Garson, G. D. (2004). The promise of digital government. In A. Pavlichev & G. Garson (Eds.), *Digital government: Principles and best practices* (pp. 2–15). Hershey, PA: IGI Global. doi:10.4018/978-1-59140-122-3.ch001

Garson, G. D. (2007). An information technology research agenda for public administration. In G. Garson (Ed.), *Modern public information technology systems: Issues and challenges* (pp. 365–392). Hershey, PA: IGI Global. doi:10.4018/978-1-59904-051-6.ch018

Gasco, M. (2007). Civil servants' resistance towards e-government development. In A. Anttiroiko & M. Malkia (Eds.), *Encyclopedia of digital government* (pp. 190–195). Hershey, PA: IGI Global. doi:10.4018/978-1-59140-789-8.ch028

Gasco, M. (2008). Civil servants' resistance towards e-government development. In A. Anttiroiko (Ed.), *Electronic government: Concepts, methodologies, tools, and applications* (pp. 2580–2588). Hershey, PA: IGI Global. doi:10.4018/978-1-59904-947-2.ch190

Gelfert, A. (2014). *A critical introduction to testimony*. London: Bloomsbury Academic.

Ghere, R. K. (2010). Accountability and information technology enactment: Implications for social empowerment. In E. Ferro, Y. Dwivedi, J. Gil-Garcia, & M. Williams (Eds.), *Handbook of research on overcoming digital divides: Constructing an equitable and competitive information society* (pp. 515–532). Hershey, PA: IGI Global. doi:10.4018/978-1-60566-699-0.ch028

Gibson, I. W. (2012). Simulation modeling of healthcare delivery. In A. Kolker & P. Story (Eds.), *Management engineering for effective healthcare delivery: Principles and applications* (pp. 69–89). Hershey, PA: IGI Global. doi:10.4018/978-1-60960-872-9.ch003

Gibson, W. (2015). Neuromancer. In *Trylogia ciągu* (P. W. Cholera, Trans.). Warszawa: Wydawnictwo MAG.

Gil-Garcia, J. R. (2007). Exploring e-government benefits and success factors. In A. Anttiroiko & M. Malkia (Eds.), *Encyclopedia of digital government* (pp. 803–811). Hershey, PA: IGI Global. doi:10.4018/978-1-59140-789-8.ch122

Gil-Garcia, J. R., & González Miranda, F. (2010). E-government and opportunities for participation: The case of the Mexican state web portals. In C. Reddick (Ed.), *Politics, democracy and e-government: Participation and service delivery* (pp. 56–74). Hershey, PA: IGI Global. doi:10.4018/978-1-61520-933-0.ch004

Glaser, B. G., & Strauss, A. L. (2009). *The discovery of grounded theory: Strategies for qualitative research*. Transaction Publishers.

Goel, L., & Shrestha, G. B. (2002). A versatile tool for electric substation reliability assessment. *Computers & Electrical Engineering an International Journal*, *28*(6), 561–576. doi:10.1016/S0045-7906(00)00064-1

Goldfinch, S. (2012). Public trust in government, trust in e-government, and use of e-government. In Z. Yan (Ed.), *Encyclopedia of cyber behavior* (pp. 987–995). Hershey, PA: IGI Global. doi:10.4018/978-1-4666-0315-8.ch081

Goodyear, M. (2012). Organizational change contributions to e-government project transitions. In S. Aikins (Ed.), *Managing e-government projects: Concepts, issues, and best practices* (pp. 1–21). Hershey, PA: IGI Global. doi:10.4018/978-1-4666-0086-7.ch001

Gordon, S., & Mulligan, P. (2003). Strategic models for the delivery of personal financial services: The role of infocracy. In S. Gordon (Ed.), *Computing information technology: The human side* (pp. 220–232). Hershey, PA: IRM Press. doi:10.4018/978-1-93177-752-0.ch014

Gordon, T. F. (2007). Legal knowledge systems. In A. Anttiroiko & M. Malkia (Eds.), *Encyclopedia of digital government* (pp. 1161–1166). Hershey, PA: IGI Global. doi:10.4018/978-1-59140-789-8.ch175

Götz, M. (Ed.). (2014). *TV-hero(in)es of boys and girls*. Frankfurt: Peter Lang. doi:10.3726/978-3-653-04304-4

Götz, M., & Lemish, D. (Eds.). (2012). *Sexy girls, heroes and funny losers*. Frankfurt: Peter Lang. doi:10.3726/978-3-653-01426-6

Götz, M., Lemish, D., Aidman, A., & Moon, H. (2005). *Media and the make-believe worlds of children*. Mahwah, N.J.: Lawrence Erlbaum.

Graham, J. E., & Semich, G. W. (2008). Integrating technology to transform pedagogy: Revisiting the progress of the three phase TUI model for faculty development. In L. Tomei (Ed.), *Adapting information and communication technologies for effective education* (pp. 1–12). Hershey, PA: IGI Global. doi:10.4018/978-1-59904-922-9.ch001

Grandinetti, L., & Pisacane, O. (2012). Web services for healthcare management. In D. Prakash Vidyarthi (Ed.), *Technologies and protocols for the future of internet design: Reinventing the web* (pp. 60–94). Hershey, PA: IGI Global. doi:10.4018/978-1-4666-0203-8.ch004

Gray, H., Gray, K., & Wegner, D. (2007). Dimensions of Mind Perception. *Science*, *315*(5812), 619. doi:10.1126/science.1134475 PMID:17272713

Groenewegen, P., & Wagenaar, F. P. (2008). VO as an alternative to hierarchy in the Dutch police sector. In G. Putnik & M. Cruz-Cunha (Eds.), *Encyclopedia of networked and virtual organizations* (pp. 1851–1857). Hershey, PA: IGI Global. doi:10.4018/978-1-59904-885-7.ch245

Gronlund, A. (2001). Building an infrastructure to manage electronic services. In S. Dasgupta (Ed.), *Managing internet and intranet technologies in organizations: Challenges and opportunities* (pp. 71–103). Hershey, PA: IGI Global. doi:10.4018/978-1-878289-95-7.ch006

Gronlund, A. (2002). Introduction to electronic government: Design, applications and management. In Å. Grönlund (Ed.), *Electronic government: Design, applications and management* (pp. 1–21). Hershey, PA: IGI Global. doi:10.4018/978-1-930708-19-8.ch001

Gross, S. (1991). Expert Evidence. *Wisconsin Law Review*, *6*, 1113–1232.

Gupta, A., Woosley, R., Crk, I., & Sarnikar, S. (2009). An information technology architecture for drug effectiveness reporting and post-marketing surveillance. In J. Tan (Ed.), *Medical informatics: Concepts, methodologies, tools, and applications* (pp. 631–646). Hershey, PA: IGI Global. doi:10.4018/978-1-60566-050-9.ch047

Gutenplan, S. (Ed.). (1994). *A companion to the philosophy of the mind.* Oxford: Blackwell.

Haas, B. (2017). Chinese man 'marries' robot he built himself. *The Guardian.* Retrieved from https://www.theguardian.com/world/2017/apr/04/chinese-man-marries-robot-built-himself

Hallin, A., & Lundevall, K. (2007). mCity: User focused development of mobile services within the city of Stockholm. In I. Kushchu (Ed.), Mobile government: An emerging direction in e-government (pp. 12-29). Hershey, PA: IGI Global. doi:10.4018/978-1-59140-884-0.ch002

Hallin, A., & Lundevall, K. (2009). mCity: User focused development of mobile services within the city of Stockholm. In D. Taniar (Ed.), Mobile computing: Concepts, methodologies, tools, and applications (pp. 3455-3467). Hershey, PA: IGI Global. doi:10.4018/978-1-60566-054-7.ch253

Hallin, A., & Lundevall, K. (2009). mCity: User focused development of mobile services within the city of Stockholm. In S. Clarke (Ed.), Evolutionary concepts in end user productivity and performance: Applications for organizational progress (pp. 268-280). Hershey, PA: IGI Global. doi:10.4018/978-1-60566-136-0.ch017

Halpern, D., Benbow, C., Geary, D., Gur, R., Hyde, J., & Gernsbacher, M. (2007). The science of sex differences in science and mathematics. *Psychological Science in the Public Interest, 8*(1), 1–51. doi:10.1111/j.1529-1006.2007.00032.x PMID:25530726

Hanson, A. (2005). Overcoming barriers in the planning of a virtual library. In M. Khosrow-Pour (Ed.), *Encyclopedia of information science and technology* (pp. 2255–2259). Hershey, PA: IGI Global. doi:10.4018/978-1-59140-553-5.ch397

Haque, A. (2008). Information technology and surveillance: Implications for public administration in a new word order. In T. Loendorf & G. Garson (Eds.), *Patriotic information systems* (pp. 177–185). Hershey, PA: IGI Global. doi:10.4018/978-1-59904-594-8.ch008

Haraway, D. (1991). Science, Technology, and Socialist-Feminism in the Late Twentieth Century. In *Simians, cyborgs and women: The reinvention of nature* (pp. 149–181). New York: Routledge.

Hashimoto, T., Kobayashi, H., Polishuk, A., & Verner, I. (2013). Elementary Science Lesson Delivered by Robot. In *Proceedings of the 8th Annual ACM/IEEE international conference on Human-robot interaction*, Japan (HRI '13) (pp. 133-134). Piscataway, NJ: IEEE Press.

Hauck, R. V., Thatcher, S. M., & Weisband, S. P. (2012). Temporal aspects of information technology use: Increasing shift work effectiveness. In J. Wang (Ed.), *Advancing the service sector with evolving technologies: Techniques and principles* (pp. 87–104). Hershey, PA: IGI Global. doi:10.4018/978-1-4666-0044-7.ch006

Haugen, V. (2015, July 3). Mann (21) drept av robot. *Tek.no.* Retrieved from http://www.tek.no/artikler/robot-drepte-mann-21/191214

Hawk, S., & Witt, T. (2006). Telecommunications courses in information systems programs. *International Journal of Information and Communication Technology Education*, *2*(1), 79–92. doi:10.4018/jicte.2006010107

Heerink, M., Albo-Canals, J., Valenti-Soler, M., Martinez-Martin, P., Zondag, J., Smits, C., & Anisuzzaman, S. (2013). Exploring Requirements and Alternative Pet Robots for Robot Assisted Therapy with Older Adults with Dementia. In *Social Robotics* (pp. 104-115).

Heidegger, M. (1975). What are poets for. In *Poetry, language, thought* (pp. 87–142). N.Y.: Harper Collins.

Heidegger, M. (1977). The question concerning technology. In *The question concerning technology and other essays* (W. Lovitt, Trans., pp. 3–35). New York: Harper and Row.

Helms, M. M., Moore, R., & Ahmadi, M. (2009). Information technology (IT) and the healthcare industry: A SWOT analysis. In J. Tan (Ed.), *Medical informatics: Concepts, methodologies, tools, and applications* (pp. 134–152). Hershey, PA: IGI Global. doi:10.4018/978-1-60566-050-9.ch012

Hendrickson, S. M., & Young, M. E. (2014). Electronic records management at a federally funded research and development center. In J. Krueger (Ed.), *Cases on electronic records and resource management implementation in diverse environments* (pp. 334–350). Hershey, PA: IGI Global. doi:10.4018/978-1-4666-4466-3.ch020

Henman, P. (2010). Social policy and information communication technologies. In J. Martin & L. Hawkins (Eds.), *Information communication technologies for human services education and delivery: Concepts and cases* (pp. 215–229). Hershey, PA: IGI Global. doi:10.4018/978-1-60566-735-5.ch014

Hernández-Espinosa, A., & Hernández-Quiroz, F. (2013). Does the Principle of Computational Equivalence Overcome the Objections against Computationalism?. In *Computing Nature* (pp. 225-233).

Hismanoglu, M. (2011). Important issues in online education: E-pedagogy and marketing. In U. Demiray & S. Sever (Eds.), *Marketing online education programs: Frameworks for promotion and communication* (pp. 184–209). Hershey, PA: IGI Global. doi:10.4018/978-1-60960-074-7.ch012

hitchBOT. (2015). hitchBOT. Retrieved from http://m.hitchbot.me; http://mir1.hitchbot.me

Ho, K. K. (2008). The e-government development, IT strategies, and portals of the Hong Kong SAR government. In A. Anttiroiko (Ed.), *Electronic government: Concepts, methodologies, tools, and applications* (pp. 715–733). Hershey, PA: IGI Global. doi:10.4018/978-1-59904-947-2.ch060

Holden, S. H. (2003). The evolution of information technology management at the federal level: Implications for public administration. In G. Garson (Ed.), *Public information technology: Policy and management issues* (pp. 53–73). Hershey, PA: IGI Global. doi:10.4018/978-1-59140-060-8.ch003

Holden, S. H. (2007). The evolution of federal information technology management literature: Does IT finally matter? In G. Garson (Ed.), *Modern public information technology systems: Issues and challenges* (pp. 17–34). Hershey, PA: IGI Global. doi:10.4018/978-1-59904-051-6.ch002

Holland, J. W. (2009). Automation of American criminal justice. In M. Khosrow-Pour (Ed.), *Encyclopedia of information science and technology* (2nd ed., pp. 300–302). Hershey, PA: IGI Global. doi:10.4018/978-1-60566-026-4.ch051

Holler, A. (2012). "He speaks Russian - just like me." How preschool programmes can support language learning. *TelevIZIon, 25*(E), 16-19.

Holler, A., & Götz, M. (2013). I like Dino Dan "because it was pure fantasy:" A reception study with preschool children. *TelevIZIon, 26*(E), 34-38.

Holler, A., Götz, M., Egerer, A., & Schwarz, J. (2012). This is the TV presenter I'd like to have. *TelevIZIon 25*/2012/E, S. 30-31.

Holloway, K. (2013). Fair use, copyright, and academic integrity in an online academic environment. In *Digital rights management: Concepts, methodologies, tools, and applications* (pp. 917–928). Hershey, PA: IGI Global. doi:10.4018/978-1-4666-2136-7.ch044

Horgan, T., Tienson, J., & Graham, G. (2004). Phenomenal Intentionality and the Brain in the Vat. In R. Schantz (Ed.), *The Externalist Challenge* (1st ed.). Berlin: De Gruyter. doi:10.1515/9783110915273.297

Horiuchi, C. (2005). E-government databases. In L. Rivero, J. Doorn, & V. Ferraggine (Eds.), *Encyclopedia of database technologies and applications* (pp. 206–210). Hershey, PA: IGI Global. doi:10.4018/978-1-59140-560-3.ch035

Horiuchi, C. (2006). Creating IS quality in government settings. In E. Duggan & J. Reichgelt (Eds.), *Measuring information systems delivery quality* (pp. 311–327). Hershey, PA: IGI Global. doi:10.4018/978-1-59140-857-4.ch014

Horst, S. (1999). Symbols and Computation: A Critique of the Computational Theory of Mind. *Minds and Machines*, *9*(3), 347–381. doi:10.1023/A:1008351818306

Hsiao, N., Chu, P., & Lee, C. (2012). Impact of e-governance on businesses: Model development and case study. In *Digital democracy: Concepts, methodologies, tools, and applications* (pp. 1407–1425). Hershey, PA: IGI Global. doi:10.4018/978-1-4666-1740-7.ch070

Huang, T., & Lee, C. (2010). Evaluating the impact of e-government on citizens: Cost-benefit analysis. In C. Reddick (Ed.), *Citizens and e-government: Evaluating policy and management* (pp. 37–52). Hershey, PA: IGI Global. doi:10.4018/978-1-61520-931-6.ch003

Huggler, J. (2015, July 02). Robot kills man at Volkswagen plant in Germany. Retrieved from http://www.telegraph.co.uk/news/worldnews/europe/germany/11712513/Robot-kills-man-at-Volkswagen-plant-in-Germany.html

Humanity+ Page. (n. d.) Retrieved from: http://humanityplus.org

Hunter, M. G., Diochon, M., Pugsley, D., & Wright, B. (2002). Unique challenges for small business adoption of information technology: The case of the Nova Scotia ten. In S. Burgess (Ed.), *Managing information technology in small business: Challenges and solutions* (pp. 98–117). Hershey, PA: IGI Global. doi:10.4018/978-1-930708-35-8.ch006

Hurskainen, J. (2003). Integration of business systems and applications in merger and alliance: Case metso automation. In T. Reponen (Ed.), *Information technology enabled global customer service* (pp. 207–225). Hershey, PA: IGI Global. doi:10.4018/978-1-59140-048-6.ch012

Hyde, W. W. (1916). The prosecution and punishment of animals and lifeless things in the middle ages and modern times. *University of Pennsylvania Law Review and American Law Register*, 696-730.

Iazzolino, G., & Pietrantonio, R. (2011). The soveria.it project: A best practice of e-government in southern Italy. In D. Piaggesi, K. Sund, & W. Castelnovo (Eds.), *Global strategy and practice of e-governance: Examples from around the world* (pp. 34–56). Hershey, PA: IGI Global. doi:10.4018/978-1-60960-489-9.ch003

IFLSCIENCE. (2015). Robot Kills Worker at Volkswagen Plant in Germany. iflscience.com. Retrieved from http://www.iflscience.com/technology/robot-kills-worker-volkswagen-plant-germany

Ihde, D. (1986). *Experimental phenomenology: An introduction.* Albany, NY: State University of New York Press.

Iliadis, A. (2013). A new individuation: Deleuze's Simondon connection. *MediaTropes eJournal, IV*(1), 83–100. Retrieved from http://www.mediatropes.com/index.php/Mediatropes/article/view/20385

Imai, K. (2013). *Needless* (Vols. 1-17). Tokyo: Shueisha.

i. Martınez, A. C. (2008). Accessing administration's information via internet in Spain. In F. Tan (Ed.), *Global information technologies: Concepts, methodologies, tools, and applications* (pp. 2558–2573). Hershey, PA: IGI Global. doi:10.4018/978-1-59904-939-7.ch186

Imran, A., & Gregor, S. (2012). A process model for successful e-government adoption in the least developed countries: A case of Bangladesh. In F. Tan (Ed.), *International comparisons of information communication technologies: Advancing applications* (pp. 321–350). Hershey, PA: IGI Global. doi:10.4018/978-1-61350-480-2.ch014

Inoue, Y., & Bell, S. T. (2005). Electronic/digital government innovation, and publishing trends with IT. In M. Khosrow-Pour (Ed.), *Encyclopedia of information science and technology* (pp. 1018–1023). Hershey, PA: IGI Global. doi:10.4018/978-1-59140-553-5.ch180

IRMA. (2010). Electronic services: Concepts, methodologies, tools and applications (3 Volumes). Hershey, PA: IGI Global. doi:10.4018/978-1-61520-967-5

IRMA. (2010). Information resources management: Concepts, methodologies, tools and applications (4 Volumes). Hershey, PA: IGI Global. doi:10.4018/978-1-61520-965-1

Ishinomori, S. (2006). Cyborg 009 (Vols. 1-10). California, LA: Tokyopop. (Original work published 1964-1981).

Islam, M. M., & Ehsan, M. (2013). Understanding e-governance: A theoretical approach. In M. Islam & M. Ehsan (Eds.), *From government to e-governance: Public administration in the digital age* (pp. 38–49). Hershey, PA: IGI Global. doi:10.4018/978-1-4666-1909-8.ch003

Jackson, F. (1982). Epiphenomenal Qualia. *The Philosophical Quarterly*, *32*(127), 127. doi:10.2307/2960077

Jacobs, J., Davis-Kean, P., Bleeker, M., Eccles, J. S., & Malanchuk, O. (2005). I can, but I don't want to: The impact of parents, interests, and activities on gender differences in math. In A. Gallagher & J. Kaufman (Eds.), Gender differences in mathematics (pp. 246-263). Cambridge, MA: Cambridge University Press.

Jaeger, B. (2009). E-government and e-democracy in the making. In M. Khosrow-Pour (Ed.), *Encyclopedia of information science and technology* (2nd ed., pp. 1318–1322). Hershey, PA: IGI Global. doi:10.4018/978-1-60566-026-4.ch208

Jain, R. B. (2007). Revamping the administrative structure and processes in India for online diplomacy. In A. Anttiroiko & M. Malkia (Eds.), *Encyclopedia of digital government* (pp. 1418–1423). Hershey, PA: IGI Global. doi:10.4018/978-1-59140-789-8.ch217

Jain, R. B. (2008). Revamping the administrative structure and processes in India for online diplomacy. In A. Anttiroiko (Ed.), *Electronic government: Concepts, methodologies, tools, and applications* (pp. 3142–3149). Hershey, PA: IGI Global. doi:10.4018/978-1-59904-947-2.ch233

James, W. (1955). The principles of psychology, authorized edition. New York, NY: Dover (Reprint of Henry Holt, 1890 edition).

James, W. (1884). What is an emotion? *Mind*, *9*(34), 188–205. doi:10.1093/mind/os-IX.34.188

Jauhiainen, J. S., & Inkinen, T. (2009). E-governance and the information society in periphery. In C. Reddick (Ed.), *Handbook of research on strategies for local e-government adoption and implementation: Comparative studies* (pp. 497–514). Hershey, PA: IGI Global. doi:10.4018/978-1-60566-282-4.ch026

Jenkins, H. (Ed.). (2009). *Confronting the challenges of participatory culture: Media education in the 21st century*. Cambridge, MA: MIT Press.

Jensen, M. J. (2009). Electronic democracy and citizen influence in government. In C. Reddick (Ed.), *Handbook of research on strategies for local e-government adoption and implementation: Comparative studies* (pp. 288–305). Hershey, PA: IGI Global. doi:10.4018/978-1-60566-282-4.ch015

Jiao, Y., Hurson, A. R., Potok, T. E., & Beckerman, B. G. (2009). Integrating mobile-based systems with healthcare databases. In J. Erickson (Ed.), *Database technologies: Concepts, methodologies, tools, and applications* (pp. 484–504). Hershey, PA: IGI Global. doi:10.4018/978-1-60566-058-5.ch031

Johnson, D. E. (2010). *Programming of life*. Sylacauga, Ala.: Big Mac.

Joia, L. A. (2002). A systematic model to integrate information technology into metabusinesses: A case study in the engineering realms. In F. Tan (Ed.), *Advanced topics in global information management* (Vol. 1, pp. 250–267). Hershey, PA: IGI Global. doi:10.4018/978-1-930708-43-3.ch016

Jones, S. (Producer). Jones, S. (Director). (2013). *Her* [Motion picture]. US: Annapurna Pictures.

Jones, T. H., & Song, I. (2000). Binary equivalents of ternary relationships in entity-relationship modeling: A logical decomposition approach. *Journal of Database Management*, *11*(2), 12–19. doi:10.4018/jdm.2000040102

Juana-Espinosa, S. D. (2007). Empirical study of the municipalitites' motivations for adopting online presence. In L. Al-Hakim (Ed.), *Global e-government: Theory, applications and benchmarking* (pp. 261–279). Hershey, PA: IGI Global. doi:10.4018/978-1-59904-027-1.ch015

Jun, K., & Weare, C. (2012). Bridging from e-government practice to e-government research: Past trends and future directions. In K. Bwalya & S. Zulu (Eds.), *Handbook of research on e-government in emerging economies: Adoption, e-participation, and legal frameworks* (pp. 263–289). Hershey, PA: IGI Global. doi:10.4018/978-1-4666-0324-0.ch013

Junqueira, A., Diniz, E. H., & Fernandez, M. (2010). Electronic government implementation projects with multiple agencies: Analysis of the electronic invoice project under PMBOK framework. In J. Cordoba-Pachon & A. Ochoa-Arias (Eds.), *Systems thinking and e-participation: ICT in the governance of society* (pp. 135–153). Hershey, PA: IGI Global. doi:10.4018/978-1-60566-860-4.ch009

Juntunen, A. (2009). Joint service development with the local authorities. In C. Reddick (Ed.), *Handbook of research on strategies for local e-government adoption and implementation: Comparative studies* (pp. 902–920). Hershey, PA: IGI Global. doi:10.4018/978-1-60566-282-4.ch047

Kamel, S. (2001). *Using DSS for crisis management*. Hershey, PA: IGI Global. doi:10.4018/978-1-87828-961-2.ch020

Kamel, S. (2006). DSS for strategic decision making. In M. Khosrow-Pour (Ed.), *Cases on information technology and organizational politics & culture* (pp. 230–246). Hershey, PA: IGI Global. doi:10.4018/978-1-59904-411-8.ch015

Kamel, S. (2009). The software industry in Egypt as a potential contributor to economic growth. In M. Khosrow-Pour (Ed.), *Encyclopedia of information science and technology* (2nd ed., pp. 3531–3537). Hershey, PA: IGI Global. doi:10.4018/978-1-60566-026-4.ch562

Kamel, S., & Hussein, M. (2008). Xceed: Pioneering the contact center industry in Egypt. *Journal of Cases on Information Technology*, *10*(1), 67–91. doi:10.4018/jcit.2008010105

Kamel, S., & Wahba, K. (2003). The use of a hybrid model in web-based education: "The Global campus project. In A. Aggarwal (Ed.), *Web-based education: Learning from experience* (pp. 331–346). Hershey, PA: IGI Global. doi:10.4018/978-1-59140-102-5.ch020

Kardaras, D. K., & Papathanassiou, E. A. (2008). An exploratory study of the e-government services in Greece. In G. Garson & M. Khosrow-Pour (Eds.), *Handbook of research on public information technology* (pp. 162–174). Hershey, PA: IGI Global. doi:10.4018/978-1-59904-857-4.ch016

Karl, A., Birbaumer, N., Lutzenberger, W., Cohen, L. G., & Flor, H. (2001). Reorganization of motor and somatosensory cortex in upper extremity amputees with phantom limb pain. *The Journal of Neuroscience*, *21*, 3609–3618. PMID:11331390

Kassahun, A. E., Molla, A., & Sarkar, P. (2012). Government process reengineering: What we know and what we need to know. In *Digital democracy: Concepts, methodologies, tools, and applications* (pp. 1730–1752). Hershey, PA: IGI Global. doi:10.4018/978-1-4666-1740-7.ch086

Katz, S. B. (1991). A computer file named Alison. *Postmodern Culture, 1*(3). Retrieved May 29, 2017 from http://muse.jhu.edu/journals/pmc/toc/pmc

Katz, S. B. (2004). The beautiful machine. *Star*line*: *Science fiction poetry association,* 27(1), 15.

Katz, P. (2014). Expert Robot: Using Artificial Intelligence to Assist Judges in Admitting Scientific Expert Testimony. *Albany Law Journal of Science & Technology*, *24*(1), 1–33. PMID:12680396

Kawagoe, J. (Director). (2011). Cyborg 009: The Cyborg Soldier [Television series]. Japan: Avenc Inc. & Japan Vistec.

Kennedy, K., Spielberg, S., & Bonnie, C. (Producers), & Spielberg S. (Director). (2001). *A.I. Artificial Intelligence* [Motion picture]. United States: DreamWorks Pictures.

Khan, B. (2005). Technological issues. In B. Khan (Ed.), *Managing e-learning strategies: Design, delivery, implementation and evaluation* (pp. 154–180). Hershey, PA: IGI Global. doi:10.4018/978-1-59140-634-1.ch004

Khasawneh, A., Bsoul, M., Obeidat, I., & Al Azzam, I. (2012). Technology fears: A study of e-commerce loyalty perception by Jordanian customers. In J. Wang (Ed.), *Advancing the service sector with evolving technologies: Techniques and principles* (pp. 158–165). Hershey, PA: IGI Global. doi:10.4018/978-1-4666-0044-7.ch010

Khatibi, V., & Montazer, G. A. (2012). E-research methodology. In A. Juan, T. Daradoumis, M. Roca, S. Grasman, & J. Faulin (Eds.), *Collaborative and distributed e-research: Innovations in technologies, strategies and applications* (pp. 62–81). Hershey, PA: IGI Global. doi:10.4018/978-1-4666-0125-3.ch003

Kidd, T. (2011). The dragon in the school's backyard: A review of literature on the uses of technology in urban schools. In L. Tomei (Ed.), *Online courses and ICT in education: Emerging practices and applications* (pp. 242–257). Hershey, PA: IGI Global. doi:10.4018/978-1-60960-150-8.ch019

Kidd, T. T. (2010). My experience tells the story: Exploring technology adoption from a qualitative perspective - A pilot study. In H. Song & T. Kidd (Eds.), *Handbook of research on human performance and instructional technology* (pp. 247–262). Hershey, PA: IGI Global. doi:10.4018/978-1-60566-782-9.ch015

Kieley, B., Lane, G., Paquet, G., & Roy, J. (2002). e-Government in Canada: Services online or public service renewal? In Å. Grönlund (Ed.), Electronic government: Design, applications and management (pp. 340-355). Hershey, PA: IGI Global. doi:10.4018/978-1-930708-19-8.ch016

Kim, J. (2007). Causation and Mental Causation. In *B. McLoughlin (Ed.), Contemporary Debates in Philosophy of Mind* (1st ed., pp. 1–22). London: Wiley-Blackwell.

Kim, P. (2012). "Stay out of the way! My kid is video blogging through a phone!": A lesson learned from math tutoring social media for children in underserved communities. In *Wireless technologies: Concepts, methodologies, tools and applications* (pp. 1415–1428). Hershey, PA: IGI Global. doi:10.4018/978-1-61350-101-6.ch517

Kirjnen, T., & Bauwen, S. (2015). *Gender and media: Representing, producing, consuming*. New York, NY: Routledge.

Kirk, R., & Squires, J. E. R. (1974). Zombies versus materialists. *Aristotelian Society*, *48*(1 Suppl.), 135–152. doi:10.1093/aristoteliansupp/48.1.135

Kirlidog, M. (2010). Financial aspects of national ICT strategies. In S. Kamel (Ed.), *E-strategies for technological diffusion and adoption: National ICT approaches for socioeconomic development* (pp. 277–292). Hershey, PA: IGI Global. doi:10.4018/978-1-60566-388-3.ch016

Kishiro, Y. (2014-ongoing). Gunnm: Mars chronicle (Vols. 1-2). Tokyo: Kodansha.

Kisielnicki, J. (2006). Transfer of information and knowledge in the project management. In E. Coakes & S. Clarke (Eds.), *Encyclopedia of communities of practice in information and knowledge management* (pp. 544–551). Hershey, PA: IGI Global. doi:10.4018/978-1-59140-556-6.ch091

Kitching, C. (2015). Inside the Japanese hotel where the front desk is staffed by ROBOTS and guests scan their faces to enter rooms. *Daily Mail*. Retrieved from http://www.dailymail.co.uk/travel/travel_news/article-3163631/Japan-s-robot-staffed-Henn-na-Hotel-opens.html

Kittner, M., & Van Slyke, C. (2006). Reorganizing information technology services in an academic environment. In M. Khosrow-Pour (Ed.), *Cases on the human side of information technology* (pp. 49–66). Hershey, PA: IGI Global. doi:10.4018/978-1-59904-405-7.ch004

Kleinginna, P. R. Jr, & Kleinginna, A. M. (1981). A categorized list of emotion definitions, with suggestions for a consensual definition. *Motivation and Emotion*, *5*(4), 345–379. doi:10.1007/BF00992553

Knobe, J. (2003). Intentional Action in Folk Psychology: An Experimental Investigation. *Philosophical Psychology*, *16*(2), 309–324. doi:10.1080/09515080307771

Knoell, H. D. (2008). Semi virtual workplaces in German financial service enterprises. In P. Zemliansky & K. St.Amant (Eds.), *Handbook of research on virtual workplaces and the new nature of business practices* (pp. 570–581). Hershey, PA: IGI Global. doi:10.4018/978-1-59904-893-2.ch041

Koh, S. L., & Maguire, S. (2009). Competing in the age of information technology in a developing economy: Experiences of an Indian bank. In S. Koh & S. Maguire (Eds.), *Information and communication technologies management in turbulent business environments* (pp. 326–350). Hershey, PA: IGI Global. doi:10.4018/978-1-60566-424-8.ch018

Kollmann, T., & Häsel, M. (2009). Competence of information technology professionals in internet-based ventures. In A. Cater-Steel (Ed.), *Information technology governance and service management: Frameworks and adaptations* (pp. 239–253). Hershey, PA: IGI Global. doi:10.4018/978-1-60566-008-0.ch013

Kollmann, T., & Häsel, M. (2009). Competence of information technology professionals in internet-based ventures. In I. Lee (Ed.), *Electronic business: Concepts, methodologies, tools, and applications* (pp. 1905–1919). Hershey, PA: IGI Global. doi:10.4018/978-1-60566-056-1.ch118

Kollmann, T., & Häsel, M. (2010). Competence of information technology professionals in internet-based ventures. In *Electronic services: Concepts, methodologies, tools and applications* (pp. 1551–1565). Hershey, PA: IGI Global. doi:10.4018/978-1-61520-967-5.ch094

Kondo, K. (2009). Can television be good for children? In N. N. Nagaraj (Ed.), Effect of television on children (pp. 129-156). Hyderabad: Icfai University Press.

Kraemer, K., & King, J. L. (2006). Information technology and administrative reform: Will e-government be different? *International Journal of Electronic Government Research*, *2*(1), 1–20. doi:10.4018/jegr.2006010101

Kraemer, K., & King, J. L. (2008). Information technology and administrative reform: Will e-government be different? In D. Norris (Ed.), *E-government research: Policy and management* (pp. 1–20). Hershey, PA: IGI Global. doi:10.4018/978-1-59904-913-7.ch001

Kripke, S. A. (1980). *Naming and necessity*. Cambridge, MA: Harvard University Press.

Kurzweil, R. (2011). *Singularity Q&A*. Retrieved from http://www.kurzweilai.net/singularity-q-a. (Original work published 2005).

Kurzweil, R. (2014). *How to create a mind: The secret of human thought revealed*. London: Duckworth.

Kyriakidou, M. (2014). Discussing robot crime interviewers for childrens forensic testimonies: A relatively new field for investigation. *AI & Society*, *31*(1), 121–126. doi:10.1007/s00146-014-0566-3

Lackey, J. (2008). *Learning from words*. Oxford: Oxford University Press. doi:10.1093/acprof:oso/9780199219162.001.0001

Lamarre, T. (March 2012). Humans and machines. *Inflexions*, 5, 30–68. Retrieved from http://www.inflexions.org/issues.html#i5

Lampathaki, F., Tsiakaliaris, C., Stasis, A., & Charalabidis, Y. (2011). National interoperability frameworks: The way forward. In Y. Charalabidis (Ed.), *Interoperability in digital public services and administration: Bridging e-government and e-business* (pp. 1–24). Hershey, PA: IGI Global. doi:10.4018/978-1-61520-887-6.ch001

Lange, C. J. (1885). *Om sindsbevaegelser: Et psyko-fysiologisk studie*. Copenhagen: Jacob Lunds.

Lan, Z., & Scott, C. R. (1996). The relative importance of computer-mediated information versus conventional non-computer-mediated information in public managerial decision making. *Information Resources Management Journal*, *9*(1), 27–0. doi:10.4018/irmj.1996010103

Latour, B. (2005). *Reassembling the social-An introduction to Actor-Network-Theory*. Oxford University Press.

Laughlin, C. D. Jr, McManus, J., & d'Aquili, E. G. (1990). *Brain, symbol and experience: Toward a neurophenomenology of consciousness*. Boston, MA: New Science Library.

Law, W. (2004). *Public sector data management in a developing economy*. Hershey, PA: IGI Global. doi:10.4018/978-1-59140-259-6.ch034

Law, W. K. (2005). Information resources development challenges in a cross-cultural environment. In M. Khosrow-Pour (Ed.), *Encyclopedia of information science and technology* (pp. 1476–1481). Hershey, PA: IGI Global. doi:10.4018/978-1-59140-553-5.ch259

Law, W. K. (2009). Cross-cultural challenges for information resources management. In M. Khosrow-Pour (Ed.), *Encyclopedia of information science and technology* (2nd ed., pp. 840–846). Hershey, PA: IGI Global. doi:10.4018/978-1-60566-026-4.ch136

Law, W. K. (2011). Cross-cultural challenges for information resources management. In *Global business: Concepts, methodologies, tools and applications* (pp. 1924–1932). Hershey, PA: IGI Global. doi:10.4018/978-1-60960-587-2.ch704

Lee, D. (2016). US opens investigation into Tesla after fatal crash. BBC. Retrieved from http://www.bbc.co.uk/news/technology-36680043

LEGO education. (2013). LEGO.com Education Home - Preschool and School - Secondary - 10+ LEGO *MINDSTORMS* Education EV3 - Instant Success - Introduction. Retrieved from http://education.lego.com/en-us/preschool-and-school/secondary/mindstorms-education-ev3/instant-success/introduction

Lemish, D. (2014). Using media to aid children in war, crisis, and vulnerable circumstances. In D. Romer & A. Jordan (Eds.), Media and the wellbeing of children and adolescents (pp. 162-171). Oxford: Oxford University Press.

Lemish, D. (2006). *Children and television: A global perspective*. New York, NY: Wiley-Blackwell.

Lemish, D. (2010). *Screening gender in children's TV: The views of producers around the world.* New York, NY: Routledge.

Lem, S. (2013). Wypadek. In *Opowieści o pilocie Pirxie* (pp. 276–311). Kraków: Wydawnictwo Literackie.

Levin, J. (2004). Functionalism. In *Stanford Encyclopedia of Philosophy*. Retrieved from http://plato.stanford.edu/entries/functionalism

Lévi-Strauss, C. (1966). *The savage mind.* Chicago, IL: University of Chicago Press.

Lewis, D. (1970). How to Define Theoretical Terms. *The Journal of Philosophy*, *67*(13), 427. doi:10.2307/2023861

Lewis, D. (1972). Psychophysical and theoretical identifications. *Australasian Journal of Philosophy*, *50*(3), 249–258. doi:10.1080/00048407212341301

Lie, M., & Sørensen, K. H. (1996). *Making technology our own? Domesticating technology into everyday life*. Scandinavian University Press Oslo.

Lii, N. Y., Leidner, D., Schiele, A., Birkenkampf, P., Pleintinger, B., & Bayer, R. (2015). Command robots from orbit with supervised autonomy: An introduction to the meteron supvis-justin experiment. In *Proceedings of the ACM/IEEE International Conference on Human-Robots Interaction* (HRI). doi:10.1145/2701973.2702022

Linebarger, D. L., Piotrowski, J., & Vaala, S. (2009). Supplementing television: What enhances or detracts from the power of television to teach. In N. N. Nagaraj (Ed.), Effect of television on children (pp. 45-71). Hyderabad: Icfai University Press.

Linebarger, D. L., & Walker, D. (2005). Infants and toddlers television viewing and language outcomes. *The American Behavioral Scientist*, *48*(5), 624–645. doi:10.1177/0002764204271505

Livingstone, S. (2007). Do the media harm children? Reflections on new approaches to an old problem. *Journal of Children and Media*, *1*(1), 5–14. doi:10.1080/17482790601005009

Livingstone, S., & Lemish, D. (2001). *Doing comparative research with children and young people*. London: London School of Economics.

Llinas, R. R. (1988). The intrinsic electrophysiological properties of mammalian neurons. *Science*, *242*(4886), 1654–1664. doi:10.1126/science.3059497 PMID:3059497

Losonczy, A., Makara, J. K., & Magee, J. C. (2008). Compartmentalized dendritic plasticity and input feature storage in neurons. *Nature*, *452*(7186), 436–440. doi:10.1038/nature06725 PMID:18368112

Luther, G. A., & Legg, J. R. Jr. (2010). Gender differences in depictions of social and physical aggression in childrens television cartoons in the U.S. *Journal of Children and Media*, *4*(2), 191–205. doi:10.1080/17482791003629651

Lutz, A., & Thompson, E. (2003). Neurophenomenology: Integrating subjective experience and brain dynamics in the neuroscience of consciousness. *Journal of Consciousness Studies*, *10*(9/10), 31–52.

Mackenzie, A. (2002). *Transductions: Bodies and machines at speed.* London, New York: Continuum.

MacLennan, B. J. (1996). Protophenomena and their Neurodynamical Correlates (Technical Report CS-96-331). University of Tennessee, Knoxville, Department of Computer Science.

MacLennan, B. J. (1997). The elements of consciousness and their neurodynamical correlates. In J. Shear (Ed.), Explaining consciousness: The hard problem (pp. 249–266). Cambridge, MA: MIT Press, .

MacLennan, B. J. (1995). The investigation of consciousness through phenomenology and neuroscience. In J. King & K. H. Pribram (Eds.), *Scale in conscious experience: Is the brain too important to be left to specialists to study?* (pp. 25–43). Hillsdale, NJ: Lawrence Erlbaum.

MacLennan, B. J. (2008). Consciousness: Natural and artificial. *Synthesis Philosophica*, *22*(2), 401–433.

MacLennan, B. J. (2009). Robots react but can they feel? A protophenomenological analysis. In J. Vallverdú & D. Casacuberta (Eds.), *Handbook of Research on Synthetic Emotions and Sociable Robotics: New Applications in Affective Computing and Artificial Intelligence* (pp. 133–153). Hershey, PA: IGI Global. doi:10.4018/978-1-60566-354-8.ch008

MacLennan, B. J. (2010). Protophenomena: The elements of consciousness and their relation to the brain. In A. Batthyány, A. Elitzur, & D. Constant (Eds.), *Irreducibly Conscious: Selected Papers on Consciousness* (pp. 189–214). Heidelberg & New York, NY: Universitätsverlag Winter.

MacLennan, B. J. (2011). Protophenomena and their physical correlates. *Journal of Cosmology*, *14*, 4516–4525. Reprinted. In R. Penrose & S. Hameroff (Eds.), *Consciousness and the universe: Quantum physics, evolution, brain & mind* (pp. 228–239). Cambridge, MA: Cosmology Science Publishers.

MacLennan, B. J. (2014). Ethical Treatment of Robots and the Hard Problem of Robot Emotions. *International Journal of Synthetic Emotions*, *5*(1), 10–18. doi:10.4018/ijse.2014010102

MacLennan, B. J. (2015). Living science: Science as an activity of living beings. *Progress in Biophysics and Molecular Biology*, *119*(3), 410–419. doi:10.1016/j.pbiomolbio.2015.08.009 PMID:26276467

Magrab, E. B. (2011). *An engineer's guide to MATLAB: With applications from mechanical, aerospace, electrical, civil, and biological systems engineering*. Upper Saddle River, N. J.: Prentice Hall.

Malcolm, C. (2000). From COGWHEELS to COGNITION: The knowledgeable robot. *Citeseer.* Retrieved from http://citeseerx.ist.psu.edu/viewdoc/summary?doi=10.1.1.194.4655

Malkia, M., & Savolainen, R. (2004). eTransformation in government, politics and society: Conceptual framework and introduction. In M. Malkia, A. Anttiroiko, & R. Savolainen (Eds.), eTransformation in governance: New directions in government and politics (pp. 1-21). Hershey, PA: IGI Global. doi:10.4018/978-1-59140-130-8.ch001

Malle, B., & Knobe, J. (1997). The Folk Concept of Intentionality. *Journal of Experimental Social Psychology*, *33*(2), 101–121. doi:10.1006/jesp.1996.1314

Malle, B., & Nelson, S. (2003). Judging mens rea: The tension between folk concepts and legal concepts of intentionality. *Behavioral Sciences & the Law*, *21*(5), 563–580. doi:10.1002/bsl.554 PMID:14502689

Maltese, A., & Tai, R. (2011). Pipeline persistence: Examining the association of educational experiences with earned degrees in STEM among U.S. students. *Science Education*, *95*(5), 877–907. doi:10.1002/sce.20441

Mandujano, S. (2011). Network manageability security. In D. Kar & M. Syed (Eds.), *Network security, administration and management: Advancing technology and practice* (pp. 158–181). Hershey, PA: IGI Global. doi:10.4018/978-1-60960-777-7.ch009

Mano, A., & Ishidori, H. (Directors). (2008). Gunslinger Girl -Il Teatrino- [Television series]. Japan: Artland.

Marciniak, M., & Klichowski, M. (2013). The paradox of commodification of the body in the society of consumption and cyborgization. *Studia Edukacyjne*, *29*, 153–167.

Mares, M. L., & Acosta, E. E. (2010). Teaching inclusiveness via TV narratives in the U.S. *Journal of Children and Media*, *4*(3), 231–247. doi:10.1080/17482798.2010.486127

Margolis, J., & Fisher, A. (2002). *Unlocking the clubhouse: Women in computing*. Cambridge, MA: MIT Press.

Marich, M. J., Schooley, B. L., & Horan, T. A. (2012). A normative enterprise architecture for guiding end-to-end emergency response decision support. In M. Jennex (Ed.), *Managing crises and disasters with emerging technologies: Advancements* (pp. 71–87). Hershey, PA: IGI Global. doi:10.4018/978-1-4666-0167-3.ch006

Markov, R., & Okujava, S. (2008). Costs, benefits, and risks of e-government portals. In G. Putnik & M. Cruz-Cunha (Eds.), *Encyclopedia of networked and virtual organizations* (pp. 354–363). Hershey, PA: IGI Global. doi:10.4018/978-1-59904-885-7.ch047

Martin, B., & Richards, E. (1995). Scientific knowledge, controversy, and public decision-making. In Handbook of science and technology studies (pp. 506-526).

Martin, N., & Rice, J. (2013). Evaluating and designing electronic government for the future: Observations and insights from Australia. In V. Weerakkody (Ed.), *E-government services design, adoption, and evaluation* (pp. 238–258). Hershey, PA: IGI Global. doi:10.4018/978-1-4666-2458-0.ch014

Marton, M. (2015). Intentional and Phenomenal Properties: How Not to Be Inseparatists. *Unpublished Article*. Retrieved from http://real.mtak.hu/27762

Maruyama, M., & Yamaki, I. (Producers), Rintaro, & Otomo K. (Directors). (2001). *Metropolis* [Motion picture]. Japan: Madhouse Studios.

Massumi, B. (2012). 'Technical mentality' revisited: Brian Massumi on Gilbert Simondon. In A. De Boever, A. Murray, J. Roffe, & A. Woodward (Eds.), *Gilbert Simondon: Being and technology* (pp. 19–36). Edinburgh: Edinburgh University Press.

Matthias, A. (2004). The responsibility gap: Ascribing responsibility for the actions of learning autonoma. *Ethics and Information Technology*, *6*(3), 175–183. doi:10.1007/s10676-004-3422-1

Maturana, H., & Varela, F. (2012). *Autopoiesis and cognition: The realization of the living*. Springer Science and Business Media. Retrieved from https://books.google.com/books?id=iOjVBQAAQBAJ

Mbarika, V. W., Meso, P. N., & Musa, P. F. (2006). A disconnect in stakeholders' perceptions from emerging realities of teledensity growth in Africa's least developed countries. In M. Hunter & F. Tan (Eds.), *Advanced topics in global information management* (Vol. 5, pp. 263–282). Hershey, PA: IGI Global. doi:10.4018/978-1-59140-923-6.ch012

Mbarika, V. W., Meso, P. N., & Musa, P. F. (2008). A disconnect in stakeholders' perceptions from emerging realities of teledensity growth in Africa's least developed countries. In F. Tan (Ed.), *Global information technologies: Concepts, methodologies, tools, and applications* (pp. 2948–2962). Hershey, PA: IGI Global. doi:10.4018/978-1-59904-939-7.ch209

McCall, R. J. (1983). *Phenomenological psychology: An introduction. With a glossary of some key Heideggerian terms*. Madison, WI: University of Wisconsin Press.

McLaughlin, J. (2016). FBI Says Apple Court Order Is Narrow, But Other Law Enforcers Hungry to Exploit It. Retrieved from https://theintercept.com/2016/02/22/fbi-says-apple-court-order-is-narrow-but-other-law-enforcers-hungry-to-exploit-it

McLeod, I. (2010). *Legal theory*. London: Palgrave.

Means, T., Olson, E., & Spooner, J. (2013). Discovering ways that don't work on the road to success: Strengths and weaknesses revealed by an active learning studio classroom project. In A. Benson, J. Moore, & S. Williams van Rooij (Eds.), *Cases on educational technology planning, design, and implementation: A project management perspective* (pp. 94–113). Hershey, PA: IGI Global. doi:10.4018/978-1-4666-4237-9.ch006

Melitski, J., Holzer, M., Kim, S., Kim, C., & Rho, S. (2008). Digital government worldwide: An e-government assessment of municipal web sites. In G. Garson & M. Khosrow-Pour (Eds.), *Handbook of research on public information technology* (pp. 790–804). Hershey, PA: IGI Global. doi:10.4018/978-1-59904-857-4.ch069

Melzack, R., & Casey, K. L. (1968). Sensory, motivational and central control determinants of chronic pain: A new conceptual model. In Kenshalo, D. R. (Ed.), *The skin senses: Proceedings of the first International Symposium on the Skin Senses, Florida State University in Tallahassee, Florida* (Ch. 20). Springfield: Charles C. Thomas.

Memmola, M., Palumbo, G., & Rossini, M. (2009). Web & RFID technology: New frontiers in costing and process management for rehabilitation medicine. In L. Al-Hakim & M. Memmola (Eds.), *Business web strategy: Design, alignment, and application* (pp. 145–169). Hershey, PA: IGI Global. doi:10.4018/978-1-60566-024-0.ch008

Meng, Z., Fahong, Z., & Lei, L. (2008). Information technology and environment. In Y. Kurihara, S. Takaya, H. Harui, & H. Kamae (Eds.), *Information technology and economic development* (pp. 201–212). Hershey, PA: IGI Global. doi:10.4018/978-1-59904-579-5.ch014

Mentzingen de Moraes, A. J., Ferneda, E., Costa, I., & Spinola, M. D. (2011). Practical approach for implementation of governance process in IT: Information technology areas. In N. Shi & G. Silvius (Eds.), *Enterprise IT governance, business value and performance measurement* (pp. 19–40). Hershey, PA: IGI Global. doi:10.4018/978-1-60566-346-3.ch002

Merwin, G. A. Jr, McDonald, J. S., & Odera, L. C. (2008). Economic development: Government's cutting edge in IT. In M. Raisinghani (Ed.), *Handbook of research on global information technology management in the digital economy* (pp. 1–37). Hershey, PA: IGI Global. doi:10.4018/978-1-59904-875-8.ch001

Meso, P. N., & Duncan, N. B. (2002). Can national information infrastructures enhance social development in the least developed countries? In F. Tan (Ed.), *Advanced topics in global information management* (Vol. 1, pp. 207–226). Hershey, PA: IGI Global. doi:10.4018/978-1-930708-43-3.ch014

Meso, P., & Duncan, N. (2002). Can national information infrastructures enhance social development in the least developed countries? An empirical investigation. In M. Dadashzadeh (Ed.), *Information technology management in developing countries* (pp. 23–51). Hershey, PA: IRM Press. doi:10.4018/978-1-931777-03-2.ch002

Middleton, M. (2008). Evaluation of e-government web sites. In G. Garson & M. Khosrow-Pour (Eds.), *Handbook of research on public information technology* (pp. 699–710). Hershey, PA: IGI Global. doi:10.4018/978-1-59904-857-4.ch063

Mills, S. (2011). FCJ-127 Concrete Software: Simondon's mechanology and the techno-social. *The Fibreculture Journal, 18*. Retrieved from http://eighteen.fibreculturejournal.org/2011/10/09/fcj-127-concrete-software-simondon%E2%80%99s-mechanology-and-the-techno-social

Mingers, J. (2010). Pluralism, realism, and truth: The keys to knowledge in information systems research. In D. Paradice (Ed.), *Emerging systems approaches in information technologies: Concepts, theories, and applications* (pp. 86–98). Hershey, PA: IGI Global. doi:10.4018/978-1-60566-976-2.ch006

Minsky, M. (1985). *Society of mind*. New York: Simon & Schuster.

Miraikan. (2014). Android: *What is Human?* National Museum of Emerging Science and Innovation. Retrieved from https://www.miraikan.jst.go.jp/en/exhibition/future/robot/android.html

Mital, K. M. (2012). ICT, unique identity and inclusive growth: An Indian perspective. In A. Manoharan & M. Holzer (Eds.), *E-governance and civic engagement: Factors and determinants of e-democracy* (pp. 584–612). Hershey, PA: IGI Global. doi:10.4018/978-1-61350-083-5.ch029

Mizell, A. P. (2008). Helping close the digital divide for financially disadvantaged seniors. In F. Tan (Ed.), *Global information technologies: Concepts, methodologies, tools, and applications* (pp. 2396–2402). Hershey, PA: IGI Global. doi:10.4018/978-1-59904-939-7.ch173

Molinari, F., Wills, C., Koumpis, A., & Moumtzi, V. (2011). A citizen-centric platform to support networking in the area of e-democracy. In H. Rahman (Ed.), *Cases on adoption, diffusion and evaluation of global e-governance systems: Impact at the grass roots* (pp. 282–302). Hershey, PA: IGI Global. doi:10.4018/978-1-61692-814-8.ch014

Molinari, F., Wills, C., Koumpis, A., & Moumtzi, V. (2013). A citizen-centric platform to support networking in the area of e-democracy. In H. Rahman (Ed.), *Cases on progressions and challenges in ICT utilization for citizen-centric governance* (pp. 265–297). Hershey, PA: IGI Global. doi:10.4018/978-1-4666-2071-1.ch013

Monteverde, F. (2010). The process of e-government public policy inclusion in the governmental agenda: A framework for assessment and case study. In J. Cordoba-Pachon & A. Ochoa-Arias (Eds.), *Systems thinking and e-participation: ICT in the governance of society* (pp. 233–245). Hershey, PA: IGI Global. doi:10.4018/978-1-60566-860-4.ch015

Moodley, S. (2008). Deconstructing the South African government's ICT for development discourse. In A. Anttiroiko (Ed.), *Electronic government: Concepts, methodologies, tools, and applications* (pp. 622–631). Hershey, PA: IGI Global. doi:10.4018/978-1-59904-947-2.ch053

Moodley, S. (2008). Deconstructing the South African government's ICT for development discourse. In C. Van Slyke (Ed.), *Information communication technologies: Concepts, methodologies, tools, and applications* (pp. 816–825). Hershey, PA: IGI Global. doi:10.4018/978-1-59904-949-6.ch052

Mora, M., Cervantes-Perez, F., Gelman-Muravchik, O., Forgionne, G. A., & Mejia-Olvera, M. (2003). DMSS implementation research: A conceptual analysis of the contributions and limitations of the factor-based and stage-based streams. In G. Forgionne, J. Gupta, & M. Mora (Eds.), *Decision-making support systems: Achievements and challenges for the new decade* (pp. 331–356). Hershey, PA: IGI Global. doi:10.4018/978-1-59140-045-5.ch020

Mori, M. (2012, June). *The Uncanny Valley*. (K. F. MacDorman & N. Kageki, Trans.). In *IEEE Robotics & Automation Magazine* (vol. 19), 98-100. Retrieved from http://ieeexplore.ieee.org/stamp/stamp.jsp?arnumber=6213238

Mori, M., MacDorman, K. F., & Kageki, N. (2012). The uncanny valley [from the field]. *Robotics and Automation Magazine, 19*(2), 98–100. doi:10.1109/MRA.2012.2192811

Morsella, E., Godwin, C. A., Jantz, T. K., Krieger, S. C., & Gazzaley, A. (2015). Homing in on consciousness in the nervous system: An action-based synthesis. In *Behavioral and Brain Sciences*. doi:10.1017/S0140525X15000643

Mörtberg, C., & Elovaara, P. (2010). Attaching people and technology: Between e and government. In S. Booth, S. Goodman, & G. Kirkup (Eds.), *Gender issues in learning and working with information technology: Social constructs and cultural contexts* (pp. 83–98). Hershey, PA: IGI Global. doi:10.4018/978-1-61520-813-5.ch005

Moss-Racusin, C., Molenda, A., & Cramer, C. (2015). Can evidence impact attitudes? Public reactions to evidence of gender bias in STEM fields. *Psychology of Women Quarterly*, *39*(2), 194–209, 39, 2, 194–209. doi:10.1177/0361684314565777

Murase, S. (Director). (2006). Ergo Proxy [Television series]. Japan: Manglobe INC.

Murphy, J., Harper, E., Devine, E. C., Burke, L. J., & Hook, M. L. (2011). Case study: Lessons learned when embedding evidence-based knowledge in a nurse care planning and documentation system. In A. Cashin & R. Cook (Eds.), *Evidence-based practice in nursing informatics: Concepts and applications* (pp. 174–190). Hershey, PA: IGI Global. doi:10.4018/978-1-60960-034-1.ch014

Mutula, S. M. (2013). E-government's role in poverty alleviation: Case study of South Africa. In H. Rahman (Ed.), *Cases on progressions and challenges in ICT utilization for citizen-centric governance* (pp. 44–68). Hershey, PA: IGI Global. doi:10.4018/978-1-4666-2071-1.ch003

Myers, K., Jahn, J., Gailliard, B., & Stoltzfus, K. (2011). Vocational anticipatory socialization (VAS): A communicative model of adolescents interests in STEM. *Management Communication Quarterly*, *25*(1), 87–120. doi:10.1177/0893318910377068

natgeobloggen.no. (2015). I Japan bygges verdens første hotell bemannet av robot-piker. Retrieved from http://www.natgeobloggen.no/2015/02/12/i-japan-bygges-verdens-forste-hotell-bemannet-av-robot-piker

Nath, R., & Angeles, R. (2005). Relationships between supply characteristics and buyer-supplier coupling in e-procurement: An empirical analysis. [IJEBR]. *International Journal of E-Business Research*, *1*(2), 40–55. doi:10.4018/jebr.2005040103

Nichols, S., & Ulatowski, J. (2007). Intuitions and individual differences: The Knobe effect revisited. *Mind & Language*, *22*(4), 346–365. doi:10.1111/j.1468-0017.2007.00312.x

Nissen, M. E. (2006). Application cases in government. In M. Nissen (Ed.), *Harnessing knowledge dynamics: Principled organizational knowing & learning* (pp. 152–181). Hershey, PA: IRM Press. doi:10.4018/978-1-59140-773-7.ch008

Norris, D. F. (2003). Leading-edge information technologies and American local governments. In G. Garson (Ed.), *Public information technology: Policy and management issues* (pp. 139–169). Hershey, PA: IGI Global. doi:10.4018/978-1-59140-060-8.ch007

Norris, D. F. (2008). Information technology among U.S. local governments. In G. Garson & M. Khosrow-Pour (Eds.), *Handbook of research on public information technology* (pp. 132–144). Hershey, PA: IGI Global. doi:10.4018/978-1-59904-857-4.ch013

Northrop, A. (1999). The challenge of teaching information technology in public administration graduate programs. In G. Garson (Ed.), *Information technology and computer applications in public administration: Issues and trends* (pp. 1–22). Hershey, PA: IGI Global. doi:10.4018/978-1-87828-952-0.ch001

Northrop, A. (2003). Information technology and public administration: The view from the profession. In G. Garson (Ed.), *Public information technology: Policy and management issues* (pp. 1–19). Hershey, PA: IGI Global. doi:10.4018/978-1-59140-060-8.ch001

Northrop, A. (2007). Lip service? How PA journals and textbooks view information technology. In G. Garson (Ed.), *Modern public information technology systems: Issues and challenges* (pp. 1–16). Hershey, PA: IGI Global. doi:10.4018/978-1-59904-051-6.ch001

Null, E. (2013). Legal and political barriers to municipal networks in the United States. In A. Abdelaal (Ed.), *Social and economic effects of community wireless networks and infrastructures* (pp. 27–56). Hershey, PA: IGI Global. doi:10.4018/978-1-4666-2997-4.ch003

O'Connor, S. (2015, July 1). A robot has killed a worker in a VW plant in Germany. ft.com/fastft/353721 [Tweet]. Retrieved from https://twittr.com/sarahoconnor_/status/616282747200479232

Ohlson, A. (2010). Anime: Fear and Anxiety in Texhnolyzed Worlds (Unpublished Thesis). Hamilton, New Zealand: University of Waikato. Retrieved from http://researchcommons.waikato.ac.nz/

Okunoye, A., Frolick, M., & Crable, E. (2006). ERP implementation in higher education: An account of pre-implementation and implementation phases. *Journal of Cases on Information Technology*, *8*(2), 110–132. doi:10.4018/jcit.2006040106

Olasina, G. (2012). A review of egovernment services in Nigeria. In A. Tella & A. Issa (Eds.), *Library and information science in developing countries: Contemporary issues* (pp. 205–221). Hershey, PA: IGI Global. doi:10.4018/978-1-61350-335-5.ch015

Orgeron, C. P. (2008). A model for reengineering IT job classes in state government. In G. Garson & M. Khosrow-Pour (Eds.), *Handbook of research on public information technology* (pp. 735–746). Hershey, PA: IGI Global. doi:10.4018/978-1-59904-857-4.ch066

Owsinski, J. W., & Pielak, A. M. (2011). Local authority websites in rural areas: Measuring quality and functionality, and assessing the role. In Z. Andreopoulou, B. Manos, N. Polman, & D. Viaggi (Eds.), *Agricultural and environmental informatics, governance and management: Emerging research applications* (pp. 39–60). Hershey, PA: IGI Global. doi:10.4018/978-1-60960-621-3.ch003

Owsiński, J. W., Pielak, A. M., Sęp, K., & Stańczak, J. (2014). Local web-based networks in rural municipalities: Extension, density, and meaning. In Z. Andreopoulou, V. Samathrakis, S. Louca, & M. Vlachopoulou (Eds.), *E-innovation for sustainable development of rural resources during global economic crisis* (pp. 126–151). Hershey, PA: IGI Global. doi:10.4018/978-1-4666-4550-9.ch011

Pagani, M., & Pasinetti, C. (2008). Technical and functional quality in the development of t-government services. In A. Anttiroiko (Ed.), *Electronic government: Concepts, methodologies, tools, and applications* (pp. 2943–2965). Hershey, PA: IGI Global. doi:10.4018/978-1-59904-947-2.ch220

Pajarinen, M., Rouvinen, P., & Ekeland, A. (2014). Computerization and the Future of Jobs in Norway. Department of Safety and Service Organization. Retrieved from http://nettsteder.regjeringen.no/fremtidensskole/files/2014/05/Computerization-and-the-Future-of-Jobs-in-Norway.pdf

Pani, A. K., & Agrahari, A. (2005). On e-markets in emerging economy: An Indian experience. In M. Khosrow-Pour (Ed.), *Advanced topics in electronic commerce* (Vol. 1, pp. 287–299). Hershey, PA: IGI Global. doi:10.4018/978-1-59140-819-2.ch015

Papadopoulos, T., Angelopoulos, S., & Kitsios, F. (2011). A strategic approach to e-health interoperability using e-government frameworks. In A. Lazakidou, K. Siassiakos, & K. Ioannou (Eds.), *Wireless technologies for ambient assisted living and healthcare: Systems and applications* (pp. 213–229). Hershey, PA: IGI Global. doi:10.4018/978-1-61520-805-0.ch012

Papadopoulos, T., Angelopoulos, S., & Kitsios, F. (2013). A strategic approach to e-health interoperability using e-government frameworks. In *User-driven healthcare: Concepts, methodologies, tools, and applications* (pp. 791–807). Hershey, PA: IGI Global. doi:10.4018/978-1-4666-2770-3.ch039

Papaleo, G., Chiarella, D., Aiello, M., & Caviglione, L. (2012). Analysis, development and deployment of statistical anomaly detection techniques for real e-mail traffic. In T. Chou (Ed.), *Information assurance and security technologies for risk assessment and threat management: Advances* (pp. 47–71). Hershey, PA: IGI Global. doi:10.4018/978-1-61350-507-6.ch003

Papert, S. (1980). *Mindstorms: Children, computers, and powerful ideas*. New York: Basic Books.

Papp, R. (2003). Information technology & FDA compliance in the pharmaceutical industry. In M. Khosrow-Pour (Ed.), *Annals of cases on information technology* (Vol. 5, pp. 262–273). Hershey, PA: IGI Global. doi:10.4018/978-1-59140-061-5.ch017

Parkin, S. (2017, June/July). Teaching robots right from wrong. *1843 Magazine*. Retrieved from https://www.1843magazine.com/features/teaching-robots-right-from-wrong

Parsons, T. W. (2007). Developing a knowledge management portal. In A. Tatnall (Ed.), *Encyclopedia of portal technologies and applications* (pp. 223–227). Hershey, PA: IGI Global. doi:10.4018/978-1-59140-989-2.ch039

Passaris, C. E. (2007). Immigration and digital government. In A. Anttiroiko & M. Malkia (Eds.), *Encyclopedia of digital government* (pp. 988–994). Hershey, PA: IGI Global. doi:10.4018/978-1-59140-789-8.ch148

Pavlichev, A. (2004). The e-government challenge for public administration. In A. Pavlichev & G. Garson (Eds.), *Digital government: Principles and best practices* (pp. 276–290). Hershey, PA: IGI Global. doi:10.4018/978-1-59140-122-3.ch018

Penrod, J. I., & Harbor, A. F. (2000). Designing and implementing a learning organization-oriented information technology planning and management process. In L. Petrides (Ed.), *Case studies on information technology in higher education: Implications for policy and practice* (pp. 7–19). Hershey, PA: IGI Global. doi:10.4018/978-1-878289-74-2.ch001

Pesce, M. (2000). *The playful world: How technology is transforming our imagination*. New York: Ballantine Books.

Pettit, D. L., Wang, S. S., Gee, K. R., & Augustine, G. J. (1997). Chemical two-photon uncaging: A novel approach to mapping glutamate receptors. *Neuron*, *19*(3), 465–471. doi:10.1016/S0896-6273(00)80361-X PMID:9331338

Pinch, T. J., & Bijker, W. E. (1984). The social construction of facts and artifacts: Or how the sociology of science and the sociology of technology might benefit each other. *Social Studies of Science*, *14*(3), 399–441. doi:10.1177/030631284014003004

Planas-Silva, M. D., & Joseph, R. C. (2011). Perspectives on the adoption of electronic resources for use in clinical trials. In M. Guah (Ed.), *Healthcare delivery reform and new technologies: Organizational initiatives* (pp. 19–28). Hershey, PA: IGI Global. doi:10.4018/978-1-60960-183-6.ch002

Plutchik, R. (2000). *Emotions in the practice of psychotherapy: Clinical implications of affect theories*. New York: American Psychological Association. doi:10.1037/10366-000

Plutchik, R. (2003). *Emotions and life: Perspectives from psychology, biology, and evolution*. New York: American Psychological Association.

Plutchik, R., & Conte, H. R. (Eds.). (1997). *Circumplex models of personality and emotions*. Washington, DC: American Psychological Association. doi:10.1037/10261-000

Pomazalová, N., & Rejman, S. (2013). The rationale behind implementation of new electronic tools for electronic public procurement. In N. Pomazalová (Ed.), *Public sector transformation processes and internet public procurement: Decision support systems* (pp. 85–117). Hershey, PA: IGI Global. doi:10.4018/978-1-4666-2665-2.ch006

Porush, D. (1984). *The soft machine: Cybernetic fiction*. London: Methuen Young Books.

Postorino, M. N. (2012). City competitiveness and airport: Information science perspective. In M. Bulu (Ed.), *City competitiveness and improving urban subsystems: Technologies and applications* (pp. 61–83). Hershey, PA: IGI Global. doi:10.4018/978-1-61350-174-0.ch004

Poupa, C. (2002). Electronic government in Switzerland: Priorities for 2001-2005 - Electronic voting and federal portal. In Å. Grönlund (Ed.), *Electronic government: Design, applications and management* (pp. 356–369). Hershey, PA: IGI Global. doi:10.4018/978-1-930708-19-8.ch017

Powell, S. R. (2010). Interdisciplinarity in telecommunications and networking. In *Networking and telecommunications: Concepts, methodologies, tools and applications* (pp. 33–40). Hershey, PA: IGI Global. doi:10.4018/978-1-60566-986-1.ch004

Prinz, J. (2006). Gut reactions: A perceptual theory of emotion. Oxford: Oxford University Press. doi:10.1093/acprof:oso/9780195309362.001.0001

Priya, P. S., & Mathiyalagan, N. (2011). A study of the implementation status of two e-governance projects in land revenue administration in India. In M. Shareef, V. Kumar, U. Kumar, & Y. Dwivedi (Eds.), *Stakeholder adoption of e-government services: Driving and resisting factors* (pp. 214–230). Hershey, PA: IGI Global. doi:10.4018/978-1-60960-601-5.ch011

Prysby, C. L., & Prysby, N. D. (2003). Electronic mail in the public workplace: Issues of privacy and public disclosure. In G. Garson (Ed.), *Public information technology: Policy and management issues* (pp. 271–298). Hershey, PA: IGI Global. doi:10.4018/978-1-59140-060-8.ch012

Prysby, C. L., & Prysby, N. D. (2007). You have mail, but who is reading it? Issues of e-mail in the public workplace. In G. Garson (Ed.), *Modern public information technology systems: Issues and challenges* (pp. 312–336). Hershey, PA: IGI Global. doi:10.4018/978-1-59904-051-6.ch016

Prysby, C., & Prysby, N. (2000). Electronic mail, employee privacy and the workplace. In L. Janczewski (Ed.), *Internet and intranet security management: Risks and solutions* (pp. 251–270). Hershey, PA: IGI Global. doi:10.4018/978-1-878289-71-1.ch009

Putnam, H. (1967). The Nature of Mental States. In W. Capitan (Ed.), *Art, mind, and religion* (1st ed., pp. 1–23). Pittsburgh: Pittsburgh University Press.

Racter. (1984). *The policeman's beard is half constructed: Computer prose and poetry by Racter.* New York: Warner.

Radkowska-Walkowicz, M. (2008). *Od Golema do Terminatora: Wizerunki sztucznego człowieka w kulturze.* Warszawa: Wydawnictwa Akademickie i Profesjonalne.

Radl, A., & Chen, Y. (2005). Computer security in electronic government: A state-local education information system. *International Journal of Electronic Government Research, 1*(1), 79–99. doi:10.4018/jegr.2005010105

Rahman, H. (2008). Information dynamics in developing countries. In C. Van Slyke (Ed.), *Information communication technologies: Concepts, methodologies, tools, and applications* (pp. 104–114). Hershey, PA: IGI Global. doi:10.4018/978-1-59904-949-6.ch008

Ramanathan, J. (2009). Adaptive IT architecture as a catalyst for network capability in government. In P. Saha (Ed.), *Advances in government enterprise architecture* (pp. 149–172). Hershey, PA: IGI Global. doi:10.4018/978-1-60566-068-4.ch007

Ramos, I., & Berry, D. M. (2006). Social construction of information technology supporting work. In M. Khosrow-Pour (Ed.), *Cases on information technology: Lessons learned* (Vol. 7, pp. 36–52). Hershey, PA: IGI Global. doi:10.4018/978-1-59140-673-0.ch003

Ray, D., Gulla, U., Gupta, M. P., & Dash, S. S. (2009). Interoperability and constituents of interoperable systems in public sector. In V. Weerakkody, M. Janssen, & Y. Dwivedi (Eds.), *Handbook of research on ICT-enabled transformational government: A global perspective* (pp. 175–195). Hershey, PA: IGI Global. doi:10.4018/978-1-60566-390-6.ch010

Reddick, C. G. (2007). E-government and creating a citizen-centric government: A study of federal government CIOs. In G. Garson (Ed.), *Modern public information technology systems: Issues and challenges* (pp. 143–165). Hershey, PA: IGI Global. doi:10.4018/978-1-59904-051-6.ch008

Reddick, C. G. (2010). Citizen-centric e-government. In C. Reddick (Ed.), *Homeland security preparedness and information systems: Strategies for managing public policy* (pp. 45–75). Hershey, PA: IGI Global. doi:10.4018/978-1-60566-834-5.ch002

Reddick, C. G. (2010). E-government and creating a citizen-centric government: A study of federal government CIOs. In C. Reddick (Ed.), *Homeland security preparedness and information systems: Strategies for managing public policy* (pp. 230–250). Hershey, PA: IGI Global. doi:10.4018/978-1-60566-834-5.ch012

Reddick, C. G. (2010). Perceived effectiveness of e-government and its usage in city governments: Survey evidence from information technology directors. In C. Reddick (Ed.), *Homeland security preparedness and information systems: Strategies for managing public policy* (pp. 213–229). Hershey, PA: IGI Global. doi:10.4018/978-1-60566-834-5.ch011

Reddick, C. G. (2012). Customer relationship management adoption in local governments in the United States. In S. Chhabra & M. Kumar (Eds.), *Strategic enterprise resource planning models for e-government: Applications and methodologies* (pp. 111–124). Hershey, PA: IGI Global. doi:10.4018/978-1-60960-863-7.ch008

Reeder, F. S., & Pandy, S. M. (2008). Identifying effective funding models for e-government. In A. Anttiroiko (Ed.), *Electronic government: Concepts, methodologies, tools, and applications* (pp. 1108–1138). Hershey, PA: IGI Global. doi:10.4018/978-1-59904-947-2.ch083

Rey, G. (1997). *Contemporary philosophy of mind*. Cambridge, Mass.: Blackwell.

Rideout, V. J., Foehr, U. G., & Roberts, D. F. (2010). Generation M2: Media in the Lives of 8- to 18-Year-Olds. Menlo Park, CA: Kaiser Family Foundation. Retrieved from http://kff.org/other/event/generation-m2-media-in-the-lives-of

Riesco, D., Acosta, E., & Montejano, G. (2003). An extension to a UML activity graph from workflow. In L. Favre (Ed.), *UML and the unified process* (pp. 294–314). Hershey, PA: IRM Press. doi:10.4018/978-1-93177-744-5.ch015

Ritzhaupt, A. D., & Gill, T. G. (2008). A hybrid and novel approach to teaching computer programming in MIS curriculum. In S. Negash, M. Whitman, A. Woszczynski, K. Hoganson, & H. Mattord (Eds.), *Handbook of distance learning for real-time and asynchronous information technology education* (pp. 259–281). Hershey, PA: IGI Global. doi:10.4018/978-1-59904-964-9.ch014

Robarts, S. (2015, July 22). New Japanese hotel has robot staff and no room keys. Robotics. *New Atlas*. Retrieved from http://newatlas.com/henn-na-hotel-robot-staff/38577

Robertson, J. (2010). Gendering humanoid robots: Robo-sexism in Japan. *Body & Society*, *16*(2), 1–36. doi:10.1177/1357034X10364767

Robitzski, D. (2017). Researchers Anticipate Problems as Cyborgs Become a Reality. *Inverse*. Retrieved from https://inverse.com/article/32271-cyborg-research-issues-looking-forward

Robot Operating System. (n. d.). Retrieved March 24, 2014 from http://www.ros.org/about-ros/

Roche, E. M. (1993). International computing and the international regime. *Journal of Global Information Management*, *1*(2), 33–44. doi:10.4018/jgim.1993040103

Rocheleau, B. (2007). Politics, accountability, and information management. In G. Garson (Ed.), *Modern public information technology systems: Issues and challenges* (pp. 35–71). Hershey, PA: IGI Global. doi:10.4018/978-1-59904-051-6.ch003

Roco, M. C., & Bainbridge, W. S. (2011). *Converging technologies for improving human performance: Nanotechnology, biotechnology, information technology and cognitive science*. Dordrecht: Springer.

Rodrigues Filho, J. (2010). E-government in Brazil: Reinforcing dominant institutions or reducing citizenship? In C. Reddick (Ed.), *Politics, democracy and e-government: Participation and service delivery* (pp. 347–362). Hershey, PA: IGI Global. doi:10.4018/978-1-61520-933-0.ch021

Rodriguez, S. R., & Thorp, D. A. (2013). eLearning for industry: A case study of the project management process. In A. Benson, J. Moore, & S. Williams van Rooij (Eds.), Cases on educational technology planning, design, and implementation: A project management perspective (pp. 319-342). Hershey, PA: IGI Global. doi:10.4018/978-1-4666-4237-9.ch017

Rodriguez, G., Aromataris, L., Donolo, M., Hernández, J., & Moitre, D. (2007). Software for calculation of electric power systems parameters. *Computers & Electrical Engineering an International Journal*, *29*(5), 643–651. doi:10.1016/S0045-7906(01)00053-2

Rolls, E. T. (2005). *Emotion explained*. Oxford, UK: Oxford University Press. doi:10.1093/acprof:oso/9780198570035.001.0001

Rolls, E. T. (2006). *Brain mechanisms of emotion and decision-making*. In International Congress Series (Vol. 1291, pp. 3–13). Amsterdam, Netherlands: Elsevier.

Rolls, E. T. (2007). A neurobiological approach to emotional intelligence. In G. Matthews, M. Zeidner, & R. D. Roberts (Eds.), *The science of emotional intelligence* (pp. 72–100). Oxford, UK: Oxford University Press.

Roman, A. V. (2013). Delineating three dimensions of e-government success: Security, functionality, and transformation. In J. Gil-Garcia (Ed.), *E-government success factors and measures: Theories, concepts, and methodologies* (pp. 171–192). Hershey, PA: IGI Global. doi:10.4018/978-1-4666-4058-0.ch010

Ross, S. C., Tyran, C. K., & Auer, D. J. (2008). Up in smoke: Rebuilding after an IT disaster. In H. Nemati (Ed.), *Information security and ethics: Concepts, methodologies, tools, and applications* (pp. 3659–3675). Hershey, PA: IGI Global. doi:10.4018/978-1-59904-937-3.ch248

Ross, S. C., Tyran, C. K., Auer, D. J., Junell, J. M., & Williams, T. G. (2005). Up in smoke: Rebuilding after an IT disaster. *Journal of Cases on Information Technology*, *7*(2), 31–49. doi:10.4018/jcit.2005040103

Roy, J. (2008). Security, sovereignty, and continental interoperability: Canada's elusive balance. In T. Loendorf & G. Garson (Eds.), *Patriotic information systems* (pp. 153–176). Hershey, PA: IGI Global. doi:10.4018/978-1-59904-594-8.ch007

Rubeck, R. F., & Miller, G. A. (2009). vGOV: Remote video access to government services. In A. Scupola (Ed.), Cases on managing e-services (pp. 253-268). Hershey, PA: IGI Global. doi:10.4018/978-1-60566-064-6.ch017

Rudrauf, D., Lutz, A., Cosmelli, D., Lachaux, J.-L., & Le Van Quyen, M. (2003). From autopoiesis to neurophenomenology: Francisco Varelas exploration of the biophysics of being. *Biological Research*, *36*(1), 27–65. doi:10.4067/S0716-97602003000100005 PMID:12795206

Russell, B. (1948). *Human knowledge*. New York: Simon and Schuster.

Saekow, A., & Boonmee, C. (2011). The challenges of implementing e-government interoperability in Thailand: Case of official electronic correspondence letters exchange across government departments. In Y. Charalabidis (Ed.), *Interoperability in digital public services and administration: Bridging e-government and e-business* (pp. 40–61). Hershey, PA: IGI Global. doi:10.4018/978-1-61520-887-6.ch003

Saekow, A., & Boonmee, C. (2012). The challenges of implementing e-government interoperability in Thailand: Case of official electronic correspondence letters exchange across government departments. In *Digital democracy: Concepts, methodologies, tools, and applications* (pp. 1883–1905). Hershey, PA: IGI Global. doi:10.4018/978-1-4666-1740-7.ch094

Sagsan, M., & Medeni, T. (2012). Understanding "knowledge management (KM) paradigms" from social media perspective: An empirical study on discussion group for KM at professional networking site. In M. Cruz-Cunha, P. Gonçalves, N. Lopes, E. Miranda, & G. Putnik (Eds.), *Handbook of research on business social networking: Organizational, managerial, and technological dimensions* (pp. 738–755). Hershey, PA: IGI Global. doi:10.4018/978-1-61350-168-9.ch039

Sahi, G., & Madan, S. (2013). Information security threats in ERP enabled e-governance: Challenges and solutions. In *Enterprise resource planning: Concepts, methodologies, tools, and applications* (pp. 825–837). Hershey, PA: IGI Global. doi:10.4018/978-1-4666-4153-2.ch048

Sakoi, M. (Director). (2009). Needless [Television series]. Japan: Madhouse Studios.

Salisbury, J. E. (2011). *The beast within: Animals in the Middle Ages*. Routledge.

Sandvik, K. (2007). The player, the controller and the amazing avatar: Extending the human body into virtual environments. København, Denmark.

Sanford, C., & Bhattacherjee, A. (2008). IT implementation in a developing country municipality: A sociocognitive analysis. *International Journal of Technology and Human Interaction*, *4*(3), 68–93. doi:10.4018/jthi.2008070104

Sayre, K. (1986). Intentionality and information processing: An alternative model for cognitive science. *Behavioral and Brain Sciences*, *9*(01), 121. doi:10.1017/S0140525X00021750

Sayre, K. (2014). *Cybernetics and the philosophy of mind*. Hoboken: Taylor & Francis.

Schelin, S. H. (2003). E-government: An overview. In G. Garson (Ed.), *Public information technology: Policy and management issues* (pp. 120–138). Hershey, PA: IGI Global. doi:10.4018/978-1-59140-060-8.ch006

Schelin, S. H. (2004). Training for digital government. In A. Pavlichev & G. Garson (Eds.), *Digital government: Principles and best practices* (pp. 263–275). Hershey, PA: IGI Global. doi:10.4018/978-1-59140-122-3.ch017

Schelin, S. H. (2007). E-government: An overview. In G. Garson (Ed.), *Modern public information technology systems: Issues and challenges* (pp. 110–126). Hershey, PA: IGI Global. doi:10.4018/978-1-59904-051-6.ch006

Schelin, S. H., & Garson, G. (2004). Theoretical justification of critical success factors. In G. Garson & S. Schelin (Eds.), *IT solutions series: Humanizing information technology: Advice from experts* (pp. 4–15). Hershey, PA: CyberTech Publishing. doi:10.4018/978-1-59140-245-9.ch002

Schlote, E. (2012). Learning how to learn. *TelevIZIon, 25*(E), 26-29.

Schlote, E. (2013). Wer begleitet Kinder in Erklär- und Wissenssendungen? Eine Medienanalyse der LernbegleiterInnen in Wissenssendungen in Deutschland, Großbritannien und den U.S.A. *TelevIZIon*, *26*(2), 44–47.

Schulz, M. (2003). Ciało do korekty - cyborgi. In P. Kletkowski & P. Marecki (Eds.), *Nowe Nawigacje II. Azja - cyberpunk - kino niezależne* (pp. 157–176). Kraków: Rabid.

Scime, A. (2002). Information systems and computer science model curricula: A comparative look. In M. Dadashzadeh, A. Saber, & S. Saber (Eds.), *Information technology education in the new millennium* (pp. 146–158). Hershey, PA: IRM Press. doi:10.4018/978-1-931777-05-6.ch018

Scime, A. (2009). Computing curriculum analysis and development. In M. Khosrow-Pour (Ed.), *Encyclopedia of information science and technology* (2nd ed., pp. 667–671). Hershey, PA: IGI Global. doi:10.4018/978-1-60566-026-4.ch108

Scime, A., & Wania, C. (2008). Computing curricula: A comparison of models. In C. Van Slyke (Ed.), *Information communication technologies: Concepts, methodologies, tools, and applications* (pp. 1270–1283). Hershey, PA: IGI Global. doi:10.4018/978-1-59904-949-6.ch088

Searle, J. (1980). Minds, brains, and programs. *Behavioral and Brain Sciences*, *3*(03), 417. doi:10.1017/S0140525X00005756

Seidman, S. B. (2009). An international perspective on professional software engineering credentials. In H. Ellis, S. Demurjian, & J. Naveda (Eds.), *Software engineering: Effective teaching and learning approaches and practices* (pp. 351–361). Hershey, PA: IGI Global. doi:10.4018/978-1-60566-102-5.ch018

Seifert, J. W. (2007). E-government act of 2002 in the United States. In A. Anttiroiko & M. Malkia (Eds.), *Encyclopedia of digital government* (pp. 476–481). Hershey, PA: IGI Global. doi:10.4018/978-1-59140-789-8.ch072

Seifert, J. W., & Relyea, H. C. (2008). E-government act of 2002 in the United States. In A. Anttiroiko (Ed.), *Electronic government: Concepts, methodologies, tools, and applications* (pp. 154–161). Hershey, PA: IGI Global. doi:10.4018/978-1-59904-947-2.ch013

Service, R. F. (2013). The cyborg era begins: Advances in flexible electronics now makes it possible to integrate circuits with tissues. *Science*, *340*(6137), 1162–1165. doi:10.1126/science.340.6137.1162 PMID:23744924

Seufert, S. (2002). E-learning business models: Framework and best practice examples. In M. Raisinghani (Ed.), *Cases on worldwide e-commerce: Theory in action* (pp. 70–94). Hershey, PA: IGI Global. doi:10.4018/978-1-930708-27-3.ch004

Shapiro, J., & Williams, A. (2012). The role of stereotype threats in undermining girls and womens performance and interest in STEM fields. *Sex Roles*, *66*(3-4), 175–183. doi:10.1007/s11199-011-0051-0

Shareef, M. A., & Archer, N. (2012). E-government initiatives: Review studies on different countries. In M. Shareef, N. Archer, & S. Dutta (Eds.), *E-government service maturity and development: Cultural, organizational and technological perspectives* (pp. 40–76). Hershey, PA: IGI Global. doi:10.4018/978-1-60960-848-4.ch003

Shareef, M. A., & Archer, N. (2012). E-government service development. In M. Shareef, N. Archer, & S. Dutta (Eds.), *E-government service maturity and development: Cultural, organizational and technological perspectives* (pp. 1–14). Hershey, PA: IGI Global. doi:10.4018/978-1-60960-848-4.ch001

Shareef, M. A., Kumar, U., & Kumar, V. (2011). E-government development: Performance evaluation parameters. In M. Shareef, V. Kumar, U. Kumar, & Y. Dwivedi (Eds.), *Stakeholder adoption of e-government services: Driving and resisting factors* (pp. 197–213). Hershey, PA: IGI Global. doi:10.4018/978-1-60960-601-5.ch010

Shareef, M. A., Kumar, U., Kumar, V., & Niktash, M. (2012). Electronic-government vision: Case studies for objectives, strategies, and initiatives. In M. Shareef, N. Archer, & S. Dutta (Eds.), *E-government service maturity and development: Cultural, organizational and technological perspectives* (pp. 15–39). Hershey, PA: IGI Global. doi:10.4018/978-1-60960-848-4.ch002

Shigematsu, S. (1999). Dimensions of desire: Sex, fantasy and fetish in Japanese comics. In J. A. Lent (Ed.), *Themes and issues in Asian cartooning: Cute, cheap, mad and sexy*. Bowling Green, OH: Bowling Green State University Popular Press.

Shukla, P., Kumar, A., & Anu Kumar, P. B.Anu Kumar P.B. (2013). Impact of national culture on business continuity management system implementation. *International Journal of Risk and Contingency Management*, *2*(3), 23–36. doi:10.4018/ijrcm.2013070102

Shulman, S. W. (2007). The federal docket management system and the prospect for digital democracy in U S rulemaking. In G. Garson (Ed.), *Modern public information technology systems: Issues and challenges* (pp. 166–184). Hershey, PA: IGI Global. doi:10.4018/978-1-59904-051-6.ch009

Silverstone, R. (2005). Domesticating domestication. Reflections on the life of a concept. In M. Hartmann (Ed.), *Domestication of media and technology* (p. 229). Open University Press.

Simondon, G. (1980). [1958]. On the mode of existence of technical objects (N. Mellamphy, Trans.). University of Western Ontario. Retrieved from https://english.duke.edu/uploads/assets/Simondon_MEOT_part_1.pdf

Simondon, G. (1992). The genesis of the individual. In *J. Crary, & S. Kwinter (Ed.) Incorporations* (pp. 297–319). Cambridge, MA: Zone Books.

Simon, H. (1977). What computers mean for man and society. *Science*, *195*(4283), 1186–1191. doi:10.1126/science.195.4283.1186 PMID:17789730

Simonovic, S. (2007). Problems of offline government in e-Serbia. In A. Anttiroiko & M. Malkia (Eds.), *Encyclopedia of digital government* (pp. 1342–1351). Hershey, PA: IGI Global. doi:10.4018/978-1-59140-789-8.ch205

Simonovic, S. (2008). Problems of offline government in e-Serbia. In A. Anttiroiko (Ed.), *Electronic government: Concepts, methodologies, tools, and applications* (pp. 2929–2942). Hershey, PA: IGI Global. doi:10.4018/978-1-59904-947-2.ch219

Singh, A. M. (2005). Information systems and technology in South Africa. In M. Khosrow-Pour (Ed.), *Encyclopedia of information science and technology* (pp. 1497–1502). Hershey, PA: IGI Global. doi:10.4018/978-1-59140-553-5.ch263

Singh, S., & Naidoo, G. (2005). Towards an e-government solution: A South African perspective. In W. Huang, K. Siau, & K. Wei (Eds.), *Electronic government strategies and implementation* (pp. 325–353). Hershey, PA: IGI Global. doi:10.4018/978-1-59140-348-7.ch014

Sjøberg, J., & Hager-Thoresen, F. (2015). Nordmenn tror ikke roboter vil ta jobbene deres. *Aftenposten*. Retrieved from http://www.aftenposten.no/article/ap-8130866.html

Smith, S. L., & Cook, C. A. (2008). *Female stereotypes: An analysis of popular film and television*. Los Angeles, CA: The Geena Davis Institute for Gender and Media.

Smith, S. T. (2006). *MATLAB: Advanced GUI development*. USA: Dog Ear Publishing.

Snoke, R., & Underwood, A. (2002). Generic attributes of IS graduates: An analysis of Australian views. In F. Tan (Ed.), *Advanced topics in global information management* (Vol. 1, pp. 370–384). Hershey, PA: IGI Global. doi:10.4018/978-1-930708-43-3.ch023

Sobol, A. (2003). Cyberprzestrzeń – elektryczny afterlife. In P. Kletkowski & P. Marecki (Eds.), *Nowe nawigacje II. Azja - cyberpunk - kino niezależne* (pp. 137–156). Kraków: Rabid.

Sommer, L. (2006). Revealing unseen organizations in higher education: A study framework and application example. In A. Metcalfe (Ed.), *Knowledge management and higher education: A critical analysis* (pp. 115–146). Hershey, PA: IGI Global. doi:10.4018/978-1-59140-509-2.ch007

Song, H., Kidd, T., & Owens, E. (2011). Examining technological disparities and instructional practices in English language arts classroom: Implications for school leadership and teacher training. In L. Tomei (Ed.), *Online courses and ICT in education: Emerging practices and applications* (pp. 258–274). Hershey, PA: IGI Global. doi:10.4018/978-1-60960-150-8.ch020

Sosa, E. (1991). Testimony and Coherence. In E. Sosa (Ed.), *Knowledge in perspective* (1st ed., pp. 215–222). Cambridge: Cambridge University Press. doi:10.1017/CBO9780511625299.013

Sosa, E. (2006). Knowledge: Testimonial and Instrumental. In J. Lackey & E. Sosa (Eds.), *The epistemology of testimony* (1st ed., pp. 116–123). Oxford: Oxford University Press. doi:10.1093/acprof:oso/9780199276011.003.0006

Speaker, P. J., & Kleist, V. F. (2003). Using information technology to meet electronic commerce and MIS education demands. In A. Aggarwal (Ed.), *Web-based education: Learning from experience* (pp. 280–291). Hershey, PA: IGI Global. doi:10.4018/978-1-59140-102-5.ch017

Spitler, V. K. (2007). Learning to use IT in the workplace: Mechanisms and masters. In M. Mahmood (Ed.), *Contemporary issues in end user computing* (pp. 292–323). Hershey, PA: IGI Global. doi:10.4018/978-1-59140-926-7.ch013

Spreen, D. (2010). *Was ver-spricht der cyborg*? Retrieved from: http://nbn-resolving.de/urn:nbn:de:0168-ssoar-207098

Stableford, B. M. (2006). Science fact and science fiction: An encyclopedia. New York: Routledge. Retrieved from https://books.google.com/books?id=uefwmdROKTAC

statista.com. (2016). Leading companies in the global industrial robot market in 2016, based on revenue from industrial robot sales (in million euros). Statista. Retrieved from http://www.statista.com/statistics/257177/global-industrial-robot-market-share-by-company

Stebbing, T. (2011). *A cybernetic view of biological growth: The Maia hypothesis*. Cambridge: Cambridge University Press.

Steinke, J., Lapinski, M., Crocker, N., Zietsman-Thomas, A., Williams, Y., Evergreen, S., & Kuchibhotla, S. (2007). Assessing media influences on middle school-aged childrens perception of women in science using the draw-a-scientist test (DAST). *Science Communication*, *29*(1), 35–64. doi:10.1177/1075547007306508

Stellefson, M. (2011). Considerations for marketing distance education courses in health education: Five important questions to examine before development. In U. Demiray & S. Sever (Eds.), *Marketing online education programs: Frameworks for promotion and communication* (pp. 222–234). Hershey, PA: IGI Global. doi:10.4018/978-1-60960-074-7.ch014

Straub, D. W., & Loch, K. D. (2006). Creating and developing a program of global research. *Journal of Global Information Management*, *14*(2), 1–28. doi:10.4018/jgim.2006040101

Straub, D. W., Loch, K. D., & Hill, C. E. (2002). Transfer of information technology to the Arab world: A test of cultural influence modeling. In M. Dadashzadeh (Ed.), *Information technology management in developing countries* (pp. 92–134). Hershey, PA: IRM Press. doi:10.4018/978-1-931777-03-2.ch005

Straub, D. W., Loch, K. D., & Hill, C. E. (2003). Transfer of information technology to the Arab world: A test of cultural influence modeling. In F. Tan (Ed.), *Advanced topics in global information management* (Vol. 2, pp. 141–172). Hershey, PA: IGI Global. doi:10.4018/978-1-59140-064-6.ch009

Strauss, A., & Corbin, J. (1994). Grounded theory methodology. In Handbook of qualitative research (pp. 273-285).

Strawson, G. (1994). *Mental reality*. Cambridge, MA: MIT Press.

Strawson, G. (2006). Realistic monism. In A. Freeman (Ed.), *Consciousness and its place in nature: Does physicalism entail panpsychism?* (pp. 3–31). Charlottesville, VA: Imprint Academic.

Strong, J. (Producer). (2017, June 2). Everyone needs a (robot) companion. *WSJ The Future of Everything*. Podcast retrieved from http://itunes.apple.com

Sturgeon, T. (1972). *More than human*. New York: Ballantine Books.

Suki, N. M., Ramayah, T., Ming, M. K., & Suki, N. M. (2013). Factors enhancing employed job seekers intentions to use social networking sites as a job search tool. In A. Mesquita (Ed.), *User perception and influencing factors of technology in everyday life* (pp. 265–281). Hershey, PA: IGI Global. doi:10.4018/978-1-4666-1954-8.ch018

Suomi, R. (2006). Introducing electronic patient records to hospitals: Innovation adoption paths. In T. Spil & R. Schuring (Eds.), *E-health systems diffusion and use: The innovation, the user and the use IT model* (pp. 128–146). Hershey, PA: IGI Global. doi:10.4018/978-1-59140-423-1.ch008

Sur, M. (2004). Rewiring cortex: Cross-modal plasticity and its implications for cortical development and function. In G. A. Calvert, C. Spence, & B. E. Stein (Eds.), *Handbook of multisensory processing* (pp. 681–694). Cambridge, MA: MIT Press.

Świech, A. (2003). Cyberpunk: kroniki w chromie. In P. Kletkowski & P. Marecki (Eds.), *Nowe Nawigacje II. Azja - cyberpunk - kino niezależne* (pp. 125–136). Kraków: Rabid.

Swim, J., & Barker, L. (2012). Pathways into a gendered occupation: Brazilian women in IT. *International Journal of Social and Organizational Dynamics in IT, 2*(4), 34–51. doi:10.4018/ijsodit.2012100103

Sytsma, J., & Machery, E. (2010). Two conceptions of subjective experience. *Philosophical Studies, 151*(2), 299–327. doi:10.1007/s11098-009-9439-x

Tarafdar, M., & Vaidya, S. D. (2006). Adoption and implementation of IT in developing nations: Experiences from two public sector enterprises in India. In M. Khosrow-Pour (Ed.), *Cases on information technology planning, design and implementation* (pp. 208–233). Hershey, PA: IGI Global. doi:10.4018/978-1-59904-408-8.ch013

Tarafdar, M., & Vaidya, S. D. (2008). Adoption and implementation of IT in developing nations: Experiences from two public sector enterprises in India. In G. Garson & M. Khosrow-Pour (Eds.), *Handbook of research on public information technology* (pp. 905–924). Hershey, PA: IGI Global. doi:10.4018/978-1-59904-857-4.ch076

Tavani, H. T. (2003). *Ethics and technology: Controversies, questions, and strategies for ethical computing*. John Wiley & Sons.

Telotte, J. P. (1995). *Replications: A robotic history of the science fiction film*. Urbana, Chicago: University of Illinois Press.

Tezuka, O. (2004). *Astro Boy* (Vols. 1-23). Milwaukie, Oregon: Dark Horse Comics. (originally purblished 1952-1968).

Tezuka, O. (2004). *Metropolis* (Vol. 1). Mozów: Wydawnictwo Arashi. (originally published 1949).

Thesing, Z. (2007). Zarina thesing, pumpkin patch. In M. Hunter (Ed.), *Contemporary chief information officers: Management experiences* (pp. 83–94). Hershey, PA: IGI Global. doi:10.4018/978-1-59904-078-3.ch007

Thielman, S. (2016). Apple v the FBI: what's the beef, how did we get here and what's at stake? *The Guardian*. Retrieved from http://www.theguardian.com/technology/2016/feb/20/apple-fbi-iphone-explainer-san-bernardino

Thomas, L. (2009, August 03). What's behind Japan's Love Affair with Robots? *Time*. Retrieved from: http://content.time.com/time/world/article/0,8599,1913913,00.html

Thomas, J. C. (2004). Public involvement in public administration in the information age: Speculations on the effects of technology. In M. Malkia, A. Anttiroiko, & R. Savolainen (Eds.), *eTransformation in governance: New directions in government and politics* (pp. 67–84). Hershey, PA: IGI Global. doi:10.4018/978-1-59140-130-8.ch004

Treiblmaier, H., & Chong, S. (2013). Trust and perceived risk of personal information as antecedents of online information disclosure: Results from three countries. In F. Tan (Ed.), *Global diffusion and adoption of technologies for knowledge and information sharing* (pp. 341–361). Hershey, PA: IGI Global. doi:10.4018/978-1-4666-2142-8.ch015

Tucker, E. (2015). How robots are helping children with autism. *The Guardian*. Retrieved from http://www.theguardian.com/lifeandstyle/2015/feb/01/how-robots-helping-children-with-autism

Ullian, J. S. (1971). Putnam Hilary. Minds and machines. The Journal of Symbolic Logic, 36(01), 177.

Unterstell, S., Götz, M., & Holler, A. (2012). Eat plenty of vegetables and drink lots of milk. *TelevIZIon 25*(E), 33-37.

Urasawa, N., & Tezuka, O. (2009). *Pluto* (Vols. 1-8). Tokyo: Shogakukan Inc.

Utz, S., Schultz, F., & Glocka, S. (2013). Crisis communication online: How medium, crisis type and emotions affected public reactions in the Fukushima Daiichi nuclear disaster. *Public Relations Review*, *39*(1), 40–46. doi:10.1016/j.pubrev.2012.09.010

van Grembergen, W., & de Haes, S. (2008). IT governance in practice: Six case studies. In W. van Grembergen & S. De Haes (Eds.), *Implementing information technology governance: Models, practices and cases* (pp. 125–237). Hershey, PA: IGI Global. doi:10.4018/978-1-59904-924-3.ch004

Van Moer, A. (2006). The Intentionality of Formal Systems. *Foundations of Science*, *11*(1-2), 81–119. doi:10.1007/s10699-004-5914-1

van Os, G., Homburg, V., & Bekkers, V. (2013). Contingencies and convergence in European social security: ICT coordination in the back office of the welfare state. In M. Cruz-Cunha, I. Miranda, & P. Gonçalves (Eds.), *Handbook of research on ICTs and management systems for improving efficiency in healthcare and social care* (pp. 268–287). Hershey, PA: IGI Global. doi:10.4018/978-1-4666-3990-4.ch013

Varela, F. J. (1997). Neurophenomenology: A methodological remedy to the hard problem. In J. Shear (Ed.), Explaining consciousness: The hard problem of consciousness (pp. 337–358). Cambridge, MA: MIT Press.

Velloso, A. B., Gassenferth, W., & Machado, M. A. (2012). Evaluating IBMEC-RJ's intranet usability using fuzzy logic. In M. Cruz-Cunha, P. Gonçalves, N. Lopes, E. Miranda, & G. Putnik (Eds.), *Handbook of research on business social networking: Organizational, managerial, and technological dimensions* (pp. 185–205). Hershey, PA: IGI Global. doi:10.4018/978-1-61350-168-9.ch010

Victor, D. (2015, August 3). Hitchhiking Robot, Safe in Several Countries, Meets Its End in Philadelphia. *The New York Times*. Retrieved from http://www.nytimes.com/2015/08/04/us/hitchhiking-robot-safe-in-several-countries-meets-its-end-in-philadelphia.html

Villablanca, A. C., Baxi, H., & Anderson, K. (2009). Novel data interface for evaluating cardiovascular outcomes in women. In A. Dwivedi (Ed.), *Handbook of research on information technology management and clinical data administration in healthcare* (pp. 34–53). Hershey, PA: IGI Global. doi:10.4018/978-1-60566-356-2.ch003

Villablanca, A. C., Baxi, H., & Anderson, K. (2011). Novel data interface for evaluating cardiovascular outcomes in women. In *Clinical technologies: Concepts, methodologies, tools and applications* (pp. 2094–2113). Hershey, PA: IGI Global. doi:10.4018/978-1-60960-561-2.ch806

Virkar, S. (2011). Information and communication technologies in administrative reform for development: Exploring the case of property tax systems in Karnataka, India. In J. Steyn, J. Van Belle, & E. Mansilla (Eds.), *ICTs for global development and sustainability: Practice and applications* (pp. 127–149). Hershey, PA: IGI Global. doi:10.4018/978-1-61520-997-2.ch006

Virkar, S. (2013). Designing and implementing e-government projects: Actors, influences, and fields of play. In S. Saeed & C. Reddick (Eds.), *Human-centered system design for electronic governance* (pp. 88–110). Hershey, PA: IGI Global. doi:10.4018/978-1-4666-3640-8.ch007

Wainwright, D., & Linebarger, D. L. (2009). Television can teach. Elements of effective educational television. In N. N. Nagaraj (Ed.), *Effect of television on children* (pp. 24–44). Hyderabad: Icfai University Press.

Wallace, A. (2009). E-justice: An Australian perspective. In A. Martínez & P. Abat (Eds.), *E-justice: Using information communication technologies in the court system* (pp. 204–228). Hershey, PA: IGI Global. doi:10.4018/978-1-59904-998-4.ch014

Walter, W. G. (1950, May 1). An imitation of life. *Scientific American, 182*(5), 42–45. doi:10.1038/scientificamerican0550-42

Wang, G. (2012). E-democratic administration and bureaucratic responsiveness: A primary study of bureaucrats' perceptions of the civil service e-mail box in Taiwan. In K. Kloby & M. D'Agostino (Eds.), *Citizen 2.0: Public and governmental interaction through web 2.0 technologies* (pp. 146–173). Hershey, PA: IGI Global. doi:10.4018/978-1-4666-0318-9.ch009

Wangpipatwong, S., Chutimaskul, W., & Papasratorn, B. (2011). Quality enhancing the continued use of e-government web sites: Evidence from e-citizens of Thailand. In V. Weerakkody (Ed.), *Applied technology integration in governmental organizations: New e-government research* (pp. 20–36). Hershey, PA: IGI Global. doi:10.4018/978-1-60960-162-1.ch002

Watt, H., & Eccles, J. (2008). *Gender and occupational outcomes: Longitudinal assessment of individual, social, and cultural Influences*. Washington, D.C.: American Psychological Association. doi:10.1037/11706-000

Wedemeijer, L. (2006). Long-term evolution of a conceptual schema at a life insurance company. In M. Khosrow-Pour (Ed.), *Cases on database technologies and applications* (pp. 202–226). Hershey, PA: IGI Global. doi:10.4018/978-1-59904-399-9.ch012

Weisberg, J. (2012). The hard problem of consciousness. *The Internet Encyclopedia of Philosophy*. Retrieved January 4, 2016 from http://www.iep.utm.edu

Weiskrantz, L., Warrington, E. K., Sanders, M. D., & Marshall, J. (1974). Visual capacity in the hemianopic field following a restricted occipital ablation. *Brain*, *97*(1), 709–728. doi:10.1093/brain/97.1.709 PMID:4434190

Wells, J. (2014). Keep Calm and Remain Human: How We Have Always Been Cyborgs and Theories on the Technological Present of Anthropology. *Revista de Antropologia*, *43*(1), 5–34.

West, C., & Zimmerman, D. H. (1987). Doing gender. *Gender & Society*, *1*(2), 125–151. doi:10.1177/0891243287001002002

West, C., & Zimmerman, D. H. (2009). Accounting for doing gender. *Gender & Society*, *23*(1), 112–122. doi:10.1177/0891243208326529

Whitehead, N. L., & Wesch, M. (Eds.). (2012). *Human no more: Digital subjectivities, unhuman subjects, and the end of anthropology* (pp. 177–198). Boulder: University Press of Colorado.

Whiting, D. (2016). *Conceptual Role Semantics | Internet Encyclopedia of Philosophy*. Retrieved from http://www.iep.utm.edu/conc-rol

Whybrow, E. (2008). Digital access, ICT fluency, and the economically disadvantages: Approaches to minimize the digital divide. In F. Tan (Ed.), *Global information technologies: Concepts, methodologies, tools, and applications* (pp. 1409–1422). Hershey, PA: IGI Global. doi:10.4018/978-1-59904-939-7.ch102

Whybrow, E. (2008). Digital access, ICT fluency, and the economically disadvantages: Approaches to minimize the digital divide. In C. Van Slyke (Ed.), *Information communication technologies: Concepts, methodologies, tools, and applications* (pp. 764–777). Hershey, PA: IGI Global. doi:10.4018/978-1-59904-949-6.ch049

Wickramasinghe, N., & Geisler, E. (2010). Key considerations for the adoption and implementation of knowledge management in healthcare operations. In M. Saito, N. Wickramasinghe, M. Fuji, & E. Geisler (Eds.), *Redesigning innovative healthcare operation and the role of knowledge management* (pp. 125–142). Hershey, PA: IGI Global. doi:10.4018/978-1-60566-284-8.ch009

Wickramasinghe, N., & Geisler, E. (2012). Key considerations for the adoption and implementation of knowledge management in healthcare operations. In *Organizational learning and knowledge: Concepts, methodologies, tools and applications* (pp. 1316–1328). Hershey, PA: IGI Global. doi:10.4018/978-1-60960-783-8.ch405

Wickramasinghe, N., & Goldberg, S. (2007). A framework for delivering m-health excellence. In L. Al-Hakim (Ed.), *Web mobile-based applications for healthcare management* (pp. 36–61). Hershey, PA: IRM Press. doi:10.4018/978-1-59140-658-7.ch002

Wickramasinghe, N., & Goldberg, S. (2008). Critical success factors for delivering m-health excellence. In N. Wickramasinghe & E. Geisler (Eds.), *Encyclopedia of healthcare information systems* (pp. 339–351). Hershey, PA: IGI Global. doi:10.4018/978-1-59904-889-5.ch045

Więckowska, K. (2011). Potomkowie Mary Shelley: Mężczyźni, kobiety i cyborgi. Literaria Copernicana, 2(8), 62-71.

Wiener, N. (1988). *The human use of human beings: Cybernetics and society*. New York: Da Capo Press.

Wig, K. (2015). Åpner verdens første robotstyrte hotell. *E24*. Retrieved from http://e24.no/article/23393633

Wolfe, C. (2009). *What is posthumanism*. Minneapolis: The University of Minnesota Press.

Wolfram, S. (2002). *A new kind of science*. Champaign, IL: Wolfram Media.

Wong, M. H. (2015, July 17). Japan's robot hotel opens its (automatic) doors. Tomorrow Transformed. *CNN*. Retrieved from http://www.cnn.com/2015/07/17/travel/japan-hotel-robot-opens/index.html

Wróblewski, M. (2011). Cyborgizacja codzienności człowieka: Stan współczesny i kilka myśli futurologicznych. Literaria Copernicana 2(8), 8-19.

Wyld, D. (2009). Radio frequency identification (RFID) technology. In J. Symonds, J. Ayoade, & D. Parry (Eds.), *Auto-identification and ubiquitous computing applications* (pp. 279–293). Hershey, PA: IGI Global. doi:10.4018/978-1-60566-298-5.ch017

Yaghmaei, F. (2010). Understanding computerised information systems usage in community health. In J. Rodrigues (Ed.), *Health information systems: Concepts, methodologies, tools, and applications* (pp. 1388–1399). Hershey, PA: IGI Global. doi:10.4018/978-1-60566-988-5.ch088

Yee, G., El-Khatib, K., Korba, L., Patrick, A. S., Song, R., & Xu, Y. (2005). Privacy and trust in e-government. In W. Huang, K. Siau, & K. Wei (Eds.), *Electronic government strategies and implementation* (pp. 145–190). Hershey, PA: IGI Global. doi:10.4018/978-1-59140-348-7.ch007

Yeh, S., & Chu, P. (2010). Evaluation of e-government services: A citizen-centric approach to citizen e-complaint services. In C. Reddick (Ed.), *Citizens and e-government: Evaluating policy and management* (pp. 400–417). Hershey, PA: IGI Global. doi:10.4018/978-1-61520-931-6.ch022

Yomiuri. (2015, August 22). Robots to help Japan's farming, fishing industries. *The Japan News/Yomiuri*. In *The Malone Telegram*. Retrieved from http://www.mymalonetelegram.com/col35/robots-to-help-japans-farming-fishing-indudstries-20150822

Young-Jin, S., & Seang-tae, K. (2008). E-government concepts, measures, and best practices. In A. Anttiroiko (Ed.), *Electronic government: Concepts, methodologies, tools, and applications* (pp. 32–57). Hershey, PA: IGI Global. doi:10.4018/978-1-59904-947-2.ch004

Yun, H. J., & Opheim, C. (2012). New technology communication in American state governments: The impact on citizen participation. In K. Bwalya & S. Zulu (Eds.), *Handbook of research on e-government in emerging economies: Adoption, e-participation, and legal frameworks* (pp. 573–590). Hershey, PA: IGI Global. doi:10.4018/978-1-4666-0324-0.ch029

Zhang, N., Guo, X., Chen, G., & Chau, P. Y. (2011). User evaluation of e-government systems: A Chinese cultural perspective. In F. Tan (Ed.), *International enterprises and global information technologies: Advancing management practices* (pp. 63–84). Hershey, PA: IGI Global. doi:10.4018/978-1-60960-605-3.ch004

zhenglimen. (Ed.). (2017). Dubai: World's first operational police robot appears at cannon fire ceremony of Ramadan. *CCTV.com*. Retrieved from http://english.cctv.com/2017/06/01/PHOAQjUwOfZMAIvgkmFHGUZM170601.shtml

Zhiguo, L., Chong, L., Jun, P., Habin, Z., & Hong, W. (2014, June 29–July 4). Motion simulation platform for a humanoid robot base on MATLAB. In *Proceedings of the 2014 11th World Congress on Intelligent Control and Automation (WCICA)* (pp. 13–18).

Zuo, Y., & Hu, W. (2011). Trust-based information risk management in a supply chain network. In J. Wang (Ed.), *Supply chain optimization, management and integration: Emerging applications* (pp. 181–196). Hershey, PA: IGI Global. doi:10.4018/978-1-60960-135-5.ch013

About the Contributors

Steven John Thompson is a researcher in digital literacy, media studies, new media, rhetorics, and in professional, science and technical communication. He has taught at Johns Hopkins University, Dartmouth College, Clemson University, Georgia Southern University, Towson University, and University of Maryland University College. Dr. Thompson was Editor of *Global issues and ethical considerations in human enhancement technologies*, published by IGI Global in 2014. He is on the Editorial Board of Academic Editors for *PeerJ* journal in Emerging Technologies, Ethical Issues, Human-Computer Interaction, Science Policy, and World Wide Web & Web Science subject areas. Thompson published pioneering quantitative research on Internet addiction and dependency in 1996, and presented his iconics theory on agency of terrorism images as virtual subscripts of artificial intelligence at University of Basel in 2009. He was plenary closing academic panel discussant at UNESCO's First International Forum on Media and Information Literacy in Fez, Morocco, in 2011. Professor Thompson's research, largely focused on the relationship between the media and terrorism, and on integration of the Internet appliance with the human body, has been presented in over a dozen countries on four continents. He holds a Ph.D. in Rhetorics, Communication, and Information Design from Clemson University, an M.S. degree in Media Arts and Science from Indiana University, and undergraduate degrees from Penn State, including a B.A. in Integrative Arts and B.A. in Media Studies, with a Minor Certificate from the Penn State College of Engineering in Science, Technology, and Society, the lens that frames his research. He may be found engaged with his lifelong avocation in fractal chaos at equashion.com.

* * *

Liudmila Baeva is Professor, Doctor of Philosophy, Dean of the Department of Social Communication, Head of the Chair of Philosophy of Astrakhan State University. She is the author of over 170 articles and six monographs. Her research interests focus on the field of axiology, philosophical anthropology, and the study of the information society issues. She is a member of the editorial board of the international journal, "*Socioloska luca: Journal of Social Anthropology, Social Demography & Social Psychology*" (Montenegro), and the editorial board of the international journal "*The Caspian Region: Economics, Politics and Culture*" (Russia). Professor Baeva has been Visiting Professor of Philosophy at Moscow State Technical University (Russia, 2005), Hainan University (China, 2007), and South Kazakhstan State University named after M. Auezov (City of Shymkent, Kazakhstan, 2011). She participated in the World Congresses of Philosophy in Istanbul, Seoul, and Athens. Her research in this volume was supported by the Russian Scientific Foundation under Grant N°16-18-10162, Sant-Petersburg State University.

Chris Chesher is senior lecturer in Digital Cultures in the Department of Media and Communications, University of Sydney. His research focuses on the cultural contexts of digital media, including computer games, Internet, mobile media and robotics.

C. Cortés Mac-Evoy was born in Santiago, Chile. He received the M.Sc. Eng. degree from Universidad de Santiago, Santiago, Chile. Since 2015, C. Cortés has been working with industrial robots as an external contributor at Volkswagen AG in Wolfsburg, Germany, on the qualification, programming, and development of software standards for the Volkswagen Group.

Natalia Dmitruk is a Ph.D. student with particular interests in transhumanistic and posthumanistic theories. She received her B.A. in Film and Theatre Studies, and her M.A. in Polish Philology. Her research interests include cyborg and android characters in Japanese popular media. She is also a member of The Association of Pop Culture Researchers and Pop-cultural Education "Trickster".

Maya Götz graduated from the Pädagogische Hochschule in Kiel, Germany, with the first state examination in education for elementary schools and the upper division of elementary schools, and with a Master of Arts degree in education. She acquired a Ph.D. degree at the University Kassel, Germany, where the theme of her thesis was "Television in everyday life of girls: Facets of media appropriation in female adolescence." From 1999 - 2002, Dr. Götz was Scientific Editor at International Central Institute for Youth and Educational Television (IZI) at the Bayerischer Rundfunk, and has been Head of the institute since 2003. Since 2006, she has been the managing director of the Prix Jeunesse International. Her main field of work is research in the area of "children/youth and television." She works and publishes in the field of gender-specific reception research, and has an extensive repertoire of empirical studies related to the televised representation of children and gender in modern society.

J. J. Johnson is a founding partner and the creative force behind Sinking Ship Entertainment. Since the company's inception in 2004, he has created and been executive producer on 15 series, including international hits *This is Daniel Cook*, *Annedroids*, and the Emmy award winning *Dino Dan*. In 2015, J. J. made Emmy history by being nominated three times in the same directing category for three different series, winning the Emmy for *Odd Squad*. J. J. is an Emmy-nominated writer with over 500 episode credits, and has been recognized by the *Hollywood Reporter* as one of their Top 20 under 35. J. J. also serves as the chair of the Youth Media Alliance in Canada.

Bruce MacLennan has a B.S. in mathematics with honors from Florida State University, and an M.S. and Ph.D. in computer science from Purdue University. MacLennan was a Senior Software Engineer with Intel Corp. from 1975 to 1979, after which he joined the Computer Science faculty of the Naval Postgraduate School in Monterey, CA as Assistant Professor (1979–83), Associate Professor (1983–87), and Acting Chair (1984–85). Since 1987, Dr. MacLennan has been a member of the Electrical Engineering and Computer Science faculty of the University of Tennessee, Knoxville. His research investigates unconventional computation and the self-organization of swarms of microscopic agents. Professor MacLennan has more than 80 refereed journal articles and book chapters, and has published three books. He has made more than 70 invited or refereed presentations. He was founding Editor-in-Chief of the *International Journal of Nanotechnology and Molecular Computation*.

Diana Iulia Nastasia is an international lecturer and scholar. In her home country of Romania, she has taught communication and media studies at the Romanian-American University, University of Bucharest, and the National University for Political Studies and Public Administration. She arrived in the US as a Fulbright Visiting Researcher, and pursued a Ph.D. in Communication and Public Discourse at the University of North Dakota, and Master's degrees in Educational Leadership and Business Administration at Southern Illinois University. She has taught communication and media studies at several universities and colleges in North Dakota, Missouri, and Illinois. Dr. Nastasia's research focuses on voices and portrayals of marginalized populations, including women, children, and immigrants. She has been awarded research project grants by the International Association of Women in Radio and Television, International Women's Media Foundation, European Institute for Gender Equality, Internationales Zentralinstitut für das Jugend- und Bildungsfernsehen based in Munich, Germany, and the Social Science Research Council. She has published chapters in edited volumes at Routledge, Palgrave, Peter Lang, Nova, and Fordham University Press, as well as articles in U.S. and European journals. She has made over 150 research presentations at national and international levels.

She has a co-authored book on media coverage of the 2011 Japan disasters, and several other publications forthcoming or in press. Dr. Nastasia has been involved in numerous university and community service activities aimed at democratizing knowledge and improving the functions of media in society. She has served on the International Communication Association Board of Directors and on the editorial boards of several journals.

Roger Andre Søraa is a Ph.D. candidate in the Department of Interdisciplinary Studies of Culture at NTNU, Norwegian University of Science and Technology, where he also received his Master's degree. He has spent several years doing fieldwork in Japan, and his research has revolved mainly around the way technology is domesticated in society and presented in the media, seeing how the sociotechnical imaginaries of robots affect their daily usage.

Claudio Urrea was born in Santiago, Chile. He received the M.Sc. Eng. and Dr. degrees from Universidad de Santiago de Chile, and the Ph.D. degree from Institut National Polytechnique de Grenoble, France. Dr. Urrea is Professor at the Department of Electrical Engineering, Universidad de Santiago de Chile, since 1998. He has developed and implemented a Robotics Laboratory, where intelligent robotic systems are development and investigated. He is Director of the Doctorate in Engineering Sciences, Major in Automation, at the Universidad de Santiago de Chile.

Billy Wheeler earned a Ph.D. in History and Philosophy of Science at the University of Cambridge. He is a Teaching Fellow in the Department for Science and Technology Studies, University College London.

Index

www.ingramcontent.com/pod-product-compliance
Lightning Source LLC
LaVergne TN
LVHW081333300926
849161LV00008B/1959
9781522529736